# Integrated Science

Volume 34

The **Integrated Science** Series aims to publish the most relevant and novel research in all areas of Formal Sciences, Physical and Chemical Sciences, Biological Sciences, Medical Sciences, and Social Sciences. We are especially focused on the research involving the integration of two of more academic fields offering an innovative view, which is one of the main focuses of Universal Scientific Education and Research Network (USERN), science without borders.

**Integrated Science** is committed to upholding the integrity of the scientific record and will follow the Committee on Publication Ethics (COPE) guidelines on how to deal with potential acts of misconduct and correcting the literature.

Jose Antonio Lara-Hernandez ·
Alessandro Melis · Paola Boarin ·
Priscila Besen
Editors

# The Dusk of Design

## Exploring Multidisciplinary Approaches and Evolutionary Biology in Architecture

Springer

*Editors*
Jose Antonio Lara-Hernandez
Huri Te Ao School of Future Environments
Auckland University of Technology
Auckland, New Zealand

Paola Boarin
School of Architecture and Planning
University of Auckland
Auckland, New Zealand

Alessandro Melis
New York Institute of Technology
New York, NY, USA

Priscila Besen
Huri Te Ao School of Future Environments
Auckland University of Technology
Auckland, New Zealand

ISSN 2662-9461 ISSN 2662-947X (electronic)
Integrated Science
ISBN 978-3-031-76866-8 ISBN 978-3-031-76867-5 (eBook)
https://doi.org/10.1007/978-3-031-76867-5



This Springer imprint is published by the registered company Springer Nature Switzerland AG
The registered company address is: Gewerbestrasse 11, 6330 Cham, Switzerland

*To the curious minds who work tirelessly in multidisciplinary environments, bridging gaps between knowledge and practice. Your commitment to collaboration and innovation shapes the future of both design and research, and this book is a testament to your enduring impact.*

# Preface

In a world increasingly defined by uncertainties, the field of architecture and urban design stands at a pivotal crossroads. *The Dusk of Design* emerges as a vital anthology that explores the relationship between form and function, between cultural heritage and forward-looking innovation. It is within the pages of this book that readers are invited to journey through realm of thought, but also practice, engaging with ideas that challenge the status quo of architectural theory and practice.

The collection of chapters curated for this volume reflects a broad spectrum of perspectives, each contributing to the discussion on the potential and maybe responsibility of architecture to respond to and shape the human condition. From the granular details of material science to the expansive urban scales of city planning, the works presented here offer a profound exploration of how we can better understand and craft our built environment in an era of unprecedented environmental, social, and technological change.

This book is rooted in the principle that design is not merely an act of creation but a complex, diverse and multi-disciplinary response to cultural, ecological, technological, and social challenges. Recognising that each piece of architecture is a statement of hope, a manifestation of how we see our place in the world and how we envision our future within it. The diverse contributions within this volume challenge us to think beyond the traditional paradigms of architecture and to consider the broader implications of our work on the world and its inhabitants.

At the core of *The Dusk of Design* is the recognition of the need for a shift from deterministic, human-centric design practices towards approaches that are more adaptive, inclusive, and evolutionary. The chapters deepen into theoretical and practical realms, exploring how we the principles of exaptation, redundancy, and diversity can foster architectural and urban forms that are capable of adapting to changing needs and unforeseen future conditions.

The book is structured to reflect this journey through layers of understanding and application. It begins with a philosophical examination of the role of design in our society and gradually transitions into practical case studies and specific design innovations that embody these principles. This structure is designed to provide a comprehensive narrative arc that mirrors the process of transformation from theory into practice, from concept into concrete reality. Although, it could seem as chaotic as evolution is too.

Moreover, the authors included in this volume bring a rich set of experiences and insights, drawing from a multitude of geographic and professional backgrounds. This diversity is not merely academic; it is a deliberate choice to reflect the varied ways in which architecture impacts and is impacted by the global community. Each chapter contributes a unique voice to the dialogue on how we can reshape our approaches to design in order to create environments that are more resilient, responsive, and reflective of the diverse world we inhabit.

As we face global challenges such as climate change, urbanisation, and social inequality, the role of architecture becomes increasingly complex and critical. Therefore, *The Dusk of Design* is a call to the architectural and planning communities to broaden their horizons and to embrace new methodologies. It urges professionals and students alike to think creatively about how design can influence and facilitate sustainable development, social cohesion, and enhanced quality of life.

In conclusion, these words seek to set the stage for a profound exploration into the future of design, offering a set of alternative lens through which we can reimagine our relationship with the natural and built environments. It is a guidepost for those who aspire to lead the charge in transforming our urban landscapes into more adaptive, inclusive, and resilient spaces. *The Dusk of Design* is not just a reflection on where architecture has been but an insightful forecast of where it must go, making it an essential read for anyone committed to the future of designing our world.

Auckland, New Zealand Jose Antonio Lara-Hernandez
September 2024

# Acknowledgements

First and foremost, we would like to express my deepest gratitude to all the contributors to this volume. Their expertise, insights, and thoughtful contributions have made this work possible. Their dedication to their respective fields and willingness to share their knowledge have enriched the quality of this book in countless ways.

Finally, we would like to thank my family and friends for their unwavering support and patience during the preparation of this book. To Sara, Barbora, Amedeo, Matheus, Isabella, Maddalena, Julieta, Adam, Manu, Elia, and Luca, your love, understanding, and encouragement have sustained us throughout this journey. This book is as much yours as it is ours.

To all those mentioned and many others whose contributions may not be specifically named, I extend my heartfelt thanks.

# Contents

# Contributors

**Lanfranco Aceti** Museum of Contemporary Cuts, Rome, Italy; Heliopolis 21 and Resilient Communities, London, UK

**Temitope M. Adebara** Department of Urban and Regional Planning, Olabisi Onabanjo University, Ogun State, Nigeria

**Fabrizio Aimar** Texas A&M University, College Station, USA

**Sofia Ayora-Talavera** Universidad Marista, Merida, Mexico

**Priscila Besen** Te Wānanga Aronui o Tāmaki Makau Rau—Auckland University of Technology, Auckland, New Zealand

**Paola Boarin** School of Architecture and Planning, University of Auckland, Auckland, New Zealand

**Andrea Brambilla** Department of Architecture Built Environment Construction Engineering, Design and Health Laboratory, Politecnico di Milano, Milan, Italy

**Stefano Capolongo** Department of Architecture Built Environment Construction Engineering, Design and Health Laboratory, Politecnico di Milano, Milan, Italy

**Antonino Di Raimo** Portsmouth School of Architecture, Art & Design, Portsmouth, UK

**Michele Dolcini** Department of Architecture Built Environment Construction Engineering, Design and Health Laboratory, Politecnico di Milano, Milan, Italy

**Alexander Gomez-Torres** University of Auckland, Auckland, New Zealand

**Mohamed Yazid Khemri** Lectutrer at Birmingham City University, Birmingham, UK

**Dongsei Kim** New York Institute of Technology, New York, USA

**Jose Antonio Lara-Hernandez** Future Environments, Auckland University of Technology, Auckland, New Zealand

**Alessandro Melis** New York Institute of Technology, New York, NY, USA

**Barbora Melis** University of Portsmouth, Portsmouth, UK

**Gregor Mews** Co-Lead Bioclimatic and Sociotechnical Cities Lab & Lecturer in Urban Design and Town Planning, University of the Sunshine Coast, Sippy Downs, Australia

**Veronica Natividade** PhD candidate, Chair Of Digital Architectonics, ETH Zürich, Zurich, Switzerland

**Christian R. Pongratz** School of Architecture and Design, New York Institute of Technology, New York, USA

**Guido Robazza** School of Architecture, Art and Design, University of Portsmouth, Portsmouth, England

**Maria Simioli** Dipartimento di Architettura, Università Degli Studi di Napoli Federico II, Napoli, Italy

**Angioletta Voghera** Politecnico di Torino, Turin, Italy;
CED PPN—European Documentation Centre on Nature Park Planning, Turin, Italy

**Dustin White** Florida Atlantic University, Boca Raton, USA

# List of Figures

**Chapter 6**



**Chapter 8**

**Chapter 9**



**Chapter 10**



**Chapter 11**



**Chapter 12**

**Chapter 13**



**Chapter 15**

**Chapter 16**



**Chapter 17**



**Chapter 18**



**Chapter 19**

**Chapter 20**

# List of Tables

# Beyond Deterministic and Human-Centric Design in Architecture

1

Alessandro Melis, Jose Antonio Lara-Hernandez, and Paola Boarin

**Abstract**

The concept of evolutionary design challenges traditional notions of intentional, purpose-driven creativity by aligning human design processes with evolutionary principles. This chapter posits that human design is an extension of biological evolution, influenced by societal categorisation and resistant to paradigm shifts. Despite awareness of environmental impacts, systemic reorientation in the construction market remains slow. This book highlights that marginalised forms of evolutionary design, often overlooked, are as prevalent as goal-oriented designs, advocating for a reinterpretation of architectural taxonomies through an ecological lens. The discussion draws parallels between architectural components and evolutionary traits, such as exaptation, where a feature evolves for one function but is repurposed for another. This concept, exemplified by the bird's wing or architectural elements like the arch, underscores the opportunistic nature of evolution. The chapter also explores the attributes of creative structures that enhance their potential for functional co-optation, emphasising redundancy, variability, and diversity. Furthermore, the book critiques the failure of contemporary architectural and urban design to address global crises, attributing this to the rigid categorisation within human civilisation. It suggests a revolutionary shift towards a non-deterministic approach to design, inspired

A. Melis · J. A. Lara-Hernandez (✉)
New York Institute of Technology, New York, USA
e-mail: Antonio.lara-hernandez@aut.ac.nz

A. Melis
e-mail: amelis@nyit.edu

P. Boarin
University of Auckland, Auckland, New Zealand
e-mail: p.boarin@auckland.ac.nz

J. A. Lara-Hernandez
Future Environments, Auckland University of Technology, Auckland, New Zealand

 
J. A. Lara-Hernandez et al. (eds.), *The Dusk of Design*, Integrated Science 34,
https://doi.org/10.1007/978-3-031-76867-5_1

by evolutionary biology. This shift advocates for a broader understanding of creativity, considering disorder, randomness, and informality as integral components. By embracing these principles, designers can create more resilient and innovative solutions, leveraging the latent creative potential of objects to adapt to changing environmental conditions. This holistic view integrates associative thinking and evolutionary adaptation, positioning creativity as a systemic, collective, and multigenerational endeavour, thereby reconnecting knowledge separated by disciplinary rigidity.

**Keywords**

Evolutionary design • Diversity • Redundancy • Variability

## 1 Evolutionary Design

Explaining the evolution of seemingly perfect organs like a bird's wing, the mimicry observed in many animal species, or an architectural component like an arch involves understanding how such features could develop without presuming a preordained design. Both human and evolutionary agencies create designs through a process akin to evolutionary refinement.

This perspective reinforces three core positions of this book:

1. Design by Humans as a Scalable Subsystem: Human design is literally a scalable subsystem of evolutionary design, a byproduct of biology. Here, design is not merely compared to evolution in metaphorical terms but is seen as an extension of evolutionary biology's adaptive potential.
2. Influence of Societal Categorization: Design has always been influenced by societal categorization, whether from a creationist perspective or an evolutionary one. The broader societal context shapes design categorization.
3. Inertia to Change: Human design categorization is influenced by discoveries or paradigm shifts, as it is primarily based on qualitative values and the extensive societal implications of such paradigm shifts impacting the built environment. For instance, the construction market is highly resistant to systemic reorientation, despite our awareness of the built environment's impact on climate crises for decades. Similarly, our understanding of adaptive design has not been sufficiently informed by the deeper significance of adaptation, which has a long history in biology.

The primary contribution of this book is not in documenting previously unrecorded design phenomena, but in emphasizing that known marginalized forms of evolutionary design are not isolated outliers. Instead, they are quantitatively equal to, if not more prevalent than, traditional goal-oriented designs. This recognition provides an opportunity to reinterpret architectural taxonomies through a more ecological lens.

The global crises humanity has faced since the onset of climate change—including the COVID-19 pandemic and the rise in inequalities—along with our obsession with mass communication via the internet and the proliferation of artificial intelligence, challenge our ability to predict future scenarios and adapt through traditional design strategies.

The failure of contemporary architectural and urban design to address complex problems is not merely a matter of inadequate design instruments. Rather, it reflects a systemic failure in the way humanity categorizes the world, including the role of design within this context.

The root problem of design's inability to respond effectively mirrors the broader failures of human civilization.

However, this challenge also offers an opportunity to rethink the categories we've used to understand the world and to reconnect pieces of knowledge that have been separated by the rigidity of current disciplinary taxonomies [7].

Since nature is the most successful designer on this planet, and adaptation—the main driver of evolution—is becoming increasingly relevant in design, we should consider rethinking design through the lens of evolution, moving beyond traditional disciplinary boundaries and professional practices.

This may suggest a revolutionary shift in design taxonomy, reminiscent of the concept of exaptation in evolutionary biology [12].

Exaptation, as defined by Stephen Jay Gould and Elisabeth Vrba in "Exaptation—a missing term in the science of form", published in "Paleobiology" in 1982, refers to a trait that originally evolved for one function but was later co-opted for a different function. Unlike adaptations, which are traits that evolved specifically for their current function, exaptations are features that initially served one use and were subsequently repurposed [3]. This concept emphasizes the opportunistic nature of evolution, where existing structures can be modified and utilized for new functions as environmental conditions and selective pressures change.

In architecture, exaptation translates to extending design towards non-deterministic approaches, providing a reservoir of possibilities for enhancing the resilience of the built environment ([5, 8]).

"The Dawn of Design" particularly focuses on the features of creative structures that maximize their potential for functional co-optation under emerging environmental pressures.

These features are redundancy, variability, and diversity, qualities that ensure the possibility of a wide range of shapes, thereby maximizing the chances of an event—functional co-optation—occurring.

Therefore, each of the three parts of this book will analyze case studies in the field of architectural design that rely on one or more of these features. These case studies will illustrate how redundancy, diversity, and variability can be harnessed to create innovative and resilient design solutions.

In this context, evolutionary design signifies a shift in determinism from the purpose of a structure to shaping its creative potential. Here, creativity is an attribute that can be applied to both humans and objects, as will be explained in the following sections. As a result, functional co-optation is serendipitous, not fortuitous, meaning that the latent creative potential of objects—such as architectures, or construction components–can be activated when environmental conditions occur.

The aim of the book is to encourage designers to utilize their full cognitive power, which involves associative thinking, through the awareness of the complexity of the environment and the limitations of using only the analytical part of our brain to describe future scenarios [17]. This will also require a more holistic understanding, in evolutionary terms, of the relationship between creativity and disorder, instinct, randomness, and informality, which are usually understood in a reductive manner. These concepts will be at the core of the following chapters.

## 2 Creativity Beyond Purpose. The Evolution of Intent

We do things for a reason, and sometimes we do things for no reason. The second part of this statement might seem like a simplification for actions we undertake without practical intentions, for more sophisticated purposes. By distinguishing initial survival needs from psychological and social needs, such as self-actualization, and later by more sophisticated categories, we attempt to explain why we do things. Without realizing it, the way we ask questions has become increasingly influenced by categories we now consider immutable, somehow making us dependent on, if not blinded by, the instruments supposed to help us understand reality.

However, certain actions we take, particularly those involving arcane qualities such as beauty, despite attempts to categorize them, seem to be sufficiently explained by the pleasure we derive from aesthetics, even though we don't fully understand the mechanisms behind them [10]. We need, then, to go back in time, beyond the boundaries of what we call the history of civilizations, to when the notion that we invent things intentionally and for a specific purpose becomes less evident. According to paleoanthropologist Heather Pringle, actions performed without a specific intention or practical use gain qualitative significance and represent the birth of creativity. This shows that the enjoyment we derive from activities is not always linked to an awareness of their purpose.

Thus, intentionality, use, and need, which we currently see as inherently linked, were not as closely associated before classical and modern categories existed. Recent advancements in science confirm that we derive pleasure from purposeful activities, whether simple or complex. Our enjoyment in doing things without a specific use is consistent with the recent paleoanthropological definition of creativity, which relies on associative thinking, as described by Heather [17], while also aligning with the evolutionary biology concept of functional cooptation, as defined by Stephen J. Gould and Richard Lewontin under the 'spandrel' metaphor [2], which emphasizes the possibilities inherent in doing things without a function.

Sometimes we discover practical uses for these things, and other times we do not [11]. This is why, in evolution, making redundant, variable, and diverse things offers more functional possibilities than relying solely on the design intentionality of individuals. More importantly, through associative thinking, we can find purposes in things done by others or by different generations, where the original creators might not have seen any or considered them useful for different reasons. The diverse possible forms of functional cooptation transform creativity into a systemic, collective, multidimensional, and multigenerational endeavor, relegating the idea of the single, talented creator of novelty to another narrative in the mythology of human behavior.

## 3 The Possibility of Redundancy, Variability and Diversity of the Creative Systems

"The Dusk of Design" argues that creativity, historically described as a mainly goal-oriented process, has proven inadequate in preparing the built world to respond to the challenges of global crises. Going back in time, we learn that humans can manipulate matter not only for specific purposes but also to imbue it successfully with a latent creative potential, depending on redundant, variable, and diverse physical morphologies. A perspective on design that attributes a creative agency to both human and objects offer a design repertoire, extending historical taxonomies towards more ecological positions [13].

Between 150,000 and 200,000 years ago, as our brains evolved in connectivity towards associative thinking, we began to enjoy doing things, likely preserving a certain level of evolutionary purpose without conscious awareness. This phenomenon, called 'creativity,' reflects a 'playful' and emergent modality that leverages the creative potential of objects' morphospace [6]. It allows humans to trade the lack of initial intention for possible functional cooptation of morphological feature or trait, when environmental conditions change.

The evolutionary advantages of creative objects reveal that creativity is a collective ability, not just an individual one, and is shared by both living and non-living agents. Our reluctance to consider categories that challenge the primacy of human agency may explain the oversight of their potential.

Neuroscience today is uncovering the mechanisms that regulate our aesthetic sense and the derivation of pleasure as a functional cooptation of the same brain regions that regulate our sense of taste, revealing a more prosaic reality—although equally fascinating—than the philosophical and religious constructs that have transformed beauty into ideals serving as a moral compass.

These mechanisms describe the creative potential of systems featuring morphological exuberance—redundancy, variability, and diversity—such as the human genome and brain. These mechanisms, therefore, form a scalable biological engine. Through functional cooptation, for example, they make our brain sensitive to exuberant expressions like beauty. They also enable humans to both design and use existing objects, harnessing aesthetics as a non-deterministic guideline. This supports the hypothesis that, once we abandon the idea that creativity is exclusively a prerogative of living agencies, multilayered creative trends permeate evolutionary processes. From the genome to any object we produce, creative systems can design other creative systems, and so on.

## 4 Adaptation Design

The Dusk of Design owes a great debt of gratitude to evolutionary biology research, particularly the multidisciplinary insights provided by Telmo Pievani, which support the notion that design is an expression of the complex, multilayered mechanisms of evolutionary biology. In fact, although adaptive design is now mainstream, designers seem to lag behind in a deep understanding of its evolutionary complexities, especially when compared with the extensive research that biologists have conducted on adaptation over the past two centuries.

According to Telmo Pievani [16], adaptation is the greatest empirical evidence of evolutionary biology, the main characteristic of natural structures, and the secret to the survival of every living being. Adaptation is what any evolutionary hypothesis must explain and, in the Darwinian view, becomes the expression of organic change that every living being must produce in response to changes and pressures in its environment. In biology, adaptation has always been considered to be in constant evolution and the result of the relentless action of natural selection acting on individual random genetic mutations.

Since the second half of the Nineteen century, the notion of adaptation has been challenging for evolutionary theories because it was deeply entrenched in the old creationist perspective of divine design and the flawless "harmony" of creation. Each organism adaptation was viewed as the proper adjustment to the eternal balances of creation. And indeed, this is how we still interpret creativity today: design, as a manifestation of creativity rooted in creationism, traditionally overlooks the evolutionary implications of adaptation and has never been subject to debate in this context. Instead, it is still imbued with an aura of reverence and mystique, which prevents it from fully embracing adaptive design perspectives.

Adaptation, as an implicit value of design, has historically been understood as the ability to make buildings and spaces suitable for the needs of their users. More explicitly, the term has conventionally referred to adapting spaces to accommodate physical limitations in their use, such as in the case of disabilities or for the elderly. Only in recent times, in response to climatic emergencies, has the term adaptation, along with the notion of sustainability, increasingly taken on the meaning of making the built environment, and thus its uses and functions, adaptable through design to rapidly changing environmental conditions.

For at least two thousand years, design has relied on immutable values, with Vitruvian principles being a prime example. Utility, beauty, and safety considered environmental variations only in terms of geographical and cultural characteristics, treating broader and deeper perspectives as constants.

Despite their apparent opposition, architectural positions, from Classicism to modern functionalism, have failed to systematically and consciously consider the environment as a fluid, dynamic presence rather than a static vacuum filled with paroxysmal inanimate objects under the banner of reified artificiality. The rapid and unexpected changes make every architectural position unstable, calling for a systemic vision of adaptive design that goes beyond merely addressing needs according to given climatic, socio-cultural, and political contexts. Even today, despite recognizing that environmental crises necessitate a higher level of adaptation, the evolutionary relationship between adaptation and the extension of design beyond deterministic mechanisms, as described in biological studies, is often overlooked.

Adaptation as a dynamic and evolving process is a pivotal lesson from biology, stemming from Darwin's recognition that adaptation is always partial and continuous, rather than a perfect, one-time solution. This insight can help us understand that architectural positions, too, cannot be viewed as flawless solutions.

The perceived perfection of classical or modern architecture as an evolutionary peak, as described in literature, reflects historians' desire to define the ultimate adaptive form. This desire has become more evident with efforts such as critical or style restoration, which aim for a purist ideal state of architecture that never truly existed (Melis & [9]).

## 5 Influence of Societal Categorization

Creationism, through the mediating concept of creativity, has been embedded in the notion of design as a central driving force in architectural literature since its inception. Whether conceived as mystical inspiration in the Middle Ages, a humanistic perspective during the Renaissance, or as an agency translating political and sociological forces into design in modern, postmodern, and contemporary contexts, human creativity has consistently occupied a central role in the taxonomy of design categorizations.

Throughout history, creativity has been the invariant feature of design, transcending different eras. However, another hypothesis could suggest that these ideas were not fundamentally different because none of them questioned a constant paradigm: the acceptance of the creationist nature of design.

Early evolutionary perspectives began to influence design concepts starting in the eighteenth century with mimetic theory, from Laugier's primitive hut to the introduction of conservation practices. The emergence of organic and radical architecture trends in the twentieth century furthered this evolution. More recent approaches, also referencing Darwin and biology and incorporating processes such as computation, generativity, autopoiesis and biomimicry, have advanced this trend more explicitly.

However, even these recent perspectives rely on orthodox Darwinism and scholastic literature when discussing evolution or adaptive design. They neglect the latest taxonomical changes in evolutionary biology and fail to challenge the concept of creativity as an individual human driving force, which translates adaptationism into a deterministic system or form of design.

Although it is widely accepted that architectural structures should continue to evolve in response to societal, technological, and environmental changes, the missing link with evolutionary biology remains. Architects still believe that further evolution in architecture can only occur along the established historical trajectory. In contrast, biologists have moved beyond the idea of linear progress [4], embracing more complex concepts such as punctuated equilibrium and niche construction [15]. In other words, classicism and modernism are not the only possible pinnacles of their eras, as the morphospace of possibilities is much larger than what has been designed and built. While acknowledging the prospect of future change, we should also explore alternative paths that could lead to different evolutionary outcomes in design to avoid the risk of an evolutionary dead end.

## 6 Non-deterministic Design: Gradualism versus Functionalism

In Chap. 6 of the 1872 edition of "On the Origin of Species," titled "Difficulties on Theory," Darwin dedicates a paragraph to "organs of extreme perfection and complexity," where he acknowledges the explanatory contradiction between gradualism and functionalism [1, 16]. According to Darwin, adaptation involves the construction of organs through a long series of continuous transformations and many intermediate stages. The difficulty of a selectionist argument lies precisely in justifying the adaptive value of these intermediate structures.

Heuristically, how can we reconcile Mannerism as a purely anti-functional drift with the harmony of Classicism, grappling with whether Mannerism represents a departure from or a continuation of classical perfection, since the inception of the debate fueled by historians like Cesare Brandi and Rudolf Wittkower [5]?

Although the final result of a design, understood in a broad sense and not limited to a single building, is easier to explain from a functional perspective, from a gradualist viewpoint, it cannot serve its ultimate function in its intermediate stages: five percent of a wing cannot be used for flying, just as a menhir cannot serve the purpose of a trilithon. According to George Mivart, the "problem of five percent of a wing" [14] is a non-issue because, as Pievani notes, such a partial structure never existed. Instead, species undergo abrupt, all-at-once changes driven by internal forces and following predefined structural plans [6].

However, Darwin could not accept such a saltationist explanation, which dismissed the gradual process of natural selection. The solution to this issue was found in the functional change hypothesis, which brought an important concept into evolutionary theory: organisms do not develop towards a specific form with future utility as a goal.

What is crucial is the continuity in differential reproductive success, which is the action of natural selection, rather than a continuous function of a single organ. The early stages of a structure must have provided a reproductive advantage to their possessors, an advantage that later evolved into a different benefit as conditions changed. If functions can change, there should not be a stable "one-to-one" correspondence between a structure and its function in evolution. Organs can have multiple functions, meaning they possess a redundant "intrinsic capacity." Different organs can perform the same function, allowing one organ to be repurposed for new uses if needed without compromising the organism's overall health (first principle of redundancy). Conversely, a single organ can perform several functions, some operational and others potential, ready to be utilized when necessary (second principle of redundancy).

In evolution, primary functions typically differ from the derived functions seen later. This means that the intermediate stages of current structures did not originally serve the functions they would eventually acquire. The function can vary radically without the form of the structure changing significantly. For example, as described by Pievani, the fin with a strong central axis, developed by some freshwater bottom-dwelling fish, was a not intentional starting point for subsequent movement onto land [6].

The continuity of morphological transformation through selection does not necessarily correspond to a continuous progression of function. Failing to acknowledge this would leave only two options: (1) denying that intermediate stages had any primary function (and thus were fixed by selection); or (2) postulating the sudden creation of a perfect final form. Both positions are equally unacceptable from an evolutionary standpoint.

While the comparison of architectural components to organisms is an accepted concept in design, the gradualistic and non-teleological nature of design represents a novel perspective in architectural literature. Accordingly, the trilithon can't be seen as a gradual structure that suddenly transitions from a metaphysical role to a technological one, and it was never designed except at a certain point in its evolution, after fulfilling other functions and continuing to do so, to become a perfect system of pillars and beams.

Cosistenly, a menhir served different purposes before evolving into a column. Even primitive dolmens and Bronze Age constructions shifted from symbolic and observational devices becoming structural systems, eventually integrating into a classical architecture “ordine". Throughout these transitions, architectures likely performed multiple functions simultaneously, often maintaining their original purpose. The transition was not necessarily abrupt; it can be assumed that the ancestral function continued to be performed even as the new function developed and eventually dominated. While the original structure effectively fulfilled its initial role, diverse adaptive pressures drove the evolution of technology in new directions, ultimately leading to the suppression of the original function.

## References

1. Darwin C (1872) On the origin of species (edition, Nov 24, 1859)
2. Gould SJ, Lewontin RC (1979) The spandrels of San Marco and the Panglossian paradigm: a critique of the adaptationist programme. Proc R Soc Lond, B 205:581–598
3. Gould SJ, Vrba E (1982) Exaptation, a missing term in the science of form. Paleobiology 8(1):4–15

4. Gould SJ (1977) Return of the hopeful monster; Eternal metaphors of palaeontology. In: Developments in palaeontology and stratigraphy, vol. 5. Elsevier, pp. 1–26.
5. Melis A (2023) Manierismo e decolonizzazione dell'ordine in architettura. Malfona and A. Crudeli, Manierismo. Pisa University Press, In L
6. Melis A, Pievani T, Lara Hernandez A (2024) Architectural expatation. Routledge
7. Melis A (2020b) La Scacchiera di Huxley e l'Architettura. In: Menichini D, Repetto D (Eds.), Panglissismo. Architetto Postpandemico Pacini Editore. https://www.pacinieditore.it/
8. Melis A, Medas B (2020) Tecnologie avanzate per la resilienza dell'architettura e della comunità. In: Bioarchitettura: Appunti per una città sostenibile Nardini Editore
9. Melis A, Melis G (1996) Architettura pisana dal Granducato lorenese all'Unita' d'Italia. ETS
10. Melis A (2023) Resilience. In De Finis G, Perin A (Ed). New words new world. Goethe Institute. Bordeaux
11. Melis A (2022) The Demiurgic position of architecture in times of environmental crises. In: De Finis G, Pecoraro C (Eds), Periferi@. Castelvecchi
12. Melis A, Pievani T (2022) Exaptation as a design strategy for resilient communities. In: Rezaei N. (eds) Transdisciplinarity. Integrated Science, vol 5. Springer, Cham
13. Melis A, Pievani T (Eds.) (2021) Architectural exaptation. Catalogo del Padiglione Italia. La XVII Biennale Venezia. Vol 1a. D Editore
14. Mivart SGJ (1871) On the genesis of species. Appleton, New York
15. Odling-Smee FJ, Laland KN, Feldman MW (1996) Niche construction. Am Nat 147(4):641–648
16. Pievani T (2024) La teoria dell'evoluzione
17. Pringle H (2013) The origins of creativity. Sci Am 308(3):36–43

# The Design of Diversity: Case Studies

# 2

Jose Antonio Lara-Hernandez, Alessandro Melis,
and Paola Boarin

**Abstract**

In the section dedicated to the Design of Diversity, "The Dusk of Design" explores the significance of incorporating diversity as a central design consideration in architectural practice. Recognizing the multifaceted benefits of diverse environments, this section highlights how diverse architectural designs can enhance social, cultural, and ecological aspects of the built environment. By incorporating diverse elements, materials, and spatial configurations, architects can create inclusive and vibrant spaces to celebrate differences and promote a sense of belonging. Through a collection of case studies from different regions of the world, this section explores architectural and urban projects that embrace diversity to foster social interactions, cultural exchanges, and ecological sustainability. By showcasing these diverse design approaches, "The Dusk of Design" encourages architects to consider diversity as an essential component of their design strategies, ultimately enriching the built environment and creating spaces that reflect the multifaceted nature of human experiences.

**Keywords**

Diverse architectural design • Urban planning • Social inclusion • Cultural exchange • Ecological sustainability

J. A. Lara-Hernandez (✉) · P. Boarin
Auckland University of Technology, New Zealand, USA
e-mail: Antonio.lara-hernandez@aut.ac.nz

P. Boarin
e-mail: p.boarin@auckland.ac.nz

A. Melis
New York Institute of Technology, New York, USA
e-mail: amelis@nyit.edu

P. Boarin
School of Architecture and Planning, University of Auckland, Auckland, New Zealand



J. A. Lara-Hernandez et al. (eds.), *The Dusk of Design*, Integrated Science 34,
https://doi.org/10.1007/978-3-031-76867-5_2

## 1 Introduction: Embracing Diversity in Architecture and Urban Design

In the intricate mosaic of global urbanism, diversity emerges not merely as a demographic characteristic, but also as a crucible for innovation in architecture and urban design. This section deeply probes the complex interplay between diversity and spatial practices, positing that embracing diversity is not solely a moral imperative but also a functional necessity for enlightened urban development [18], Chap. 1). Through an amalgamation of diverse scholarly perspectives and detailed case studies, this introduction meticulously sets the stage for a nuanced exploration of how diversity influences and is reciprocally shaped by the disciplines of architectural and urban planning.

Urban landscapes, marked by their sprawling diversity, serve as fertile grounds where sociocultural, economic, and environmental differences catalyse the evolution of space and community dynamics. There are several examples across the globe in which cities exemplify a rich spectrum of human interaction, where diversity is more than just an incidental backdrop, rather it is a dynamic force moulding the built environment. Thus, the vibrancy and resilience of urban life are defined. Avoiding the monofunctional or over-designed of urban spaces is increasingly recognized as crucial for creating environments that are not only physically accessible, but also culturally resonant and socially inclusive [12].

The conscious integration of diversity in architectural and urban design challenges conventional planning paradigms, urging a shift towards innovative methodologies that reflect the complex realities of urban populations. Such an approach requires a departure from traditional design orthodoxies, advocating for creative and inclusive strategies that accommodate and celebrate the multifaceted nature of urban life [4]. This shift is strategic and essential, enabling cities to adapt to the dynamic flux of urban demographics and evolving needs of their inhabitants.

Furthermore, embracing diversity is pivotal for addressing some of the pressing contemporary challenges in urban development, such as social inequality, spatial justice, and ecological sustainability [3]. It offers a robust framework through which cities can be designed (and/or redesigned) to foster environmental resilience and social equity, ensuring that urban development is inclusive and beneficial to all societal segments. Here, diversity functions as a bridge linking disciplines to foster a holistic approach to urban challenges, rich in innovation, and effective in execution.

As we go into the intricate relationship between diversity and urban design, it becomes evident that the successful melding of varied cultural, economic, social, and ecological elements into architectural practice is not just advantageous, but essential. History has shown that environments crafted under this paradigm are richer in experiential quality and more efficacious in functionality, exemplifying how a diversified approach can lead to more comprehensive and enduring urban solutions [18], Chap. 4).

This section introduces endeavours to unravel the intricate nature that showcases the multifaceted, multidisciplinary, and multiscalar role of diversity in sculpting contemporary urban landscapes. An in-depth examination of global case studies and theoretical discussions illuminates the transformative potential that diversity holds within the architectural and urban planning professions. By doing so, it aims to inspire a reimagined vision of urban development that is deeply informed by and actively engages with the rich tapestry of human diversity.

## 2 Historical Overview

Architecture and urban planning have traditionally fluctuated between the adoption of global solutions and the endorsement of culturally specific local expressions [6], in most cases at the expense of fully embracing diversity. For instance, the modernist movement of the 20th-century has irrefutably shown a preference for uniformity and predictability, standardizing urban landscapes and even whole cities. Such historical inclinations have regularly marginalised the chaotic yet vibrant realities of human diversity [8, 9, 16], favouring semblance of control over the multifaceted nature of what used to be urban life.

This section aims to illustrate examples moving from top-down and hegemonic practices that often ignore diverse cultural expressions to a nuanced and fragmented understanding of urban space. These case studies reflect the broader socio-political shifts brought about by global migration patterns and evolving multicultural tapestries of some contemporary cities. It explores how significant historical movements, particularly postcolonial urban developments and the ascendance of global cities, have acted as catalysts to bring diversity issues to the forefront of the architectural and urban planning discourse. For instance, taking inspiration of Barad's work [2], Greg Mews in Chap. XX introduces the novel term Flux Cities as a conceptual tool aiming to pinpoint sporadic bursts of activities that reveal adaptive design capacities within cities, where fragments contribute to the ongoing production of the 'real' in cities and different thinking about our spatial/material cultural practice pro-life is being tested.

In doing so, it becomes apparent how deeply ingrained practices in architecture and planning have evolved. Ancient cities, with their rigid hierarchies and structured layouts, often mirrored prevailing power dynamics, leaving little room for the incorporation of diverse perspectives. In contrast, the upheavals of the twentieth century—marked by decolonisation and the subsequent reconfiguration of global power structures—introduced new dynamics into urban development. These changes challenged the established norms and opened up avenues for diverse and previously marginalised [13, 17], voices that influence urban planning and architectural design.

Furthermore, this section highlights the role of significant urban theorists and practitioners, who have been instrumental in advocating for a more inclusive approach to urban design. Their work has often focused on how urban spaces can be reimagined to better reflect the diversity of their populations, not just in terms of cultural background, but also in accommodating varying socioeconomic statuses, gender identities, and even age groups.

By charting these historical shifts, this narrative underscores the importance of acknowledging and incorporating diversity as a central component of architectural and urban planning. The evolution from a monocultural approach to the recognition of multicultural realities illustrates a broader trend towards inclusivity and sustainability in shaping future urban environments. This historical perspective not only enriches our understanding of past practices but also provides a critical foundation for anticipating and crafting future developments in this field. Thus, the examination of historical trajectories in architecture and urban planning is not merely academic, but serves as a vital tool for practitioners aiming to create more equitable and vibrant urban spaces.

## 3 Conceptualising Diversity

Defining diversity within the realms of architecture and urban design necessitates adopting a broad perspective that includes sociocultural, economic, and ecological dimensions. Within this context, diversity is perceived not merely as an indicator of demographic variation, as is commonly understood, but as a pivotal element guiding the philosophy of design [18], Chap. 1). This section outlines frameworks from a diverse range of theoretical perspectives, advocating for an inclusive understanding of urbanism.

These frameworks are helpful in understanding the intricacies involved in designing (and redesigning) environments that are culturally diverse environments, as in the case of global cities. They aim to offer architects, urban designers, and planners the opportunity to create spaces that are functional, aesthetically appealing, and culturally sensitive. By incorporating these diverse perspectives, urban spaces can be transformed to support a wider range of social interactions and to foster a sense of community among various population groups.

Moreover, a discussion on these frameworks is necessary; therefore, we aim to highlight the critical role that diversity plays in driving innovation within the field of urban design. It posits that by embracing varied cultural insights and drawing from a wide spectrum of human experiences, designers can break new ground by creating flexible, adaptive urban spaces that respond dynamically to the needs of their inhabitants. Such spaces are better equipped to withstand socioeconomic shifts and environmental challenges, thereby enhancing their sustainability and resilience.

This exploration of theoretical frameworks enriches our understanding of the crucial role of diversity in shaping the constantly changing urban landscapes. It encourages both practitioners and scholars to reconsider traditional approaches to urban design and architecture through the lens of exaptation, ensuring that the cities of tomorrow are built on principles of diversity, equity, and environmental stewardship.

## 4 Current Practices

The practical application of diversity in architecture and urban design offers fertile ground for in-depth analysis. This section illustrates contemporary practices in which diversity has catalyzed innovative design solutions, drawing upon examples from the book Chapters of this section, which illustrate how various regions have integrated cultural, economic, and ecological diversity into their urban planning processes.

For instance, Chap. 10 "Enabling Flexibility and Residents' Agency in Shaping Social Housing: Case Studies from South America" by Priscila Besen examines social housing projects in South America, critiquing the traditional inflexibility of design, and showcasing alternative approaches that offer adaptability and resident agency. It highlights projects such as incremental housing by Elemental in Chile and progressive housing in Brazil, illustrating how these initiatives empower residents and enable the personalisation of living spaces over time.

Chapter 4 "The City Through Her Eyes" by Barbora Melis explores the intersection of gender and urban design, emphasizing the need for a gender-sensitive approach to urban planning. Pointing out how urbanism affects women's experiences in urban spaces, advocates for designs that increase safety, accessibility, and inclusivity.

Similarly, in the context of public spaces, Adebara in Chap. 18 utilised evolutionary theories to understand people's underlying motivations and preferences when interacting with their environment. Identifying the traditional uses of public squares, such as rituals and gatherings for ceremonies, reflects long-standing needs for social cohesion and well-being in Nigerian communities.

Following the same vein of informality but more towards housing, Veronica Natividade, this Chaper explores the unique context of the favelas in Brazil through the lens of exaptation. Highlighting them as an example of self-organization, and unsupervised in impoverished architecture and urban fabrics, which underscores the urgent need to rethink material flows within our socioeconomic system.

Lastly, Sofia Ayora in Chap. 5 "The Mayan Solar" illustrating the Mayan domestic space as a cultural blending, an exaptation that, in addition to spatial and functional aspects, includes cultural expression in a new form. Thus, an example of human creativity promotes the preservation of material, cultural, and symbolic structures without necessarily seeking perfection. This illustrates the Mayan dwellings as an architectural artifact of great resilience to cultural challenges over time, both in the colonial period and in modernity.

## 5 Challenges and Barriers

Despite the growing recognition of its importance, integrating diversity into architectural practices poses substantial challenges. First, institutional biases, economic constraints, and sociopolitical and cultural resistance profoundly shape the landscape of diverse urban development. The Chapters in this section elucidate some of these barriers and illuminate how they manifest across different cultural and economic contexts.

Institutional biases often stem from long-standing practices and norms within architecture and urban planning professions [11]. These biases can subtly influence decision-making processes, favoring traditional designs and resisting innovative approaches that embrace diversity. For instance, zoning laws and building codes in many cities may inadvertently reinforce segregation by economic class, ethnicity, and stifling efforts to create more integrated and diverse communities [1, 7, 10].

Economic constraints also play a critical role in shaping urban development. The high cost of land in urban centres can limit opportunities for developing diverse, inclusive spaces, pushing lower-income populations to the periphery [20, 21]. This economic barrier not only affects residential architecture, but also limits the diversity of uses within public spaces, often prioritizing commercial interests over communal or cultural uses [14, 15].

Sociopolitical resistance to diversity can manifest in various forms, from community opposition to specific developments to broader political or policy environments that do not prioritize diversity. This resistance can be particularly potent in contexts in which there is a lack of understanding or appreciation of the benefits of diverse urban environments [19]. Political reluctance or hostility towards policies promoting social integration in housing and public spaces further complicates the realization of diverse urban forms.

Policy and regulatory frameworks are crucial in either facilitating or hindering the development of diverse urban forms. For instance, well-conceived policies can support diversity initiatives by incentivizing inclusive housing developments or mandating diverse community engagement in the planning process [5]. Conversely, outdated or narrowly focused policies can significantly obstruct these efforts by embedding exclusionary practices into the fabric of urban development.

Additionally, this section explores examples from various global contexts to illustrate these practical challenges. For instance, in some European cities, heritage preservation policies may unintentionally hinder the development of culturally diverse urban environments by overly prioritizing historical aesthetics over contemporary needs and inclusivity. Similarly, in the North American context, the legacy of redlining and suburban sprawl continues to influence urban planning decisions, often to the detriment of creating more diverse and integrated urban spaces.

## 6 The Role of Policy and Governance

Urban policy and governance significantly influence the integration of diversity into architectural outcomes, as underscored by the various Chapters in this book. By examining the interaction between policy initiatives and spatial design, we can understand how government policies support and impede the development of diverse urban environments. Key themes, such as zoning laws, public housing policies, and urban renewal initiatives, are explored in this context, highlighting their critical role in shaping the inclusivity of urban spaces.

For instance, as described in Chap. 10, zoning laws can profoundly affect urban diversity. Besen's analysis of incremental housing projects, such as Elemental in Chile, illustrates how flexible zoning policies that accommodate growth and personalization can foster community resilience and empowerment. These projects exemplify how thoughtful zoning can facilitate housing that adapts over time to meet the changing needs of residents, thus supporting a diverse and dynamic urban fabric.

Chap. 6 by Veronica Natividade, focused on progressive housing models in Brazil, demonstrates how public housing policies can foster inclusivity or perpetuate segregation. By allowing for expansion and personalization, these models challenge traditional rigid approaches to social housing. They showed that public housing can be a tool for social integration if designed with adaptability and resident involvement in mind, directly contrasting with more restrictive policies that limit residents' ability to shape their living environments.

If we turn now to urban renewals, Barbora Melis, in Chap. 4, touches on the dual nature of urban renewal initiatives. It explores how these projects, while potentially revitalizing neighbourhoods, can also lead to gentrification, displacing long-established communities, and reducing diversity. This highlights the need for governance that is acutely aware of and responsive to the socioeconomic dynamics at play, ensuring that renewal efforts enhance rather than diminish urban diversity.

Therefore, the roles of public policy and urban governance in shaping diverse urban architectural outcomes are profound. Ideally, policies should be thoughtfully considered and inclusively formulated to ensure that they contribute positively to the diversity of urban environments. Despite the uniqueness of each urban environment, urban planners and policymakers can better design interventions that promote inclusivity and diversity by learning from successful and less-successful policies, as highlighted in the Chapters. This understanding is crucial for developing urban spaces that are not only diverse, but also vibrant and resilient, reflecting the wide array of human experiences and cultures that populate modern cities.

## 7 Conclusion

As the intricacies of urban dynamics continue to evolve, the urgency to weave diversity into the fabric of our cities rises. This section offers insights from a variety of historical, theoretical, and empirical perspectives, making a compelling case for a shift towards more inclusive and versatile architectural practices. By embracing the multifaceted challenges and opportunities presented by diversity, architects, urban designers, and planners are uniquely positioned to not only enhance the aesthetic and functional attributes of their projects, but also drive the creation of equitable and sustainable urban futures.

The glance presented here underscores the transformative power of diversity as a quinte-essential feature in shaping urban spaces. By integrating diverse perspectives from the outset, urban projects can achieve deeper relevance to the communities they serve, fostering environments in which all inhabitants can thrive.

Moreover, the discussion highlighted the crucial role of policy and governance in facilitating or impeding the incorporation of diversity into urban design. Without supportive frameworks and proactive governance, efforts to integrate diversity can be significantly hampered. Therefore, it is imperative for policymakers, alongside architects and planners, to collaboratively dismantle existing barriers and create environments that enable diversity to flourish.

Looking forward, the path towards inclusive urban development demands not only creativity and innovation but also a steadfast commitment to embracing diversity at all levels of planning and execution. As urban planners and architects look to the future, they must continuously seek and apply lessons learned from around the globe, adapting and evolving practices to meet the dynamic needs of diverse populations.

In summary, the call to action is clear: to cultivate cities that reflect and respect the varied mosaic of human experience. This requires dedicated and sustained efforts to challenge conventional norms and champion diversity, as an essential principle of urban development. The vision articulated in the present book aims to inspire professionals across the urban realm spectrum to imagine and construct spaces that advocate for the diverse societies of the future. This is not just an opportunity to transform urban landscapes but also a profound responsibility to foster communities that are robust, vibrant, and inclusive for generations to come.

## References

1. Adelman RM, Mele C (Eds.) (2015) Race, Space, And Exclusion: Segregation And Beyond In Metropolitan America. Routledge
2. Barad K (2007) Meeting the Universe Halfway: quantum physics and the entanglement of matter and meaning (Kindle ver). Duke University Press. https://doi.org/10.2307/j.ctv12101zq
3. Beebeejaun Y (2022) Whose diversity? race, space and the European city. J Urban Affairs. https://doi.org/10.1080/07352166.2022.207526
4. Bodnar J (2015) Reclaiming public space. Urban Studies 52(12):2090–2104. https://doi.org/10.1177/0042098015583626
5. Boros L, Fabula S, Horváth D, Kovács Z (2016) Urban diversity and the production of public space in Budapest. Hung Geogr Bull 65(3):209–224. https://doi.org/10.15201/hungeobull.65.3.1
6. Brenner N (2000) The urban question as a scale question: reflections on Henri Lefebvre, urban theory and the politics of scale. Int J Urban Reg Res 24(2):361–378. https://doi.org/10.1111/1468-2427.00234
7. Caldeira TPR (1996) Building up walls: the new pattern of spatial segregation in Sao Paulo. Int Soc Sci J 48:55
8. Gehl J (2011) Life between buildings. Island Press
9. Jacobs J (1961) The death and life of great American cities. Random House
10. Johnston R, Forrest J, Poulsen M (2002) Are there ethnic enclaves/ghettos in English cities? Urban Stud 39(4):591–618. https://doi.org/10.1080/0042098022011948

11. Khayesi M, Monheim H, Nebe JM (2010) Negotiating "Streets for All" in Urban transport planning: the case for pedestrians, cyclists and street vendors in nairobi, Kenya. Antipode 42(1):103–126. https://doi.org/10.1111/j.1467-8330.2009.00733.x
12. Lara-Hernandez JA (2022) Cities : how urban design can make people less likely to use public spaces. The Conversation, 1–5. https://theconversation.com/cities-how-urban-design-can-make-people-less-likely-to-use-public-spaces-184079
13. Lara-Hernandez JA (2023) Temporary appropriation of public spaces as an indicator of social resilience. In: Melis A, Lehmann S (Eds.), Comunità Resilienti: Una Raccolta di Brevi Saggi Sulla Capacita' Adattiva delle Citta 2:163–173. D Editore. https://www.comunitaresilienti.com/
14. Leary-Owhin M (2015) A fresh look at lefebvre's spatial triad and differential space: a central place in planning theory? 2nd Planning Theory Conference University of the West of England, pp 1–8
15. Low SM (2006) The erosion of public space and the public realm: paranoia, surveillance and privatization in New York city. City and Soc 18(1):43–49. https://doi.org/10.1525/city.2006.18.1.43
16. Madanipour A (2014) Urban design, space and society. Palgrave Macmillan
17. Melis A, Pievani T (2022) Exaptation as a design strategy for resilient communities. In: Rezaei N (ed) Integrated science: transdisciplinarity across the different disciplines. Springer Nature, pp 307–327
18. Melis A, Pievani T, Lara-Hernandez JA (2024) Architectural Exaptation: When Function Follows Form. Routledge
19. Orsman B (2013) Not in my back yard. New Zealand Herald. https://www.nzherald.co.nz/nz/not-in-my-back-yard/FHLY4UFR23ZTATLNK5KUN5FZMA/
20. Ricci M (2021) The periphery does not exist: or about the need to be radical in architecture. In: Carta M, Perbellini M, Lara-Hernandez JA (Eds.). Resilient Communities and the Peccioli Charter - Towards the possibility of an Italian Charter for Resilient Communities. Springer Nature
21. Southworth M, Owens PM (1993) The evolving metropolis–studies of community, neighborhood, and street form at the urban edge. J Am Plann Assoc 59(3):271–287. https://doi.org/10.1080/01944369308975880

# On Designing Exteriority from within Through Flux Cities in Public Space

3

Gregor Mews

**Abstract**

Our evolution is an ongoing story that contains diverse spatial practices and dynamic everyday life relational connected far beyond the human. It has reached a decisive point where the metabolic complexities of cities with their underlying interdependencies are at an evolutionary turning point in their becoming. On the one hand, there is an emerging collective awareness about the state of the human and more-than-human world(s) and the impacts of our cumulative spatial practices. This is reflected in popular rhetoric and actions to counteract biodiversity collapse, resource grabbing, economic uncertainty, housing unaffordability, the increasing global greenhouse gas emissions at the planetary scale, displacement in light of the conflict and the rise of technological innovation related to artificial computation (Anderson et al. in Dialogues Hum Geogr 2:212–215, 2012; Davis (2022), Ernstson and Swyngedouw (2019) Politicizing the environment in the Urban Century. In Ernstson and Swyngedouw (eds), Urban Political Ecology in the Anthropo-obscene: Interruptions and Possibilities pp 4. Routledge.; Fry in remaking cities: an introduction to urban metrofitting, Bloomsbury Publishing, 2017; Morton in Environ Plan D Soc Space 33:831–849, 2015; Morton (2016). Dark ecology: For a logic of future coexistence. Columbia University Press.; Morton (2017) Humankind: Solidarity with non-human people. Verso Books.; Sasaki in cities 27:S3–S9, 2010; Swyngedouw New Geogr 1:20–27, 2017). On the other hand, less conscious practices are prevalent in diverse activities within cities, for example, when considering the perception of safety, small everyday life deeds, spatial practices of underprivileged population groups, or playful encounters in public spaces that we often take for granted. They impact people and other animate objects, such as insects or birds, live in cities and shape

G. Mews (✉)
Co-Lead Bioclimatic and Sociotechnical Cities Lab & Lecturer in Urban Design and Town Planning, University of the Sunshine Coast, Sippy Downs, Australia
e-mail: gmews@usc.edu.au

 
J. A. Lara-Hernandez et al. (eds.), *The Dusk of Design*, Integrated Science 34,
https://doi.org/10.1007/978-3-031-76867-5_3

the production of the 'real' in unforeseen and unpredictable ways. However, both aspects are two sides of the same coin that cast a shadow that relationally entangles them. From an ecological perspective, (Gould and Eldredge in Models Paleobiol 1972:82–115, 1972; Gould and Eldredge in Nature 366:223–227, 1993) suggest that the evolution of life occurs in swift spurts instead of following the slow and steady path that Darwin initially introduced. Long periods of stasis with little activity in terms of extinctions or the emergence of new species are interrupted by intermittent bursts of activity; they called it a punctuated equilibrium (Gould and Eldredge in Models Paleobiol 1972:82–115, 1972; Gould and Eldredge Nat 366:223–227, 1993). To pinpoint sporadic bursts of activities that hold adaptive capacities within cities, this chapter follows a line of flight where fragments of these casted shadows contribute to the ongoing production of the 'real' in cities. The idea of Flux Cities will function as the method to search for possible emerging turning points that bring together different accounts of events related to public space. The chapter concludes with a reflection on the role of working with Flux Cities from a post-structural perspective to enable different ways of creating knowledge related to punctuated equilibrium, which mainstream social science and urban disciplines have largely ignored.

**Keywords**

Flux cities • Spatial practices • Urban dynamics • Punctuated equilibrium • Ecological awareness

## 1 Introduction

Our planetary health is at a decisive turning point as the human impact on the earth system is becoming increasingly evident and rendered as a new geological epoch—the Anthropocene [35, 44]. A meta-crisis associated with our collective impact at the planetary scale is increasingly recognised within urban policy discourse. Rhetoric leaves traces across our societies and focuses on solutions thinking to counteract biodiversity collapse, pandemics, resource grabbing, economic uncertainty, increasing global greenhouse gas emissions at the planetary scale, mental health crisis and the rise of technological innovation related to artificial computation in diverse ways [2, 11, 14, 16, 28–30, 34, 37]. They all demand a change in how humans design the environment in the broadest sense, seek to exist, and can continue to exist [38]. At the same time, more than ever in human history, people choose to live in urban areas as a way of life [39]. For the purpose of this argument, there are two fundamental underlying assumptions. First, those urban areas are predicted to continue to grow in the foreseeable future, meaning that population growth will continue equally as the need for resource grabbing to sustain spatial practices designed to accommodate the human needs within such areas in diverse, dynamic and unequal ways. Second, popular rhetoric regarding urbanisation in most Western Societies continues to favour dualist and reductionist thinking of decision-makers, focusing on reactive and politically motivated short-term solutions to maintain a modus operandi that potentially undermines the intent of good holistic public policy. For example, the production of nature under neoliberal paradigms aimed at human habitation and their spatial material affects are uneven, problematic, and evident in urban experiences across the globe. After all, we are the only species that consistently exceeds their resource capacity, commonly called Earth Overshoot Day, highlighting that we consume our ecological capital faster than what is available to sustain a balanced earth system. In 2023, our limit was reached on the 2nd of August compared to the 25th of December in the year 1971 [21]. In addition, cities addiction to economic growth led to a doubling since 2000 to US$66.2 trillion in 2010 and increased to US$84.7 trillion in 2020 [43] while at the same time, urban inequality increased by more than 70 per cent of the world's population. [40] latest World Cities Report stresses that we are facing a twin crisis that includes climate change and the loss of global biodiversity, threatening the futures of cities, posing the question: Why are we obsessed with rescuing capital instead of nurturing life at all scales? Fortunately, scholars such as [16] or Morton remind us that anthropocentric superiority thinking needs to become redundant in order to sustain life for humans on Earth's surface. At the same time, our technology and research methods to understand the dynamic impacts of such spatial practices is becoming more refined and nuanced. However, despite the new approaches, we continue to downplay the hyper-complexity, ethical dilemmas and decision-making processes concerning stewardship of life. To date, contemporary Australian spatial practices and the production process of cities have not fully reconciled that humans are only one part of nature and depend on it. At the same time, the indigenous relationship with the country remains undervalued and underappreciated. Subsequently, how we currently modify external materiality and design our cities demonstrates a distressing level of cognitive dissonance about our adaptive capacity and the co-production of the 'real'.

With such a level of awareness, we cannot ignore research on human emotions embedded in the underlying ethical practices to realise a duty of care and a moral obligation to rethink, undo and generate new ways we design environments for pro-life. We are again at a decisive crisis point of rapid and random transformation after a prolonged period of stasis? From an ecological perspective, [19, 20]) originally suggested that the evolution of life occurs in swift spurts instead of following the slow and steady path that Darwin initially introduced, they called it a punctuated equilibrium [19, 20]. To identify evolutionary and transformative pro-life activities within our urban areas that hold the potential to overcome the meta-crisis, I shall explore intermittent bursts of seemingly random activities that emerge from long periods of stasis and are found in the public spaces in cities.

For the purpose of this chapter, I shall borrow the concept as a starting point for a journey aimed at pinpointing sporadic bursts of activities that reveal adaptive design capacities within cities. Hence, this chapter follows a line of flight where fragments contribute to the ongoing production of the 'real' in cities and different thinking about our spatial/material cultural practice pro-life is being tested. The novel concept of 'Flux Cities' will function as the method for pinpointing sporadic bursts or emerging turning points found in different accounts of events in public spaces of cities. I call them metaphorically 'shadow casting' of the 'real'. This approach is a conceptual tool that instead of overly relying on modification efforts of materiality in the exteriority considers equally the 'exteriority from within' [5]. In other words, the reader actively participates in the ongoing production process of the 'real' by engaging with this text while being acutely aware of their own ethico-onto-epistemological position [5, 10]. Emergent concepts such as Flux Cities, have a role to play to advance understandings of the 'real' in cities by examining approaches that embrace complexity embedded in a larger circle of creativity after crisis. Creativity is central to this inquiry for two reasons. First, it does not prioritise established research approaches, which may dimmish our capacity for creation. Second, even in the process of creation through words and things remains always part of a larger assemblage [12] that contributes to the production of the 'real'. For [12], 'thoughts lag behind nature' suggests that nature and culture are deeply entangled within the material world and cannot be understood as two fully separate realms. Creative concepts such as Flux Cities embrace relationality with dynamic, ever-changing, self-regulating and open systems at heart that can be found in all life forms, including people, trees, insects, birds and materiality in cities that shape the production of the 'real' in unforeseen and unpredictable ways. Following the shadow casting of two very different events in public space, the chapter concludes with a reflection on the role of working with Flux Cities as a post-structural method to enable different ways of creating new knowledge to pinpoint punctuated equilibrium in the evolution of cities, which mainstream social science and urban disciplines have largely ignored.

## 2 Flux Cities

Flux Cities is inclusive of multiplicities of more-than-human 'realities' and conducive to ontological perspectives of readers that seek to challenge ways of structuring urban systems aimed at creating order, certainty, control or security while contesting the moving notion of reality, characteristics and parameters that co-produce our contemporary 'cities'. In its most simplistic form, Flux Cities is a consciously ambiguous milieu[1] of matter and an ongoing form-giving process. As a process, it encompasses multiplicities of space in an ever-changing state of flux. It contains coherently emerging properties of a loosely energic state of entanglement in mo(ve)ment directed towards what our human minds conceive as a 'city'. Flux Cities differ from established definitions of the 'city' precisely as it continues to escape concrete and reducible forms of articulation. At the same time, it integrates existing definitions of 'cities' as a naturally occurring and directly

[1] A Milieu can be understood as the middle of a relational field [12, 25], p.164, /2021, p.22).

observable fact established through human activities in specific social, environmental, economic, and cultural functions. In this context, Flux Cities is different to metaphysical processes. Within the ambiguous milieu of matter mo(ve)ment is a key form as it functions as the opening into Flux Cities. Here, movement and moment come together with one event, reminding us that perceivable matter is situated in a milieu. If we pay attention to what moves in any given moment (ongoing change), new insights and appreciation of the world(s) becoming can be obtained. The performative aspect of Flux Cities becomes evident in the present when mo(ve)ment is evident in events produced by contradictions, juxtapositions, or ambiguities within inseparable interactions of objects and social structures of cities. Events that have mo(ve)ment are the openings I am looking for when applying Flux Cities. The concept emphasises that each moment is casting a fragmented shadow through movement. These shadows can be captured and traced through an ever-growing array of research methods. Under a state of coherence—the shadows can be perceived as 'stasis' and produce the illusion of certainty in what we might conceive as 'real'. Nevertheless, they remain relational entangled even under the state of coherence. Rationale and formal conclusions drawn from research of shadows are valid as long as they acknowledge situatedness due to their reductionist approach to 'reality' and the ongoing and iterative open-ended nature of worlding [18]. After all, matter in its core is non-reducible, interconnected and holds indeterminacy as it is always more than the sum of its parts [41]. Gamble et al. [18] highlight that matter is never inherently an active (subject) or a passive agent (object), but rather passive and active at the same time, confirming that its unevenness is always relational and immanently self-caused [18], 127). Flux Cities is a connecting point between the internal and external worlds that Barad [5] described in her work as 'exteriority from within'.

## 3 Mo(ve)ment Events in Cities

The early post-structuralist [41] embraced the term 'events', pointing towards an endless cascade of events comprising the discursive material affects of nature and culture that produce matter in which perceptions of human history and feelings are included. As cities are marvellous products of events in the broader story of human civilisation, they also face complex challenges associated with crisis, evident in the impacts of rapidly changing climate conditions, biodiversity collapse, resource grabbing, economic uncertainty, housing unaffordability, and displacement through geopolitical conflict at the planetary scale. Readily available planning and design instruments are geared towards working with control and predictability to ensure stability, seemingly unable to resolve the challenge as nature remains unpredictable and random. Here, Flux Cities comes into play as it shifts the paradigm towards working with rather against the unpredictability and randomness of nature. For example, more-than-human techniques and methods enrich the theoretical discourse by decentring the anthropocentric perspectives. This type of decentred approach to human activities acknowledges the role of non-human elements in shaping urban environments [37] and accommodates the possibility of embracing an ecological awareness that is underpinned by a new duty of care. Such duty of care is found in a new cognitive architecture in which conceptual division of nature and culture on life in urban places are obsolete [6, 9, 32]. The following accounts of two different events in public spaces in Berlin, Germany, and Melbourne, Australia, focus on a more-than-human perspective by applying Flux Cities as a method to pinpoint sporadic bursts of activities under the state of life 'crisis' that hold clues about adaptive capacities at different scales.

## 4 Crisis Event One: Climate Activism in Public Space

Transgressive forms of public space appropriation are one type of observable spatial practice that includes conscious acts aimed at shifting the perception of publics in relation to distinct human activities, the dominating body politics within a specific context [26, 36]. Such public space appropriations can be temporal [24] in nature and, therefore, align with the mo(ve)ment of Flux Cities. They generate a coherent entanglement in spatial material practices with human cognition that can even be disconnected from place. Through such bursts of activities, the stasis of 'reality' as an external product is being raptured, and an opening pointing towards punctuated equilibrium in our process of becoming is generated. The following event captures fragmented shadows of a milieu of matter associated with climate activism in Berlin, Germany, on the 21st of December during the pre-Christmas celebration period in 2022. In response to the climate emergency and the lack of meaningful political commitment to significantly slow down the release of greenhouse gas emissions into the atmosphere, a climate activist group called Last Generation was generated by a selected group from the incoming generation of young people. In this case, their task was to cut off the top of the Christmas tree positioned at the iconic Pariser Platz in front of the Brandenburg Gate Deutsche [13]. Instead of appropriating the public space over a prolonged period, the group decided to engage in an informal activity designed to modify materiality that has much attached emotional value to shift discourse as part of people's everyday life in cities. The Pariser Platz is a prominent historically significant public space in Berlin featuring the iconic Brandenburg Gate, the American and French Embassy, the Academy of the Arts, Banks and the Adlon Hotel all near the Reichstag, the German parliament. In this instance, two activists came on a transportable lifting platform around 9.20 am to cut off the top of the 15-m-tall, decorated Nordmann fir tree (see Fig. 1).

To underpin their political motivation, they revealed a banner with the following message: "This is just the tip of the Christmas tree." Pointing at the dilemma that Lilli Gomez, highlighted in formal statement: "*So far we're seeing only the tip of the underlying disaster in Germany.*" (Deutsche [13]). Further, the activists noted, "*While all of Germany spends the week getting the best gifts from the biggest stores, others are wondering where they will get their water to drink after droughts*

CLIMATE | GERMANY

**Climate activists cut top off iconic Berlin Christmas tree**

12/21/2022

**Last Generation climate activists calling for tougher government action to combat the global climate crisis sawed the top off a Christmas tree at the Brandenburg Gate. Police said they attended the scene and took action.**



**Fig. 1** Decapitated Christmas tree at the Pariser Platz in Berlin (Source: https://www.dw.com/en/climate-activists-cut-top-off-iconic-berlin-christmas-tree/a-64175003)

*and floods have destroyed their crops. Climate collapse in Germany is just around the corner and the federal government is making no effort to protect us.*" Deutsche [13]. It is an ancient German tradition to erect pine trees to celebrate Christmas. This form of climate activism in public space impacted diverse groups of the German public in the increasingly ubiquitous and individualised realm of the internet. Figure 2 contains accounts of the rich emotional/rational spectrum responses to this transgressive public space appropriation.

It highlights that movement in cognitive behaviour is entangled with emotion as well as facts-driven rationale, while the moment even created ripple effects in the internet social media space that shapes and therefore impacts much of the social production process of the 'real'. The more-than-human, in this case, digital technology, functions as a tool that, in dynamic, random and unforeseen ways, reaches diverse users. The fragmented shadows casted in Fig. 2 reveals contradictions and highlights that the desired impact was achieved. In a sense, what matters here is that conceptual dualism was overridden, and this uncomfortable burst of activities demonstrates the current capacity and (ability) of diverse publics to question our own spatial practices, critically interrogate their own ontology and ethics of acting differently to enact a sense of stewardship pro-life. Here, the design intent of materiality comes into play, as collective awareness of the cultural richness of such public space in the capital city of Berlin functions as a foundational catalyst to enable the impact of such activity when materiality is being modified to challenge unconscious everyday life practices.

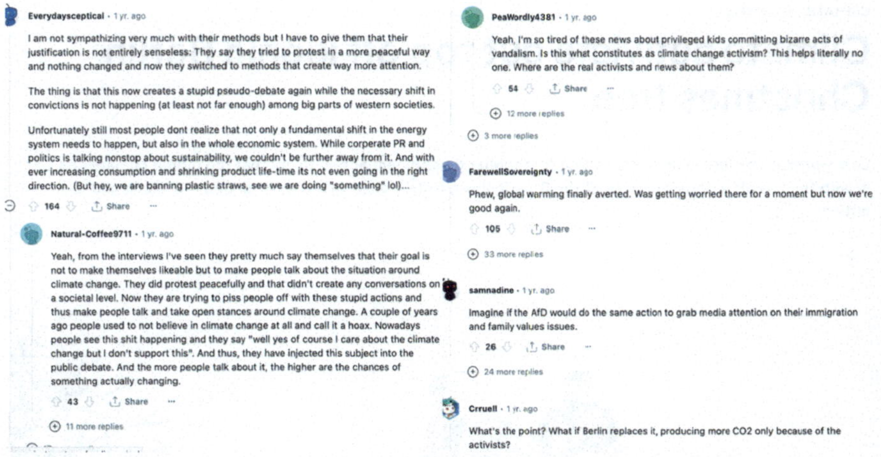


**Fig. 2** Emotional social media response to the activity (Source: Reddit [r/europe])

## 5 Crisis Event Two: Play and Risk in Public Space

Even in a resource-rich society such as Australia, everyday playfulness is subject to spatial restriction, often carefully designed, curated, and even subjugated [25]. Within such culture play is widely socially accepted for children, even though play occurs in adults as well as other living beings Huizinga [8, 27]. At the same time, play is fundamental to our physical and mental health and can, as an activity, occur in all public spaces [27]. Alexander, Frohnlich and Fusco [4], as well as [33], highlight that play is mostly instrumentalised as a means to promote physical health, even though it does so much more while it continues to escape reductionism [42]. Informal and non-instrumental forms of public space appropriations, such as play, have much potency to generate diverse opportunities for social cohesion in highly urbanised areas where social isolation and loneliness are prevalent [3]. Within the urban policy discourse, they often need to be more appreciated [27]. These impacts are evident regarding policies and design related to risk at various scales. For example, at the playground scale, [17] points out that the culture of risk aversion and the level of control reduces and even compromises children's ability to develop into capable and healthy adults. Subsequently, many playgrounds in Australia lack the sense of challenge that children can develop a sense of risk competency through cognitive, social and emotional capacity, which in turn builds internal resilience, and they can deal with life's challenges in creative ways. At the same time, the fear of injury claims from caregivers, councils across Australia often react and aim to reduce risks regarding playground design, management, and policies that impact children in diverse ways (reference). The following accounts relate to a crisis within the conceptual assemblage of risk perception, health and social well-being in response to dominant spatial practices and cultural norms.

The event is about a novel spatial material design approach related to the public playground design of "Rocks on Wheels" in Naarm (Melbourne). At the same time a piece of conceptual art that questions the status quo and challenges the use across all ages. The physical aspects, such as the material used, the process of creation as well as the use of such space, contribute to mo(ve)ment in the user who decides what risky element to use and what level of perceived risk is being challenged.

The playground, designed by Mike Hewson, contains 24 rough-hewn bluestone boulders located at the Southbank Boulevard in the heart of the city next to a green space (see Fig. 3).

In addition, slides, robes, swings, and monkey bars connect the boulders of a seemingly self-assembled structure. Intriguing small features such as miniature animals, upcycled toy cars, garden hose and other items enrich the diverse sequences and experiences with a sense of adventure and discovery. The fallout surfaces around the boulders are an important safety feature, but in this case were designed unconventionally. Robes seemingly hold monkey bars together; bluestone appear pavers but are, in fact, soft—fall mats. These are important optical absurdities that have consciously been integrated into the composition in order to challenge risk perceptions. As shown in Fig. 4, despite the real look and feel, the boulders appear to be lightweight as they sit on reinforced dollies underneath them.

In November 2022, the key features sparked much media attention and concern around Australia and internationally as the playground was perceived as too risky.

**Fig. 3** Rocks on dolleys and small design features (Source image: Mike Hewson https://mikehewson.co.nz/2022/11/rocks-on-wheels)

**Fig. 4** Buckets on the rocks (Source image: Mike Hewson https://mikehewson.co.nz/2022/11/rocks-on-wheels)

In actuality, the design process was rigorous engineered and detailed so that the playground is safe and satisfies the Australian Safety Standards. The design is a testimony to Melbourne's reputation of being a design and arts city open to experimentation. The affects that the spatial, material and behavioural assemblage generated at the moment of opening cast diverse shadows and impacted users and designers in dynamic and unpredictable ways. The urban design, art and landscape architecture disciplines praised the project as a much-needed human response to raise awareness of the risk-averse culture and the sensitive integration of natural elements (Fig. 5). The accounts of such material composition and spatial practices can evoke a level of discomfort when facing the obvious contradictions, such as large rocks on seemingly movable dolleys that could hardly hold the structural composition together (Fig. 6). In contrast to standard product playgrounds and dominant cultural norms, this design is an obvious juxtaposition.

Hewson's artistic eye and skill played with ambiguities of what elements may hold together or break when approached (Fig. 6). Caregivers are inevitably internally challenged regarding the level of control over their children in this place. The only way forward is by developing trust and surrendering to the soundness of the design. Each moment when the user engages with such assemblage, a risky life-invocating process takes place. Diverse emotional responses can touch other users and, therefore, entangle them temporarily with each other. They remind me that we are inevitably always part of a larger whole where interactions of objects and social structures of cities remain arguably inseparable from each other.

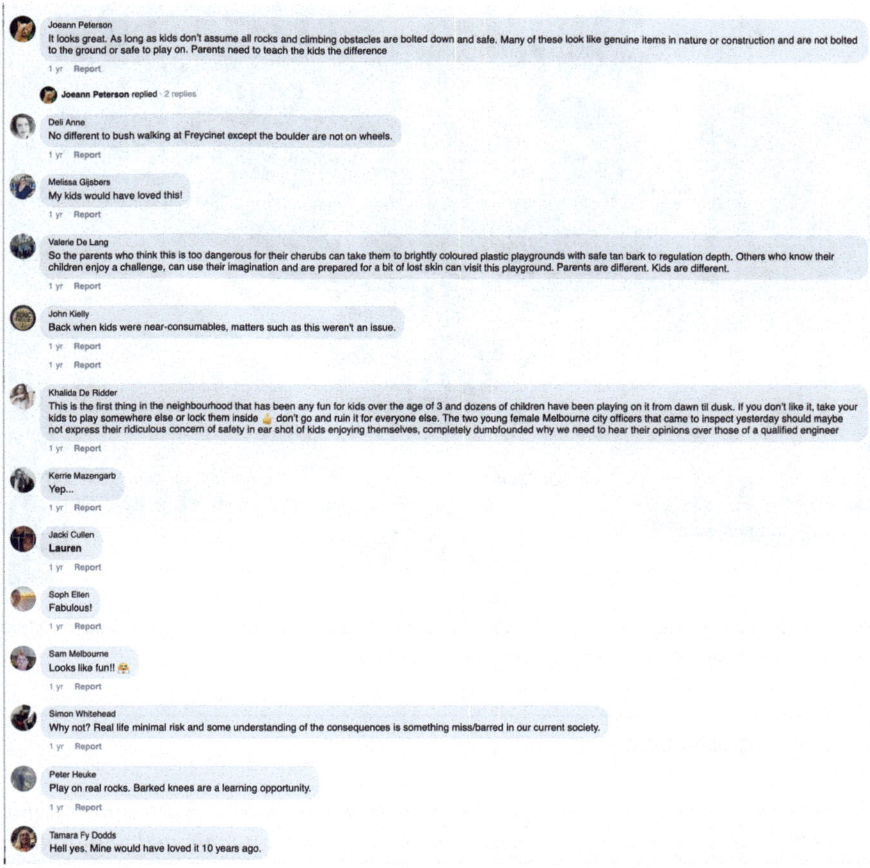


**Fig. 5** Social media responses on playground design and risk (Source: ABC News https://m.facebook.com/72924719987/posts/this-new-playground-on-wheels-has-raised-eyebrows-among-parents-but-the-local-go/10163885206539988/ Accessed on 23/12/2023)

**Fig. 6** Children climbing at Rock on Wheels playground in Melbourne (Source image: Mike Hewson https://mikehewson.co.nz/2022/11/rocks-on-wheels)

## 6 Conclusion

*"Life does not merely operate at the level of individual organism (a presumed stable being) but emerges from the self-organising capacities of organic and non-organic materials to co-create novel formations, a continuous and fluid state of becoming".* [25], p.71.

Both events outlined accounts related to spatial material practices in public spaces of two cities. From a post-structural perspective, Flux Cities was applied as a creative method to pinpoint accounts of sporadic bursts of activities that hold adaptive capacities for the becoming of our cities conducive to life, human and non-human.

Lester's insights about everyday play indicate that such bursts cannot be found at an individual level in static isolation but must be understood in the fluid state of becoming. Therefore, starting a social inquiry with a concept rather than established research methods is worthwhile. Furthermore, Flux Cities is useful as it does not attempt to represent events fully, as there are no methods available that generate 100 per cent truth. Instead, openness, fluidity and complexity are embraced. They imply that social 'realities' continue to rely on creative narratives to produce the 'real' that considers 'exteriority from within' [5]. Working with Flux Cities, such complex assemblages, when journeying into both events, invite the reader to venture into a creative cognitive space. The sporadic form-giving process of the events, in this case, the transgressive public space appropriation of climate activities in Berlin, happened in various planes where self-organising capacities and sporadic form-giving processes are in mo(ve)ment at any given moment. At a meta-level, there is something inherently interconnected at play, in this case, social media related to place.

The Melbourne event had a different focus. The complex assemblage between organic (human and non-human life) and non-organic materials, nature and spatial practice was closely examined. Again, the fluid state of becoming was evident in the event. Organic actors interact with non-organic material, taking positions on social media (Fig. 5). Both are equally needed in order to open us up to the idea of embracing a different cognitive architecture in design. Being confronted with perhaps uncomfortable situations either in Berlin or in Melbourne event, contributes to an increased level of awareness to question cultural norms and, therefore, our relationship with the nature of things at large while casting shadows into this moment. Even the process of presenting both events together in Flux Cities contribution creates and binds previously unrelated crisis events together in a hyper-complex production process of the 'real'. While accurate situatedness and description of boundary conditions are necessary to render a coherent journey, the notion of control over nature through cultural practices remains embedded within an irreducible part of a larger whole that is and continues to be in mo(ve)ment. We must acknowledge that emotion and evidence-based facts contribute equally to the production of the 'real', no matter if it is part of the pitching of an idea to realise such risky playground art in Melbourne or the impact diverse opinions on the internet can have about a small transgressive intervention in a place that is attached to high emotional value during a time of year during a season that is all about the emotion and lived values geared towards compassion and love towards the living. Both events are testimony to a new cognitive design architecture at work. The open-ended possibility to redesign our cognitive, relational architecture due to produced facts related to the state of global climate or child health and risk harnesses the adaptive and imaginary capacities of the human mind in random and unforeseen ways. Even if they bring out the uncomfortable feelings in the human condition, they are necessary as we urgently need to activate new ways of perceiving and acting in a more-than-human paradigm, what Haraway [22] calls worlding. Such worlding integrates the nurturing capacity of randomness, also found 'in' punctuated equilibrium, and 'of' nature through us at any given mo(ve)ment. James Bride suggests that *"Randomness assigns value to everything and everyone it touches by giving each participant equal weight: everything is equally valuable [...] Each and every thing matters; everyone matters."* [7], p. 249). I find that his observation transfers agency to the cognitive design architecture of Flux Cities as it prescribes value to randomness and value to everything related, it might generate uncomfortable situations but produces a capacity for mo(ve)ment that can be turned towards pro-life when curiosity and criticality are imbued.

## References

1. ABC News (2023). This new playground on wheels has raised eyebrows among parents—but the local government says it's an example of a "risk play park", which is actually "very safe". https://m.facebook.com/72924719987/posts/this-new-playground-on-wheels-has-raised-eyebrows-among-parents-but-the-local-go/10163885206539988/
2. Anderson B, Kearnes M, McFarlane C, Swanton D (2012) Materialism and the politics of assemblage. Dialogues Human Geogr 2(2):212–215
3. Aelbrecht P, Stevens Q (2019) Public space design and social cohesion: an international comparison. Routledge
4. Alexander SA, Frohlich KL, Fusco C (2014) Playing for health? Revisiting health promotion to examine the emerging public health position on children's play. Health Promot Int 29(1):155–164. https://doi.org/10.1093/heapro/das042
5. Barad K (2007) Meeting the universe halfway: quantum physics and the entanglement of matter and meaning. Duke University Press
6. Bennett J (2010) Vibrant matter: a political ecology of things. Duke University Press
7. Bridle J (2022) Ways of being: animals, plants, machines: the search for a planetary intelligence. Penguin UK
8. Brown S, Vaughan C (2010) Play: how it shapes the brain, opens the imagination, and invigorates the soul. Scribe, Vol 1
9. Coole D, Frost S (2010) Introducing the new materialisms. New materialisms: Ontology, agency, and politics, 1–43
10. Cozza M, Gherardi S (2023) Posthuman feminism and feminist new materialism: towards an ethico-onto-epistemology in research practices. In: Handbook of feminist research methodologies in management and organization studies. Edward Elgar Publishing, pp 55–71
11. Davis J (2022) The caring city: ethics of urban design. Policy Press
12. Deleuze G, Guattari F (1987) A thousand plateaus. Bloomsbury (first published by University of Minnesota Press)
13. Deutsche W (2022) Climate activists cut top off iconic Berlin Christmas tree. Deutsche Welle
14. Ernstson H, Swyngedouw E (2019) Politicizing the environment in the Urban Century. In: Ernstson H Swyngedouw E (eds), Urban political ecology in the anthropo-obscene: interruptions and possibilities. Routledge, pp 4
15. Fry, T. (2012). Becoming human by design. A&C Black.
16. Fry T (2017) Remaking cities: an introduction to urban metrofitting. Bloomsbury Publishing
17. Gill T (2007) No fear: growing up in a risk averse society. Calouste Gulbenkain Foundation
18. Gamble CN, Hanan JS, Nail T (2019) What is new materialism? Angelaki 24(6):111–134. https://doi.org/10.1080/0969725X.2019.1684704
19. Gould SJ, Eldredge N (1972) Punctuated equilibria: an alternative to phyletic gradualism. Models Paleobiol 1972:82–115
20. Gould SJ, Eldredge N (1993) Punctuated equilibrium comes of age. Nature 366(6452):223–227
21. Global Footprint Network (2023) Earth overshoot day. National Footprint and Biocapacity Accounts 2023 Edition. Retrieved 19/12/2023 from https://www.overshootday.org/newsroom/past-earth-overshoot-days/
22. Haraway DJ (2016) Staying with the trouble. Duke University Press
23. Hewson M (2023) Rocks on wheels. Mike Hewson Melbourne. https://mikehewson.co.nz/2022/11/rocks-on-wheels
24. Lara-Hernandez JA, Coulter CM, Melis A (2020) Temporary appropriation and urban informality : Exploring the subtle distinction. Cities 99:102626
25. Lester S (2020) Everyday playfulness: a new approach to children's play and adult responses to it. Jessica Kingsley Publishers
26. McAuliffe C, Iveson K (2011) Art and crime (and other things besides…): Conceptualising graffiti in the city. Geogr Compass 5(3):128–143

27. Mews GH (2022) Transforming public space through play. Routledge
28. Morton AD (2015) The warp of the world: Geographies of space and time in the Border Trilogy by Cormac McCarthy. Environ Plan D Soc Space 33(5):831–849. https://doi.org/10.1177/0263775815604919
29. Morton T (2016) Dark ecology: for a logic of future coexistence. Columbia University Press
30. Morton T (2017) Humankind: solidarity with non-human people. Verso Books
31. Reddit [r/europe] (2022) Iconic Berlin Christmas tree decapitated by climate activists [social media post]. politico.eu. https://www.reddit.com/r/europe/comments/zshqkb/iconic_berlin_christmas_tree_decapitated_by/?rdt=51680
32. Robertson SA (2018) Rethinking relational ideas of place in more-than-human cities. Geogr Compass 12(4):e12367
33. Ryall E, Russell W, MacLean M (2013) The philosophy of play. Routledge
34. Sasaki M (2010) Urban regeneration through cultural creativity and social inclusion: rethinking creative city theory through a Japanese case study. Cities 27:S3–S9. https://doi.org/10.1016/j.cities.2010.03.002
35. Steffen W, Grinevald J, Crutzen P, McNeill J (2011) The Anthropocene: conceptual and historical perspectives. Philos Trans R Soc Math Phys Eng Sci 369(1938):842–867
36. Stevens Q, Dovey K (2020) Appropriating the spectacle. In: Activating Urban Waterfronts. Routledge. pp 115–131
37. Swyngedouw E (2017) More-than-human constellations as immuno-biopolitical fantasy in the urbicene. New Geogr 1:20–27
38. United Nations (2017) New Urban Agenda Secretariat HI, (ed) [Outcome document]. UN-Habitat. http://habitat3.org/wp-content/uploads/NUA-English.pdf
39. United Nations. (2019). World Urbanization Prospects: The 2018 Revision.https://population.un.org/wup/Publications/Files/WUP2018-Report.pdf
40. UN-Habitat (2022) World cities report 2022. Retrieved 19/12/2023 from https://unhabitat.org/sites/default/files/2022/06/wcr_2022.pdf
41. Whitehead AN (1942) Adventures of ideas. CUP Archive
42. Wittgenstein L (2009) Philosophical investigations Hacker PMS, Schulte J, Trans.; 4th (ed). John Wiley & Sons
43. World Bank (2020) GDP (current US$). World Bank Retrieved 19/12/2023 from https://data.worldbank.org/indicator/NY.GDP.MKTP.CD
44. Zywert K (2017) Human health and social-ecological systems change: rethinking health in the Anthropocene. Anthropocene Rev 4(3):216–238. https://doi.org/10.1177/2053019617739640

# Enhancing Women's Experiences in Urban Public Spaces

4

Barbora Melis

**Abstract**

Dusk, the transitional phase between day and night, symbolically connects to women's experiences and needs in urban public spaces, especially in design. In the urban landscape, dusk often marks the shift from bustling daytime activities to a quieter, sometimes dimly lit environment. For women, this temporal juncture can capture a complex intersection of safety concerns, empowerment, and the search for inclusive spaces. As the sun sets and the city transforms, this transition can be both liberating and constraining for women in public spaces. On one hand, the relative anonymity provided by the dimming light may offer a sense of freedom, allowing women to move through the urban environment with a degree of privacy. On the other hand, diminishing visibility can also amplify feelings of vulnerability, raising concerns about personal safety. In the context of women's experiences, dusk extends beyond the literal fading of sunlight, becoming a metaphor for the nuanced challenges and opportunities in the urban environment. Design is decisive in the ongoing evolution of urban spaces, adapting to better suit community needs. This includes creating spaces that address the specific requirements and experiences of diverse groups, including women. Densely populated, with a diverse community and strong cultural identity, the South Bronx is an example of a shifting urban context from a predominantly industrial and deprived area, segregated from wealthy Manhattan, through recent urban and social regeneration processes to a new urban paradigm within the city's urban fabric. Considering the present urban dynamics, it represents therefore an appropriate context for the exploration of the women's perspectives on the public spaces in the neighbourhood analysing their ability to favour (or not) women's inclusion. The metaphorical dusk invites us to reconsider the narratives surrounding women's presence in public spaces challenging

B. Melis (✉)
University of Portsmouth, Portsmouth, UK
e-mail: barbora.melis@port.ac.uk

 
J. A. Lara-Hernandez et al. (eds.), *The Dusk of Design*, Integrated Science 34,
https://doi.org/10.1007/978-3-031-76867-5_4

traditional norms and stereotypes, and advocating for a society where women can navigate urban landscapes with the same sense of ownership and agency as their male counterparts. Design becomes a key tool in shaping a city that embraces the diversity of its inhabitants, ensuring that the urban environment reflects the needs and aspirations of all, regardless of gender.

**Keywords**

Urban design • Safety • Inclusivity • Gender equity • Public spaces

## 1 Introduction

Dusk, the transitional phase between day and night, symbolically connects to women's experiences and needs in urban public spaces, particularly in urban design. In the urban landscape, dusk often marks the shift from bustling daytime activities to a quieter, sometimes dimly lit environment, encapsulating a complex intersection of safety concerns, feelings of empowerment, access to amenities, and the search for inclusive spaces for women.

As the sun sets and the city transforms, this transition can be both liberating and constraining for women. On one hand, the relative anonymity provided by the dimming light may offer a sense of freedom, allowing women to move through the urban environment with a degree of privacy. On the other hand, diminishing visibility can also amplify feelings of vulnerability, raising concerns about personal safety.

In the context of women's experiences, dusk extends beyond the literal fading of sunlight, becoming a metaphor for the nuanced challenges and opportunities in the urban environment. Urban design plays a decisive role in transforming public spaces, adapting to better suit community needs. As dusk signifies a transition, it mirrors the phase where new methodologies and tools must be developed to assess diverse needs and address inclusivity issues in urban design. Despite growing awareness, there remains a gap in understanding how design choices through a gender lens can foster innovative approaches that accommodate the nuances shaping women's interactions with their surroundings.

To bridge this gap, the contribution will explore specific urban contexts that exemplify the complexities of inclusivity in public spaces. One such context is the South Bronx, a densely populated neighbourhood in New York City with a diverse community and strong cultural identity. Historically, the South Bronx has undergone significant transformations, presenting an opportunity to explore women's perspectives on public spaces within the neighbourhood and analyse their ability to foster inclusivity.

Building on these insights, a methodological framework is proposed that reflects women's perspectives and creative agency. Based on literature and preliminary fieldwork in the case study area, this contribution suggests a methodology that can be employed in further research, combining observation, photo-elicitation interviews and focus groups with women, which can drive innovation, particularly in urban inner-city threshold areas. This methodological approach aligns with the principle of "Leave no one behind", a call for collective action to address systemic inequalities and create a fairer society. This principle is crucial in achieving the United Nations' Sustainable Development Goals (UN SDGs), which emphasise inclusivity and equal access to opportunities and resources for all individuals, regardless of their background or circumstances [28].

The metaphorical dusk invites us to reconsider narratives surrounding women's presence in public spaces, challenging traditional norms and advocating for a society where women can navigate urban landscapes with the same sense of ownership and agency as their male counterparts. Urban design thus becomes a crucial tool in shaping cities that embrace the diversity of their inhabitants, ensuring that public spaces as integrated components of the urban fabric reflecting the needs and aspirations of all, regardless of gender [9].

The urban experiences of women and how the built environment conditions these experiences are often overlooked by professionals in the field. This lack of diversity in the voices shaping urban design contributes to and sustains gender inequity. Recognising and valuing the diverse needs, experiences, and challenges faced by different genders is essential. Incorporating a wide range of perspectives, particularly those of women, into urban design is imperative for fostering inclusive and equitable cities [2, 19]. Ensuring these diverse voices are heard and integrated into planning processes helps create public spaces that reflect and serve the entire population.

This contribution highlights the need for creativity as a methodological tool to measure urban resilience, emphasising the ability to adapt to unexpected change during global crises and unprecedented urbanisation growth as an essential prerogative for developing future urban planning strategies [31]. Drawing on scholarly and practice-based experiences, it demonstrates how thoughtful design can address the critical issues of safety and inclusivity for women in urban spaces.

## 2 The Symbolism of Dusk

Dusk, the brief period between sunset and nightfall, represents more than just a transition in the sky; it symbolises a broader shift in the urban environment and its impact on daily life. As the natural light fades, the dynamics of public spaces change dramatically. This transition period encapsulates a mixture of beauty and apprehension, creating a unique atmosphere that profoundly influences human behaviour and perception.

For women, dusk holds particular significance as it often marks the end of the typical workday and the beginning of evening activities, when many navigate public spaces on their way home to social gatherings or other engagements. This duality of dusk, evoking tranquillity and unease, mirrors the complexities of women's experiences in urban environments. Women face unique challenges while navigating cities in their multifaceted roles as caregivers, workers, and community members. The diversity of their experiences informs how they perceive and use cities, creating disparities between the physical layout of urban spaces and the gendered perspectives of safety and navigation [1, 30].

As dusk settles, urban spaces undergo a noticeable transformation. The decreasing visibility can turn familiar paths and public areas into shadowy, ambiguous environments. For women, this change often heightens awareness and caution. The interplay of light and shadow can obscure potential threats, making it challenging to estimate one's surroundings accurately. Street harassment, assaults, and other safety concerns are not uncommon, particularly in poorly lit or less populated areas [27].

Studies have shown that women are more likely than men to alter their behaviour in response to safety concerns. This includes avoiding certain areas, modifying routes, or travelling in groups. The fear of violence or harassment during dusk can significantly impact women's mobility and freedom, constraining their ability to engage fully in urban life. The psychological toll of constant vigilance and the need for self-protection strategies contribute to an ongoing sense of unease [8].

Conversely, dusk can also offer a sense of empowerment and anonymity. The diminishing light can provide a degree of privacy, allowing women to navigate urban spaces without the scrutiny that daylight often brings. This relative anonymity can be liberating, offering a respite from societal expectations and judgments. For some women, dusk represents a time of increased autonomy and confidence. The ability to blend into the semi-darkness can foster a sense of control and independence, encouraging exploration and engagement with the city [15].

The duality of the experience, the simultaneous presence of freedom and vulnerability, highlights the complex interplay of factors influencing women's experiences during dusk. Understanding the symbolism of dusk and its impact on women's experiences is pivotal for developing inclusive urban design strategies that prioritise safety and empowerment, fostering more welcoming and inclusive environments for all residents.

Including women's perspectives represents a starting point for a fairer urban planning process, facilitating a paradigm shift that transcends gender issues [14, 32]. In this context, "dusk" serves as both a metaphor and a repository of knowledge, encapsulating the nuanced challenges and opportunities inherent in designing public spaces and inspiring the exploration of new methodologies to address diverse needs and implement inclusive design measures, thereby paving the way for a more equitable urban landscape.

## 3 The Role of Urban Design

Urban design has historically shaped the experiences of city dwellers, affecting daily routines, work habits, mobility, and recreation. The interconnection between planning and design can perpetuate existing inequities, as these fields have historically been male-dominated, resulting in a one-sided perspective on the urban realm [11, 34]. Traditional urban design focused on facilitating commerce, transportation, and industrial growth, frequently overlooking the specific needs of women and leading to environments prioritising efficiency and functionality over safety and inclusivity, highlighting the disparity between women's daily experiences and our built environments [5, 17]. This historical context is crucial to understanding women's challenges in urban public spaces today, particularly during dusk.

Despite urban planning advancements, many cities struggle with inclusivity and safety in public spaces, particularly for women. Women frequently report feeling unsafe in poorly lit areas, inadequately maintained parks, and isolated pathways, challenges exacerbated during dusk when reduced visibility heightens vulnerability. Urban spaces often lack amenities catering to women's needs, such as accessible restrooms, nursing facilities, and safe transportation options. This contributes to a pervasive sense of exclusion and insecurity that limits their access to and participation in urban life.

The urban interactions of women in various roles and responsibilities inform how the city is perceived as a whole and in specific components like public space. Falu' & Sassen [7] outline how urban public spaces fail to meet women's needs, citing time fragmentation due to their diverse daily duties. They argue that women's use of public space, often with children, disabled individuals, and seniors during daylight hours, exposes neglect in recognising and addressing women's reproductive and productive roles in urban planning. This oversight translates into fewer rights, especially for poorer women, regarding differing women's needs for housing, employment, safety, political participation, autonomy, work access [6].

Moreover, studies in Canada and the UK identify common design factors that make women feel unsafe. These include entrapment risks along pedestrian routes, isolation in multi-storey parking garages, reduced visibility due to inadequate lighting and signage, and "blind corners." Neglecting signs such as poor maintenance and graffiti further mark locations as vulnerable to assault [33].

Applying a gender lens in urban design involves recognising and addressing women's unique experiences and needs. This approach advocates for inclusive design principles that consider safety, accessibility, and comfort for all genders. Incorporating gender perspectives can lead to innovative solutions that enhance the overall quality of urban spaces. For instance, well-lit streets, open sightlines, and diverse transportation options can significantly improve women's sense of safety and freedom of movement. Integrating public amenities such as childcare facilities and secure waiting areas can make urban environments more welcoming and inclusive Micklow et al. [20].

UN-Habitat states that the lack of gender-disaggregated data in urban transformation processes substantially impacts the lives of women and girls, affecting their

"Right to the city." Achieving gender equality in urban planning and empowering women and girls to contribute to sustainable development fully are pivotal steps for improving human health, fostering resilience, and protecting the environment [29]. Robust methodologies, such as photo-elicitation interviews and focus groups, are essential for effectively incorporating gender considerations into urban design.

Photo-elicitation interviews involve using photographs to prompt discussions and deeper insights from participants [18]. This method allows women to visually document their experiences and highlight specific areas of concern or appreciation in their urban environment. The visual nature of this approach helps uncover subtle aspects of daily life that might be overlooked in traditional interviews. Participants can share stories and emotions connected to the images, providing richer and more nuanced data for urban designers to consider [24].

Focus groups bring together small groups of women to discuss their experiences and perspectives on urban spaces. These discussions can reveal common themes and diverse viewpoints, offering a comprehensive understanding of the issues. Focus groups encourage interaction and dialogue, enabling participants to build on each other's ideas and generate new insights. This collaborative process can uncover practical solutions and design recommendations that might not emerge from individual interviews.

Combining photo-elicitation interviews with focus groups can provide a holistic view of women's experiences in urban spaces. This integrated approach allows for individual reflections and collective discussions, enriching the data and offering multiple layers of insight. Urban designers can use these methodologies to gather detailed information on specific areas, identify patterns, and develop targeted interventions that address women's needs and concerns.

Applying these methodologies in real-world contexts is crucial for validating their effectiveness and generating actionable insights. For instance, conducting a case study in the South Bronx, an area with diverse demographics and unique urban challenges, can provide valuable lessons in designing inclusive public spaces. By engaging women in the community through photo-elicitation interviews and focus groups, urban designers can better understand local needs and develop tailored solutions that enhance safety and inclusivity during dusk. According to Richard [26], this technique results in empowerment and promotes participants' agency in the interviewing process, making the interview seem more participant-directed.

The South Bronx is one such area where this methodological approach can be effectively applied. With its diverse community and strong cultural identity, this densely populated neighbourhood in New York City provides a rich context for exploring women's perspectives on public spaces. Historically, the South Bronx has undergone significant transformations, presenting a unique opportunity to analyse how these changes impact inclusivity and safety for women in urban environments.

## 4 Case Study—South Bronx

Located in New York City, the South Bronx is a densely populated neighbourhood with a rich cultural heritage and a history of resilience and hardship. It is home to a diverse population, including a significant proportion of Hispanic and African American residents, contributing to a strong sense of community identity and presenting unique challenges in addressing the varied needs of its inhabitants.

The South Bronx has undergone significant urban and social regeneration processes, evolving from a historically low-profile and deprived area once segregated from wealthy Manhattan [13]. Formerly synonymous with urban decay and crime, it now exemplifies shifting urban dynamics within New York City [23]. Known for its vibrant culture and diverse community, the South Bronx provides an ideal context for examining the complexities of inclusivity in urban public spaces.

Despite extensive redevelopment efforts to revitalise the neighbourhood, challenges persist in shedding its rugged image amidst industrial lots and housing projects lacking essential amenities [21], and ensuring safety, accessibility, and inclusivity, particularly for women. The South Bronx exemplifies urban revitalisation efforts to overcome the stigma of social and economic marginalisation. The area's ongoing redevelopment offers a valuable perspective on the intersection of urban design and women's experiences. Historically marked by economic and social changes, recent transformations have improved infrastructure and amenities while highlighting ongoing inequalities and safety concerns [3]. These dynamics underscore the necessity of actively involving the community in all phases of neighbourhood redevelopment and transformation processes.

From the initial fieldwork activities, the relevance of the visual impact immediately emerged, highlighting missed opportunities in the design process to make the South Bronx more responsive to the community's needs. Initial discussions with both community members and experts clearly indicated the necessity of finding a common language to discuss women's perspectives that can be both visual and textual. Utilising photo-elicitation interviews and focus groups, the research offers the opportunity to examine women's experiences in the case study area.

Utilising photo-elicitation interviews and focus groups, the research will offer the opportunity to examine women's experiences in the South Bronx consistently with the aim of gathering detailed insights into local spatial conditions and their impact on women, informing more inclusive urban planning guidelines and tools. Considering the paradigmatic case of the South Bronx, these combined methodologies can also offer valuable insights into transferable urban planning practices, fostering more equitable, inclusive, and socially conscious approaches to urban development processes.

## 5 Methodologies for Inclusive Urban Design

Developing inclusive urban design requires a deep understanding of the lived experiences and perspectives of women who navigate these spaces daily. Traditional design approaches often overlook these nuances, creating environments that may inadvertently perpetuate inequality and discomfort. As mentioned in the previous paragraph, this section explores innovative methodologies prioritising women's voices, providing valuable insights to drive more inclusive and responsive urban design, as emerged from initial activities carried out in the South Bronx.

In fact, photo-elicitation interviews enhance our understanding of women's experiences in urban public spaces, with results later verified through focus groups. These findings are then identified and categorised using photographs as potential indicators affecting women's experiences in these spaces and organised into a photographic report. This method is used as a visual research tool to structure "articulated visual statements" of the community and to generate "explanatory models" for analysis and interpretation during the following interviews [22].

Photo-elicitation interviews are a powerful tool for capturing women's experiences in urban spaces. This methodology involves showing researcher-generated photographs of the public space of the case study area and then discussing these images in interviews. The visual component helps to elicit deeper reflections and emotions that might not emerge through verbal descriptions alone [12].

The researcher shows a selection of photographs that capture different aspects of the public space. These can include areas they find empowering and safe or places they avoid due to safety concerns. The pictures are prompts during the interviews, guiding discussions about their experiences and perceptions. The visual nature of this method allows for a more nuanced understanding of how women interact with and perceive their environments. It also empowers participants by giving them prompts over what aspects of their surroundings to highlight. This can lead to richer, more detailed data that captures the complexity of their experiences. In urban design, insights gained from photo-elicitation interviews can inform the development of physical and psychological spaces attuned to women's needs. For example, understanding the specific elements that make a space feel safe or unsafe can guide the placement of lighting, seating, and pathways.

Complementing photo-elicitation, focus groups provide a collaborative environment where women can share their experiences and ideas about urban spaces. This method fosters a sense of community and allows participants to build on each other's insights, leading to a deeper and more comprehensive understanding of the issues at hand.

Focus groups typically consist of 6–10 participants and are facilitated by a researcher. The discussions can be structured around specific themes, such as safety during dusk, access to amenities, or overall feelings of empowerment in public spaces. Facilitators can use prompts and scenarios to stimulate conversation and ensure that all voices are heard. The group dynamic encourages participants to articulate their thoughts and feelings in ways that might not emerge in one-on-one interviews [10, 25]. It also provides a platform for women to validate each other's experiences, which can be particularly powerful when discussing sensitive issues like safety and harassment. According to Longhurst [16], data and results from the photo-elicitation interviews discussed in the focus group can highlight common themes and divergent perspectives, offering a broad view of women's experiences. This information is invaluable for urban designers looking to create inclusive environments. For instance, if a common concern is inadequate lighting in certain areas, this can be addressed in design plans.

During the data analysis phase, key themes, patterns, and discrepancies in women's experiences. Can be identified. Such analysis unveils underlying issues and potential solutions that may not be readily apparent through traditional research methods alone.

Exploring women's interactions with public space requires a comprehensive approach aligned with the triangulation methodology in qualitative research, which integrates multiple methods or data sources, such as observation, photo-elicitation interviews, and focus groups, to understand the phenomenon thoroughly Creswell [4]. By triangulating these methods, researchers cross-verify data from diverse sources, thereby enhancing the validity and robustness of their analysis. Specifically, combining photo-elicitation interviews and focus groups within specific urban contexts allows for a nuanced examination of women's experiences. A case study approach further enriches this approach, providing detailed insights into particular context-specific areas that are broadly applicable to urban settings. Integrating these research methods is pivotal in capturing the complexity of women's experiences in public spaces, enabling researchers to gather multifaceted data that captures both individual perspectives and collective experiences. This holistic approach deepens our understanding of how specific design elements influence women's perceptions and behaviours within urban environments.

## 6 Data Analysis

For this study, the South Bronx has emerged as an ideal site due to its richly diverse community and significant history of urban transformation. In selecting case study sites, choosing diverse urban areas that reflect varying demographic, cultural, and socio-economic backgrounds is crucial in ensuring a comprehensive range of insights.

Regardless of improvements, safety remains a critical issue in the South Bronx, particularly at dusk. Many women feel vulnerable in poorly lit or secluded areas, which affects their mobility and access to public spaces. Access to essential amenities, such as public transportation, parks, and community centres, varies widely across the South Bronx. This uneven distribution can exacerbate feelings of exclusion and limit opportunities for social interaction and engagement. Efforts to involve residents in the planning and development have had mixed success. While some initiatives have successfully incorporated community input, others have failed to address women's needs and concerns.

The present research has consistently employed a combination of photo-elicitation interviews and focus groups to gain a deeper understanding of women's experiences in the South Bronx. These methodologies have provided rich, qualitative data highlighting the complexities of navigating urban spaces. Photographs of different aspects of the public space served as the basis for in-depth interviews, allowing participants to articulate their experiences and perceptions in detail. Complementing the individual interviews, focus groups provided a collaborative platform for women to discuss their experiences and share insights. These discussions revealed common themes and unique perspectives, offering a holistic view of the challenges and opportunities in the South Bronx.

The data from photo-elicitation interviews and focus groups reveals key themes and insights for understanding women's experiences in the South Bronx.

1. **Safety concerns**: A predominant theme across all participants was the concern for personal safety. Women highlighted specific areas where inadequate lighting, lack of visibility, and limited foot traffic contributed to feelings of vulnerability.
2. **Empowerment and community**: Despite these challenges, many women expressed a strong sense of community and empowerment in certain parts of the South Bronx. Public spaces that were well-lit, maintained, and actively used by the community were often considered safer and more welcoming.
3. **Design preferences**: Participants provided valuable feedback on design elements that could enhance their sense of safety and inclusion. Suggestions included better lighting, increased presence of security personnel, more accessible public amenities, and spaces designed to encourage social interaction.

The insights gained from this case study highlight the importance of incorporating women's perspectives into urban design processes. By addressing the specific safety concerns and preferences identified by participants, urban planners can create physically safer and more psychologically welcoming environments.

4. **Lighting and visibility**: Improving lighting in key areas can significantly enhance women's sense of safety and mobility. Strategic placement of streetlights, along with the use of brighter and more energy-efficient lighting, can make a substantial difference.
5. **Community spaces**: Designing public spaces that encourage community interaction and engagement can foster a sense of belonging and empowerment. This includes creating parks, plazas, and recreational areas that are accessible, well-maintained, and actively used by the community.
6. **Inclusive design**: Incorporating feedback from diverse groups, particularly women, into the design process ensures that urban spaces reflect the needs and aspirations of all residents. This can involve participatory planning workshops, surveys, and ongoing dialogue with the community.

The case study of the South Bronx underscores the critical role of inclusive urban design in enhancing women's experiences in public spaces during dusk. By utilising methodologies that prioritise women's voices, urban planners can develop targeted strategies to address safety concerns and promote inclusivity. The lessons from the South Bronx can be applied to other urban contexts, contributing to the broader goal of creating safe, welcoming, and inclusive cities for all.

## 7 Discussion: Designing Inclusive Urban Public Spaces

Inclusive urban design aims to create spaces that are accessible, welcoming, and safe for everyone, regardless of gender, age, or background. Drawing on the insights gained from previous chapters, this chapter explores practical strategies and design principles that can enhance the inclusivity of urban public spaces. By prioritising women's experiences, urban designers can address safety concerns and foster a sense of belonging and empowerment.

Guided by several key principles, inclusive design ensures that urban spaces cater to the diverse needs of their inhabitants. These principles include accessibility, safety, visibility, and community engagement. Ensuring that public spaces are easily accessible to all individuals is a fundamental aspect of inclusive design. This includes considering physical accessibility for people with disabilities and ensuring that amenities are within easy reach for all residents.

Designing for safety involves more than just addressing physical threats; it also includes creating environments where people feel psychologically secure. This can be achieved through thoughtful lighting, clear sightlines, and the presence of security measures. Enhancing visibility in public spaces can significantly improve safety and comfort, particularly for women. This includes physical visibility, such as well-lit pathways, and social visibility, where the community actively uses and monitors public spaces.

Involving community members in the design process ensures that the spaces reflect their needs and preferences. This participatory approach fosters a sense of ownership and investment in the public realm.

Safety is a critical concern for women in urban public spaces, especially during dusk and darkness. Implementing strategies that address safety can significantly improve the inclusivity of these environments.

**Improved Lighting**: Ensuring that public spaces are well-lit is one of the most effective ways to enhance safety. This includes main thoroughfares and side streets, parks, and transit stops. The use of energy-efficient, bright lighting can help reduce shadows and increase visibility.

**Clear Sightlines**: Designing public spaces with clear sightlines allows individuals to see and be seen, reducing opportunities for concealment and increasing the sense of security. This can be achieved by strategically placing pathways, seating, and landscaping.

**Active Spaces**: Creating spaces that encourage active use by the community can deter crime and enhance safety. This includes designing areas for social interaction, recreation, and cultural activities. The presence of people naturally increases surveillance and deters criminal activity.

**Surveillance and Security**: The presence of security personnel and surveillance cameras can further enhance safety. However, it is important to balance these measures with the need for privacy and to avoid creating an environment that feels overly policed.

Beyond safety, inclusive urban design aims to foster a sense of belonging and empowerment among all users by creating welcoming, comfortable spaces that reflect the community's diversity. Providing amenities that cater to diverse needs, such as seating, public restrooms, and recreational facilities, can make public spaces more welcoming. These amenities should be designed with input from the community to ensure they meet local needs.

Incorporating local culture and history elements into the design of public spaces can foster a sense of pride and belonging. This can include public art, cultural markers, and spaces for community events. Designing areas that encourage social interaction and community building, such as plazas, parks, and communal seating areas, can further enhance the sense of belonging by providing places where people can gather and connect.

Additionally, ensuring that public spaces are aesthetically pleasing and comfortable can make them more inviting. This involves using materials and design elements that are both functional and beautiful, creating environments that people enjoy spending time in.

Designing inclusive urban public spaces requires a comprehensive approach considering all users' diverse needs and experiences. By prioritising safety, accessibility, visibility, and community engagement, urban designers can create welcoming and empowering environments for women, particularly during dusk. The strategies and case studies presented in this chapter provide practical insights for developing inclusive urban spaces that reflect the aspirations of their communities and promote a sense of belonging and security.

## 8 Conclusion: Creativity as an Emergency Mode in Urban Design

This contribution explores the critical role of creativity in urban design, particularly in addressing challenges faced by women in public spaces during dusk. Examining the metaphorical and practical significance of dusk and understanding women's experiences through innovative methodologies emphasises the necessity of inclusive design principles. The case study of the South Bronx serves as a practical example of applying these methodologies to create safer, more welcoming urban environments.

In urban planning and design, innovating methodologies tailored for inclusivity is crucial. A gender-sensitive perspective significantly contributes to understanding urban environments and their impact on women's interactions, enabling the creation of spaces that accommodate diverse experiences and needs.

Listen to everybody's voice to ensure safety in public spaces, especially during dusk, which is essential for everyone, with specific implications for women. In this context, crime and fear of crime are viewed through a gendered lens in urban design, which can mitigate these issues through thoughtful planning and design interventions.

Threshold urban areas undergoing socio-economic changes, like the South Bronx, present unique challenges and opportunities. Understanding their socio-economic, cultural, and historical contexts is vital for fostering inclusive and sustainable development strategies that benefit all residents.

Achieving Sustainable Development Goal 11 (Sustainable Cities and Communities) requires transformative urban policy and investment shifts. Addressing local disparities through equitable urban resilience strategies involves collaborative efforts between designers, communities, and local governments.

Looking forward, the future of urban resilience depends on translating community needs into scalable planning solutions. This approach integrates indigenous wisdom with innovative urban strategies, fostering equitable environments through collective action and integrated planning efforts.

## References

1. Bondi L (1992) Gender symbols and urban landscapes. Prog Human Geogr 16(2):157–170. https://doi.org/10.1177/030913259201600201
2. Borch C, Kornberger M (2015) Urban Commons: Rethinking the City, Space (eds), Materiality and the Normative. Abingdon, New York, NY: Routledge. https://doi.org/10.4324/9781315780597
3. Ciampa F (2018) 'Participatory design approach for the Bronx waterfront', Proceedings of The 6th International virtual conference on advanced scientific results, 6, pp 202–206. https://doi.org/10.18638/scieconf.2018.6.1.499
4. Creswell JW, Creswell JD (2018) Research design. qualitative, quantitative and mixed methods approaches. 5th Editio. Thousand Oaks, CA: SAGE Publications
5. Darke J (1996) The Man-Shaped City. In: Booth C, Darke J, Yeandle S (eds) Changing places: women's lives in the city. P. Chapman, London, pp 183–207
6. Fainstein SS, Servon LJ (eds) (2005) Gender and planning. New Brunswick, NJ; London: Rutgers University Press
7. Falú A, Sassen S (2017) 'Women and the city: reclaiming the streets to impose equal rights', The conversation [Preprint]. Available at: https://theconversation.com/women-and-the-city-reclaiming-the-streets-to-impose-equal-rights-88279
8. Franck KA, Paxson L (1989) 'Women and urban public space', Public Places and Spaces, pp 121–146. https://doi.org/10.1007/978-1-4684-5601-1_6
9. Gehl J (2010) Cities for people. Island Press, Washington
10. Gibbs A (1997) 'Focus groups', social research update, Department of Sociology, University of Surrey, (19), pp 1–7. Available at: https://openlab.citytech.cuny.edu/her-macdonaldsbs2000fall2015b/files/2011/06/Focus-Groups_Anita-Gibbs.pdf

11. Greed CH (1994) Women and planning: creating gendered realities. Routledge, London, New York
12. Gube J (2022) Photo-eliciting technique as an integrated meaning-making practice: an introspective look at diversity identities and voice-giving. Vis Stud 37(1–2):116–127. https://doi.org/10.1080/1472586X.2020.1854053
13. Hughes CJ (2021) 'In the bronx, Mott haven suddenly gets a skyline', The New York Times, (Okt. 17), p 12. Available at: https://www.nytimes.com/2021/10/15/realestate/mott-haven-bronx-new-construction.html
14. Hutson M (2020) 'Time for women leaders', In: Scientific American (ed) Women: Why equality, health and safety matter to everyone. New York, p 171
15. Kern L (2020) Feminist city. Claiming space in a man-made world. Verso, London, New York
16. Longhurst R (2003) Semi Structured Interviews and Focus Groups. In: Clifford N et al (eds) Key methods in geography, 3rd edn. SAGE Publications Ltd., London, pp 103–115
17. Matrix (1984) Making space: women and the man made environment. Matrix. London, Sydney: Pluto Press Ltd
18. McIntosh MJ, Morse JM (2015) 'Situating and constructing diversity "Diversity" in semi-structured interviews.' Global Qual Nurs Res. https://doi.org/10.1177/2333393615597674
19. Melis A, Melis B (2021) 'Community resilience through exaptation: notes for a transposition of the notions of exaptation into a design practice to promote diversity and resilience as an alternative to planning determinism during crisis', Forum A+P, 22(January), pp 70–77. Available at: https://forumap.org/portfolio/forum-ap-vol-22/
20. Micklow A, Kancilia E, Warner M (2015) 'The need to plan for women', APA Planning & Women, Planning w(November), p 15. Available at: http://s3.amazonaws.com/mildredwarner.org/attachments/000/000/533/original/f8ffecaf94ed7e8f5f700ffe9bfbd391
21. Mokha K (2011) 'Mott haven tries to rekindle momentum', Wall Street Journal, pp 1–5. Available at: https://online.wsj.com/article/SB10001424052748704099704576289213167219704.html
22. Norman WR Jr (1991) 'Photography as a research tool.' Vis Anthropol. https://doi.org/10.1080/08949468.1991.9966560
23. NYU Furman Center (n.d.) Neighborhood profiles: Mott Haven/Melrose BX01. Available at: https://furmancenter.org/neighborhoods/view/mott-haven-melrose (Accessed: 23 October 2021)
24. Patton MQ (2001) Qualitative reserach & evaluation methods. rdh. Thousand Oaks, CA: Sage Publications, Inc.
25. Powell RA, Single M (1996) Methodology matters: focus groups. Int J Quality Health Care 8(5):499–504
26. Richard VM, Lahman MKE (2015) Photo-elicitation: reflexivity on method, analysis, and graphic portraits. Int J Res Method Educ 38(1):3–22. https://doi.org/10.1080/1743727X.2013.843073
27. Tacoli C, Satterthwaite D (2013) Gender and urban change. Environ Urbanization 25(1):3–8. https://doi.org/10.1177/0956247813479086
28. UN (2015) Transforming our world: the 2030 Agenda for Sustainable Development. https://doi.org/10.1201/b20466-7
29. UN Habitat III (2017) New Urban Agenda. Available at: www.habitat3.org
30. UN Women (2017) 'Safe cities and safe public spaces: Global results report', pp 1–24. Available at: http://www.unwomen.org/-/media/headquarters/attachments/sections/library/publications/2015/unwomen-safecities-brief_us-web(1).pdf?la=en&vs=953
31. UNDESA (2019) World urbanization prospects. The 2018 revision, United Nations News. New York. Available at: https://population.un.org/wup/Publications/ (Accessed: 5 October 2021)
32. Vitale G (2021) Shaping the female city, Arup. Available at: https://www.arup.com/perspectives/shaping-the-female-city (Accessed: 20 June 2021)

33. Whitzman C (1995) 'What do you want to do? Pave parks? Urban planning and the prevention of violence', In: Eichler M (ed) Change of plans: towards a non-sexist sustainable city. Toronto, Ont.: Garamont Press, pp 89–110. https://doi.org/10.3138/j.ctt2ttv77.10
34. World Bank (2020) Handbook for gender-inclusive urban planning design

# The Mayan Solar

5

Sofia Ayora-Talavera

**Abstract**

The reduction in the number of settlements and the congregation of the population in towns and republics of natives was a Spanish strategy to control the original inhabitants and their habitable spaces. In the case of the domestic space, these policies confined and constricted it in lots or land perfectly delimited, located next to each other and around a square. This with the intention of regulating and monitoring what was happening within of it and modify the Mayan way of living. In cultural and spatial terms, this event signified a violent change in the environment and context. On the other hand, the shape and characteristics of the space or plot assigned by the Spaniards possessed the conditions for a "Western" way of living; however, the cultural and spatial structures of the Maya were employed and readapted to the new circumstances, consequently creating a new housing model: the Mayan solar. In this sense, the solar can be considered an exaptation that, in addition to spatial and functional aspects, includes cultural expression in a new form. This is an example of how human creativity promoted change while simultaneously preserving material, cultural, and symbolic structures without necessarily seeking perfection. Thus, the Maya domestic space becomes a guardian of dwelling that has demonstrated great resilience to cultural challenges over time, both in the colonial period and subsequently in modernity. So much so, that currently, we can recognize some of the spatial characteristics derived from ancient Maya dwelling.

**Keywords**

Mayan solar • Mayan spatial configuration • Mayan dwelling • Architectural exaptation • Resilience

S. Ayora-Talavera (✉)
Universidad Marista, Merida, Mexico
e-mail: sayora@marista.edu.mx



J. A. Lara-Hernandez et al. (eds.), *The Dusk of Design*, Integrated Science 34,
https://doi.org/10.1007/978-3-031-76867-5_5

## 1 Introduction

Natural evolution is a complex process that involves a variety, perhaps still undetermined, of principles and mechanisms through which organisms undergo transformation over time and space. Building upon Darwin's theory of natural selection, Gould and Vrba [6] introduce new perspectives to biological design by proposing that it involves two complementary mechanisms: adaptation and exaptation. Exaptation enters the world devoid of an evolutionary history and simply affects it (Borrero and Borrazzo, [1], p. 10). Therefore, evolution finds in this mechanism the ability for the emergence and existence of multiple dimensions, directions, and rhythms beyond adaptation, which positions the organism in a passive role to adjust to external conditions and variations. This perspective grants "living systems an active, creative, innovative role, exploring new possibilities in development and coexistence with other organisms" (Muñoz, [13], p. 56).

Viewing the evolutionary process in this way allows for a parallel with architectural design to be established. This correspondence has been described by some authors [11], who, exploring concepts associated with evolutionary biology, have recognized some of their relationships and effects in architecture. Such is the case of exaptation, which operates through two mechanisms: change and functional co-option. In line with biological design, it enables the transformation of architectural space, also in multiple dimensions, directions, and rhythms, thus opening up a myriad of possibilities to find resilient, sustainable, and innovative solutions [10].

To exemplify the above, this chapter explores the case of the Maya domestic space that emerged during the colonial period in Yucatán, Mexico: the solar. As will be seen later, it is a spatial system that can be considered highly exaptive because, even today, it is the result of a spontaneous praxis (poietic), inscribed in the specific space–time of its inhabitants, without the specific intervention of specialists (architects or engineers) for its design and construction. Therefore, it is something unique in its own way, forged almost unintentionally and built as it is inhabited, understanding the concept of dwelling as the inherent ways of being and existing in the world [7]. These ways of dwelling manifest in the spatial characteristics and spatiality and transform over time at various rhythms and in ways that are not necessarily linear, intentional, ordered, perfect, rational, or progressive.

Juhani Pallasmaa [17] points out that there is a very close relationship between dwelling and space: for him, humans exist to the extent that they dwell; therefore, dwelling reveals the different ways of being in the world and relating to it, establishing, on one hand, an exchange between the inhabitant situated in the place and the space that is situated in their consciousness, and on the other hand, an extension, as the place becomes an externalization and an extension of their being. We then understand that architecture is not just a matter of form and function, and that space is not necessarily or exclusively physical, but also includes how humans construct their corporeality and their relationship with the universe. Social space is broader and more complex than architectural physical space because it encompasses, in addition to materiality, mental constructs and abstractions that become concrete through human practices.

Therefore, to understand the Maya space, it is necessary to move away from classical ecological determinism and consider the importance of its cosmological, aesthetic, and subjective dimensions that, in the landscape, combine geographical sense with the sense of a subjectively perceived space [3]. Unlike the Western view, which separates and delineates very clear boundaries between what humans produce and what grows in nature, the Mayan perspective manifests a dialectical relationship between nature and culture.

> The inhabited territory [Maya] is a coherent whole within which men, animals, plants, stars, geographical orientation, and temporal flow play a specific function and in which all are interdependent; nature possesses an order with the human being at its centre. Likewise, the intimate relationship between humans and their environment is assumed as a religious duty; all elements composing the world are believed to be "alive," meaning they possess a "heart" and a divine essence. The human population coexists with the plant, animal, mineral, and meteoric population in a biological exchange expressed through specific myths and rituals. (Morales, [12], p. 298)

The above implies conceiving, representing, and using space in a particular way. To understand this, we will first address the spatial characteristics of pre-Columbian Maya territory as a reference for the conception and use of space, which was transformed through the reorganization measures employed for the domination of indigenous people in the region. One of the main innovations was the emergence of the traditional domestic space known as the Mayan solar. This can be considered an exaptation that, in addition to spatial and functional aspects, includes cultural expression in a new form. It is an example of how human creativity promoted change while simultaneously conserving material, cultural, and symbolic structures without necessarily seeking perfection.

Subsequently, we will analyze the spatial and formal structures that can still be found in the Mayan solar, demonstrating the connection of this archetype with the concepts of redundancy, variability, and diversity, which, within biology, are associated with resilience and sustainability.

## 2 Pre-hispanic Maya Space

> *The towns now inhabited by indigenous people have no form, nor can they have, of streets because the houses are made of wood covered with thatch, and thus they appear, depending on the town, large or small, a congregation of huts. But there are very good, peaceful houses to live in, although with no security, due to the danger of burning down, as they often do. Stone houses are not suitable for Indians because they fall ill and die in them, and for their way of life and nature, straw houses are more useful to them.* (*Palomar,* [18]*[1579], p.71*)

The socio-political and spatial organization of the Maya before the arrival of the Spanish was not formed by continuous extensions with defined and physically marked boundaries, as previously believed [14]. The dispersed pattern with seemingly random arrangements that the conquistadors found, described by Palomar in the previous quote, although it did not correspond to Western organizational and orderly canons, was undoubtedly a response to the natural environment and the complex network of social, familial, political, economic, production, symbolic, and religious relationships that characterized Maya dwelling.

Such organization was based on the variable and flexible interconnection of the basic unit, the "*cah*," which, according to various conditions, formed structures more or less complex [14, 16] through repetition, addition, or redundancy. The minimal cell was formed by a group of twelve or more extended families related and situated in territorial extensions of variable dimensions, depending on the forest occupied by their domestic spaces (rooms) and their crops: "The lands, for now, are common, and so, the first who occupies them owns them. They sow in many parts, so if one fails, the other will supplement it." (Landa, [9] [1566], p. 46).

The union of these units was achieved through alliances promoted by representatives seeking greater benefits for the family groups; therefore, the grouping of several "*cah*," which formed a "*batabil*" or lordship, was flexible in dimension, form, and spatial occupation, as it transformed according to the achievement or rupture of agreements. The "*batabil*" was headed by a "*batab*", the lord who represented the union of all members and had the power to dispose, in a certain way, of the people and resources of the "*cah*," which was called "*cah-cuchteel*" when integrated into a "*batabil*" [14].

The third organizational structure was the "*cúuchcabal*," which, under the same scheme and showing spatial characteristics similar to those described, united a "*batabil*" with other analogous entities. The "*Halach Uinic*" was the representative of the "*cúuchcabal*," and its place of residence, named "*noh cah*" (main town), was recognized as the political capital (Fig. 1).

Due to their origin and as described, in the complex structure or spatial formation of the "batabil" and the "cúuchcabal," the location where the representative resided was not important; rather, it was the extent of their power. Therefore, the "cah-cuchteelob" of a "batabil" or the "batabilob" of a "cúuchcabal" were dispersed in a non-geographically continuous territory, meaning they were not necessarily contiguous or adjacent. Indeed, one or several sets from another lordship could be closer than one's own [15]. Thus, it is appreciated that the cohesion of the group was not based on the possession of material goods, but rather the system of redistribution and reciprocity, as well as various cycles of rituals and ceremonies, wove together and defined the relationships of the members of the lordship, thus their identity in belonging to it. This entire set of elements—cultural, social, economic, and religious—validated the existence of the "batabil," hence its corporate nature. (Okoshi, [14], p.18).

This spatial model based on the "*cahob*," "*batabilob*" and "*cúuchcabalob*" shows, on one hand, the close link between nature and culture in the Maya conception and how this is expressed in the construction of space. On the other hand, it

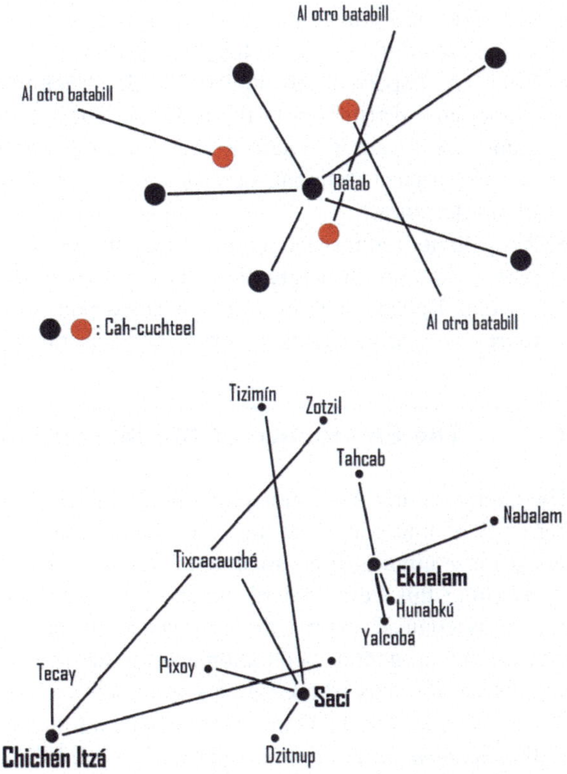


**Fig. 1** They illustrate the absence of linear boundaries and the spatial complexity derived of their organisations [15]. *Source* Tsubasa Okoshi Harada. Okoshi, "Las entidades políticas de las tierras bajas del norte al tiempo de la invasión española", pp. 5 & 7

allows us to find certain spatial characteristics that, due to their dialectical relationship with nature, align with the concepts of variability, redundancy, and diversity, and can be considered suitable for promoting sustainability and resilience.

In summary, this seemingly spontaneous design begins with a germinal element (*cah*) that, as a spatial system, operates under certain family criteria and possesses various components (physical and intangible). This element repeats to form more complex and diverse structures, leading to redundancy, since when it joins with others, it does not lose its independence or essential characteristics: it remains itself while simultaneously becoming part of other structures.

Both the "*cah*," the "*batabil*" and the "*cúuchcabal*" maintain the following characteristics.

- They do not have linearly drawn boundaries or borders; therefore, they do not determine continuous zones and are very flexible: they expand, or contract as needed.
- The growth of each of the organizational structures results from the addition of new members. Alliances materialize in places without requiring a relationship of proximity or contiguity to be concretized and, nonetheless, create and hierarchize existential centers. In this way, the territory extends from the intimate

space of the room to the domestic space of the extended family, to the "*cah*," to the "*batabil*" and up to the "*cúuchcabal*."
- They are dispersed and highly flexible structures that constantly vary in size, shape, and distribution in the territory. These transformations occur in an "organic" or "natural" manner, that is, away from what is linear, intentional, orderly, perfect, rational, or progressive. The structures grow by the addition of similar ones.
- They do not require geographic contiguity or spatial proximity to be considered part of the same structure; thus, they cannot be spatially represented as zones or regions. Rather, they are interconnected points; therefore, their spatial graphic representation generates patterns that seem random and unordered.

## 3 The Emergence of the Mayan Solar

The Conquest and the socio-spatial reorganization carried out by the Spanish during the Colonial period allowed for domination, the introduction of Catholicism, and a clear ethnic differentiation in Yucatán. The Spanish enterprises aimed at establishing this new "order" included the creation of towns or cities for white people as control centers, the emergence of the encomienda system which distributed the indigenous population among the Spanish, and the congregation of the population in towns whose regions were defined linearly (Fig. 2).

For this mission, the *Ordenanzas* played a fundamental role. For example, those of the Spanish judge Tomás López Medel (1552–1553), known through López Cogolludo [14, 21], established the following for the congregation of towns: build houses close together with a layout similar to Spanish towns, with wide roads and removing vegetation so they appeared clean. In the blocks, the houses had to be adjacent to each other, have linear boundaries, and be arranged according to the alignment of the street.

In spatial terms, these and other massive transformations at the settlement level, in contrast to the sociopolitical organization based on the "*cahoob*," "*batabiloob*," and "*cúuchcabaloob*," promoted the fragmentation of space and the disappearance of the dispersed and discontinuous spatial organization. They also hindered organic growth and addition, making it impossible to appropriate the land and its dimensional variation, as the physical and linearly drawn boundaries defined fixed particular territories. In this way, spatial fragmentation becomes an element that dilutes the bond with the land and clearly separates the built space from the natural one, establishing an asymmetrical relationship between them, prioritizing human production and considering nature only as a source of resources.

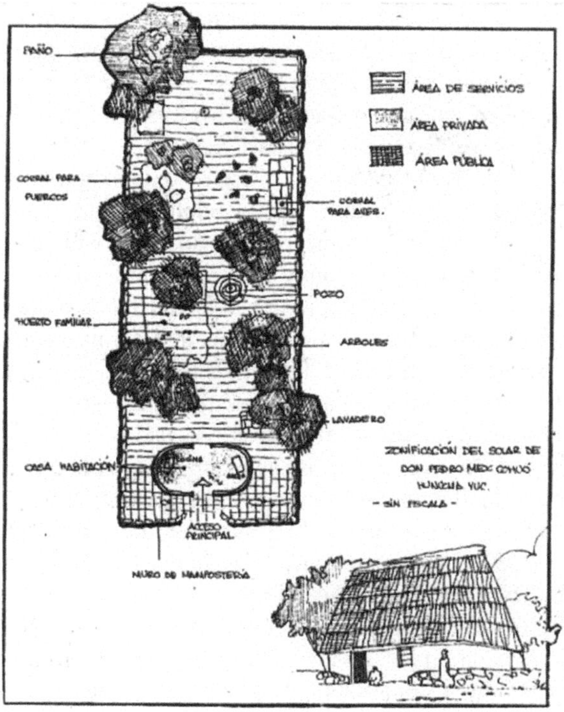


**Fig. 2** The Yucatecan Solar [24]. *Source* Lucía Tello Peón, "La vivienda en Yucatán: su espacialidad y esencia", p. 8

## 4 Spatial Characteristics of the Mayan Solar

To contextualize the characteristics of the domestic space described in this section, it is important to make two considerations: the first is that the various Spanish strategies for socio-spatial reorganization gave rise to a range of Mayan Solares that vary according to their origin. Thus, we can mention those that emerged in indigenous towns and those created around agricultural and livestock production (ranches and livestock-corn estates) or agro-industrial production (henequen estates). On the other hand, it is also important to mention that, in the current rural environment of Yucatán, the characteristics of domestic space and its spatiality manifest a heterogeneity of stages of change-permanence in the modes of inhabiting concerning traditional Maya habitation, promoted by Westernization, modernity, and urbanization. Therefore, since we cannot speak of homogeneity in the characteristics of the "*solares*" in Yucatán, those described in this section are general features, many of them observed in fieldwork and others described in specialized bibliography.

The solar, the Mayan dwelling of the Yucatán Peninsula, is a very spacious residential complex. It consists of several buildings: one or more houses accompanied by a kitchen and sometimes a storage room or a granary; the exterior spaces also serve as the site for numerous activities. It accommodates one or more nuclear families, each having its own house. If these families cooperate in domestic tasks and share the annexes, they form what is known as a domestic group; otherwise, if the families lead independent daily lives, they will have their own annexes (Pierreborg, [19], p. 235).

When entering a traditional Mayan solar, the ensemble may give the impression of being somewhat disordered (Fig. 3). As suggested, Maya spatial order does not necessarily adhere to Western or geometric schemes, but rather to organic ones, linked to nature, family relations, and symbolic aspects. Even symmetry results from the cosmic order associated with the cardinal points and not as a parameter to achieve unity in the ensemble. Thus, the morphological, spatial, and organizational configuration can be attributed to the way of inhabiting which, like pre-Columbian socio-spatial structures, forms a complex and flexible system interconnected with others. Recalling the dialectical relationship between nature and culture, this system has the possibility to establish links not only with the jungle through agricultural production but also with the cosmos and its elements. However, it also contains subsystems: the Mayan house and its groupings, which derive from family composition that still prioritizes the extended family today.

The organization of the solar could be compared to what happened among the "*cahob*", "*batabilo*b" and "*cúuchcabalob*" as small groups of traditional rooms and other spaces make up larger and more complex structures. These elements are organized in an organic and dispersed manner on the ground plane, as if they were natural pieces whose placement contributes to their life cycle. The dispersion maintains a communion between forms and spaces; likewise, it allows for the exploitation of their characteristics to achieve suitable orientations in terms of ventilation and sunlight exposure. The components do not make physical contact with each other, they do not touch, they are not contiguous, and yet they can be recognized as part of the same ensemble: while the buildings of a family nucleus may have nearby locations (within the same area and sharing some space like the kitchen), there may also be annexes, such as animal cages or washing areas, which move away from the assigned area and may even be grouped with others of similar use and from another family.

Similarly, as with the elements and structures of the cahob, batabilob, and cúuchcabalob, in the Mayan residential complex buildings are added as needed [19]. This means that functional multiplication is not carried out by spatial division, but by the addition of new cells or rooms that are created through repetition or redundancy. These cells house new members or functions and have a certain degree of independence to act autonomously while being integrated into the rest of the complex. However, if needed, they also can change function or take on additional roles without altering their form (Fig. 4).

The dialectical and parity relationship that the Mayans establish between nature and the built, from a functional perspective, reduces the importance that might be given to the size of covered spaces. Conceptually and in practice, both roofed and outdoor spaces jointly meet the pragmatic and existential requirements of the architectural program. This allows the traditional Mayan house to be considered a vital space that extends across the entire plot, rather than merely a 30 $m^2$ construction.

**Fig. 3** Mayan Solar parts A. Naj, B. Kitchen, C. Troje (wooden structure for corn storage), D. Ka'anche', E. Pigpen, F. Henhouse, G. Well, H. Laundry trough, I. Outhouse or place for evacuation [4]. *Source* Atlas de Procesos Territoriales de Yucatan, p. 239

This complementarity makes the solar a highly flexible and variable system: the space transforms to adapt to various activities and uses, as these constantly vary according to family needs. In other words, the forms and their organization do not have a permanent or fixed character in terms of presence, location, or number. Both the exterior space (patio) and the interior space (rooms) are multifunctional, allowing for multiple activities to be carried out simultaneously and, therefore, capable of continuously renewing themselves and being used for new activities or purposes.

The patio is used for raising and caring for animals and planting, but also for washing and hanging clothes to dry, for cooking, for playing, and even for some family gatherings. On the other hand, in the house, activities such as playing, cooking, gathering, sewing, or eating can take place during the day, and at night, it transforms to accommodate rest and sleep, so the free space that allows for a variety of activities during the day is filled with hammocks that are set up at night,

**Fig. 4** Spatial characteristics of the Solar. The absence of physical contact (contiguity) between elements, dispersion, organic and additive growth, as well as the flexibility and multifunctionality of the patio space and the lack of physical boundaries between domestic lots can be observed

taking on completely different characteristics and configurations. This shows that, depending on the circumstances, there are activities that can be done both inside and outside the rooms, and therefore, life flows smoothly between these two spaces (Fig. 5).

The solar acquires value for its flexibility, variability, diversity, and multifunctionality, not for the quantity or dimensions of its built and permanent structures. In fact, if it is necessary to set up a framework for a certain task—corrals, sheds, places to wash and hang clothes, or sit—ephemeral elements are generally used, built with perishable, easy-to-dismantle, or recyclable materials; that is, those that can change position or use according to family requirements. Something similar happens with the furniture inside the rooms; it is always flexible and multifunctional, accommodating the constant metamorphosis of the space. This allows us to understand the solar and the traditional house as living organisms, as their structures, forms, and spaces, in their condition of being useful and mobile, meet the conditions to adequately carry out daily tasks thanks to their power of transformation.

The spatial characteristics described exhibit a conception of space distinct from the Western one. From this perspective, for example, it's impossible to understand the interior as something different and opposed to the exterior. Nor is the existence (conceptual and physical) of specific containers for activities with permanent functions conceived. Likewise, the idea of fixed barriers or limits that define these containers as closed and impenetrable, or that separate spaces, does not exist in the Mayan vision. This is closer to the Heideggerian perspective, which considers the limit not as where something ends but where something begins. Thus, one inhabits fluidly between inside and outside, dissipating boundaries into successive ambiguous layers, whose function is not to contain but to connect [25]. In this

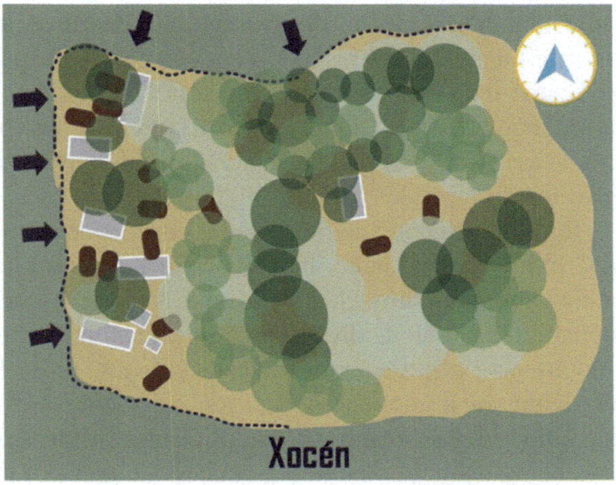

**Fig. 5** Schematic configuration of interior and exterior spaces in an extended family yard in the town of Xocén, Yucatán. It can be observed that there is a fluid transition between natural and man-made elements, without distinction, allowing for the development of activities without the existence of an interior opposed to an exterior. Design by the author

sense, Hoppan [8] mentions that the Mayan language does not have a word that signifies a border in the sense of a separating limit, only a close concept (*táan*) that means crossing, passage place, or meeting point.

## 5 Conclusion

As seen in this chapter, the violent and punctual changes that occurred during the colonial period led to the emergence of the "solar." It became a formally, spatially, and functionally innovative structure whose design was able to adapt and, therefore, preserve the cultural essence of the Maya, their way of inhabiting, and their spatial conceptions.

Thus, if we want to talk about resilient, sustainable, and adaptable design, it is necessary to mention the Maya domestic space. The "solar" has become a guardian of the living habits of this culture for over 500 years and has demonstrated, over time, great resilience to cultural and environmental challenges, both in the colonial period and subsequently in modernity. In this way, we can still recognize in it some of the spatial characteristics derived from ancestral Mayan living.

Similarly, the traditional Maya dwelling is an architectural element whose characteristics have granted it great resilience: it has been able to adapt efficiently from pre-Columbian times to the present day. Based on archaeological evidence, such as dwellings with over 3000 years of age found by Hammond in Cuello, Belize [22], and the descriptions by Landa in "Las Relaciones de Yucatán," it is recognized that the Mayan house has preserved over time many of its formal and spatial characteristics, construction materials, and construction system, aspects that also give it a sustainable character.

Both the structure and the spatial conception of the Maya possess highly exaptive traits. On one hand, they generate flexible characteristics, as we have seen, allowing space, through its use and activities, to continuously and organically transform itself along diverse paths in an autopoietic manner. However, these innovations also acquire familiar characteristics, those belonging to the Maya way of living, in a continuum between exaptation and adaptation. We believe this has allowed the solar to promote change while simultaneously preserving and enhancing resilience, both in practices and ways of living.

It is important to note that the constant in this review of the Mayan solar is the spatial manifestation of variability, redundancy, and diversity. These concepts open paths and opportunities to seek resilience and sustainability, as they render structures, forms, and spaces more flexible, allowing for their continuous and seemingly random transformation; that is to say, removed from perfection, rationality, determination, organization, and intentionality. In these terms, the characteristics of Mayan space and spatiality such as dispersion, discontinuity, organic and additive growth, the absence of boundaries or their constant variability, and the multifunctionality of systems can be considered within this category.

However, we must not forget that there are other aspects mentioned here that are part of the Mayan conception of nature. Granting nature a value similar to everything produced by humans and considering that there is a dialectical relationship between the two emphasizes renewal, the use of natural materials, their circular employment, and, of course, recycling. It also highlights the positive aspects of the perishable, mobile, and variable nature of architectural systems, the complementarity of indoor and outdoor spaces, and the possibility of physical and environmental transformation of spaces according to activities. These attributes bring this archetype closer to sustainability, resilience, and adaptability.

Undoubtedly, much of what has been discussed alludes to architectural conceptions that diverge from the known; however, we believe it is worth revisiting, reclaiming, and reinterpreting them to seek better alternatives for ourselves and our planet. The solutions and ways of using Mayan space, in fact, run counter to contemporary pursuits such as permanence, security, and stability, derived from human craving for control. We hope that these reflections promote the idea that architectural objects can emancipate themselves from pre-established values and encourage creative exploration that liberates architecture from exclusively technological and aesthetic intentions to discover lateral connections with an environmental and social context that often is not the work of anyone, much less of academic design and planning [2].

## References


1. Borrero LA, Borrazzo K (2013) Exaptaciones, cambio y oportunismo en Arqueología", *Cazadores-Recolectores del Cono Sur*, Universidad Nacional de Mar del Plata. Facultad de Humanidades, pp 9–29. http://fh.mdp.edu.ar/revistas/index.php/cazrec/issue/viewIssue/102/39
2. Cienfuentes QCA (2023) La arquitectura como sistema natural. El pensamiento informacional, la mutación ontológica del objeto y la biologización de la arquitectura", Academia XXII 14(28), Universidad Nacional Autónoma de México, 2023, pp 30–57. https://doi.org/10.22201/fa.2007252Xp.2023.14.28.87235
3. DESCOLA P (2003) El paisaje maya y su historia: comentario", en: Alain Breton, Aurore Monod Becquelin y Mario Humberto Ruz, (edit.), *Espacios Mayas. Usos, representaciones, creencias.* México, Universidad Nacional Autónoma de México y Centro Francés de Estudios Mexicanos y Centroamericanos, pp 220–201
4. García De Fuentes A, Alfredo Alonso Aguilar ROL et. al (1999) (coord.), Atlas de Procesos territoriales de Yucatán, Mérida, Universidad Autónoma de Yucatán
5. García QAyASS (2014) Introducción a las historias de la Maya *naj*", en Aurelio Sánchez Suárez y Alejandra García Quintanilla, (edit.), *La casa de los mayas de la Península de Yucatán: historias de la maya Naj*, Mérida, Facultad de Arquitectura de la Universidad Autónoma de Yucatán, pp 9–18
6. Gould SJ, Vrba ES (1982) Exaptation-A Missing Term in the Science of Form", *Paleobiology* 8(1), pp 4–15. https://doi.org/10.1017/S0094837300004310
7. Heidegger M (1994) *Conferencias y artículos*, Barcelona, Ediciones del Serbal
8. Hoppan J-M (2014) Miradas epigráficas sobre la vivienda maya", en Nah, Otoch. *Concepción, factura y atributos de la morada maya*, Mérida, Yucatán, Secretaría de Educación del estado de Yucatán y Universidad Autónoma de México, pp 23–24
9. LANDA, Diego de, Relación de las cosas de Yucatán [1566], Yucatán, Consejo editorial de Yucatán A.C., 1984.
10. Melis A, Lara-Hernández José Antonio and Melis Barbora, Exaptive Design: A Revolutionary Approach to Resilient Architecture in Times of Crises
11. Melis A, Pievani T (2022) Exaptation as a design strategy for resilient communities", In Rezaei N (Ed.), *Integrated Science: Transdisciplinarity Across the Different Disciplines*, pp 307–327
12. Morales DMA (2010) Territorio sagrado: cuerpo humano y naturaleza en el pensamiento maya", en *Cuicuilco* 17(48) disponible: http://www.scielo.org.mx/scielo.php?script=sci_arttext&pid=S0185-16592010000100014, pp 279–298
13. Muñoz RJ (2019) La revolución en Biología como cuestionamiento de las relaciones esencia-apariencia (una crítica al inmanentismo neodarwinista)", en Muñoz, Rubio Julio (Coord.), *La Biología Evolutiva Contemporánea: ¿Una revolución más en la ciencia?*, Universidad Nacional Autónoma de México (Centro de Investigaciones Interdisciplinarias en Ciencias y Humanidades), pp 23–70
14. Okoshi HT (2018) De lo ajeno impuesto a lo nuestro fundado: el proceso de la "domesticación" del espacio de los pueblos de indios de la gobernación de Yucatán", en Tsubasa Okoshi, Julien Machault y Alberto Tepoxtecatl, (coord.), *Recorriendo el lindero, trazando la frontera. Estudios interdisciplinarios sobre el espacio y las fronteras en las sociedades indígenas*, México, Universidad Nacional Autónoma de México
15. Okoshi HT (2012) Las entidades políticas de las tierras bajas del norte al tiempo de la invasión española", en *Revista Dgital Universitaria*, vol 13, Número 12, México, Universidad Autónoma de México. http://www.revista.unam.mx/vol.13/num12/art119/art119.pdf
16. Okoshi HT, John F. Chuchiak, editores (2009) *Texto y contexto: La literatura maya yucateca en perspectiva diacrónica*, Alemania, Shaker Verlag
17. Pallasmaa J (2016) *Habitar*, Barcelona, Gustavo Gili

18. Palomar De M (2008) Relación de la ciudad de Mérida" [1579], en Mercedes de la Garza, directora, Relaciones histórico-geográficas de la Gobernación de Yucatán, Tomo I, México, 2008, Universidad Nacional Autónoma de México
19. Pierrebourg De F (2003) La vivienda maya, entorno natural y mundo natural: un enfoque etnoarqueológico", en Alain Breton, Aurore Monod Becquelin y Mario Humberto Ruz (edit.), *Espacios Mayas. Usos representaciones y creencias*, México, Universidad Nacional Autónoma de México y Centro Francés de Estudios Mexicanos y Centroamericanos, pp 235–260
20. Rivas GD, La choza maya (2012) Cuna y custodia de los grandes misterios de la sabiduría de una cultura que sigue viva, Yucatán, Universidad Autónoma de Yucatán
21. Ruz MH (2014) Nombrar para habitar: la morada maya en las grafías coloniales, en Fabienne de Pierrebourg y Mario Humberto Ruz, (coord.), *Nah, Otoch. Concepción, factura y atributos de la morada maya*, Mérida, Yucatán, Secretaría de Educación del estado de Yucatán y Universidad Autónoma de México, pp 65–122
22. Sánchez y Alejandra García Quintanilla (Editores) (2014) La casa de los mayas de la Península de Yucatán: historias de la maya Naj, Mérida, Facultad de Arquitectura de la Universidad Autónoma de Yucatán
23. Sánchez SA (2006) La casa maya contemporánea. Usos, costumbres y configuración espacial", en Península, vol 1, no 2, México, Universidad Nacional Autónoma de México
24. Tello PL (1992) La vivienda en Yucatán: su espacialidad y esencia", en Cuadernos de Arquitectura 5, Mérida Yucatán, Facultad de Arquitectura de la UADY, pp 7–14
25. Vasileva NN (2017) Sistema de objetos del habitar japonés. Acciones y enseres en el espacio doméstico durante el período de modernización (tesis doctoral), Madrid, Universidad Politécnica de Madrid, 2017, disponible en: https://docplayer.es/138630595-Sistema-de-objetos-del-habitar-japones.html

# Favelas, Exaptation and Self-organised Maps

# 6

Veronica Natividade

**Abstract**

In this chapter, we delve into the unique context of favelas, Brazil's version of slums, as a case study to explore the concepts of exaptation, self-organisation, and unsupervised learning – a specific form of artificial intelligence—in impoverished architecture and urban fabrics. We blend biological, philosophical, and architectural perspectives to illuminate key urban milestones: the global shift towards urbanisation and the 2020 tipping point where anthropogenic mass surpassed living mass. This underscores the urgent need to rethink material flows within our socioeconomic system. We highlight the overwhelming prevalence of buildings and infrastructure as the dominant human-made mass category, illuminating the urban footprint and architectural foundation. Our chapter is structured into five parts: critically examining exaptation from biological and architectural standpoints, presenting a case study exhibited at the Venice Biennale 2023, and proposing a material-based approach to exaptation. We prioritise primary sources, particularly in biology, to ensure accuracy and avoid deterministic or moralistic interpretations, especially when integrating current information technologies.

**Keywords**

Favelas • Exaptation • Self-organised maps • Unsupervised learning • Venice Biennale 2023

V. Natividade (✉)
PhD candidate, Chair Of Digital Architectonics, ETH Zürich, Zurich, Switzerland
e-mail: vgomes@ethz.ch



J. A. Lara-Hernandez et al. (eds.), *The Dusk of Design*, Integrated Science 34,
https://doi.org/10.1007/978-3-031-76867-5_6

## 1 Introduction

In this chapter, we elected the context of favelas—the Brazilian version of *slums* – as our case scenario to explore the concepts of exaptation, self-organisation and unsupervised learning—a specific case of artificial intelligence—in impoverished architecture and urban fabrics, weaving biological, philosophical, and architectural perspectives.

We based our decision on two critical urban milestones. First, in 2008, the urban exceeded the rural population. Although the shift towards urbanisation is a global trend, urban informality is often the primary mode of 21st-century metropolitan space production [44], making informal housing builders the largest constructors of the 21st Century. The second significant development is the tipping point reached in 2020, as anthropogenic mass exceeded living mass [13]. This revelation underscores the pressing need to rethink the material flows of our socioeconomic system, referred to as the "socioeconomic metabolism" [13]. By drawing attention to the material composition, stocks, and flows on both local and global scales, Elhacham's group calls upon researchers to delve into the material crystallisation of society's cultural underpinnings in our cities. Notably, the study sheds light on the overwhelming prevalence of buildings and infrastructure as the dominant human-made mass category, spotlighting the urban footprint and architectural foundation.

The paper has five parts. In "**Creative Redundancies**", we examine the concept of exaptation and its importance to evolution from the biological point of view. Next, in the **"Exaptation in Architecture"** session**,** we offer a critical view of the term, contrasting architectural and biological perspectives. In the **"Rendered Parasitism"** session, we will present our case study—an artefact produced for the collateral event "Students as Researchers" hosted by the NYIT at the Venice Biennale 2023. The piece expresses the discussion around the topics addressed in the paper, percolated by the philosophical concepts of "parasite" from Michel Serres [46] and "aesthetics of the smooth" from Byung Chu Han [20]. We end the paper with a **"Discussion"** session followed by the **"Final Remarks",** proposing a material-based approach to exaptation.

As much as possible, we prioritised primary sources, especially regarding concepts derived from Biology. We did so in response to our equal preoccupation with accuracy when using biological analogies and metaphors, as well as in avoiding false determinisms and moralism, especially when current information technologies are mixed in. Listening to Ludwig von Bertalanffy's alerts: *"[…] unlike the phylogeny that is determined by organic regularities only, it is up to man himself whether he directs the development that is determined by his peculiarity towards progress or towards self-annihilation" (Bertalanffy apud* Hofkirchner, [22]*, as cited in Auf den Pfaden des Lebens, 1951, p. 214—Hofkirchner's translation).*

## 2 Creative Redundancies

*"We have seen that man by selection can certainly produce great results, and can adapt organic beings to his own uses, through the accumulation of slight but useful variations, given to him by the hand of Nature. But Natural Selection, we shall hereafter see, is a power incessantly ready for action, and is as immeasurably superior to man's feeble efforts, as the works of Nature are to those of Art." Darwin,* [8]

*"Exaptive possibilities define the "internal" contribution that organisms make to their own evolutionary future." Gould e Vrba,* [19]

*"Natural selection merely modified, while redundancy created." Susumu Ohno,* [40].

We are all familiar with Charles Darwin's "On the Origin of Species" [8] and his introduction to the term 'adaptation'. While beautifully pictured in affable examples, such as how a polar bear's fur is an adaptation that helps it stay warm in cold climates, Darwing also demonstrated nature's crude side. Always at work, "silently and insensibly" [8], nature's motto is adapt or perish. In the vernacular and sciences other than evolutionary biology, all the meanings for the word adaptation are consistent with the etymology of *ad* ('to) + *aptus* ('fitted'), giving us"towards a fit" or "for a particular role". To be fitted for a role, however, should not be confounded with teleology, not even at the molecular level. Darwin knew it back then.

In a spectacular demonstration of political incorrectness for the contemporary sensibilities, Alfred Russel Wallace—a strict adaptationist born in 1823, argued that although '*savages*' were equipped with a brain as powerful as his ('ours', he said...), they kept themselves in a primitive culture [19]. Wallace dismissed the interpretation of natural selection as built for immediate use, pointing out that those 'living primitives' were not using their full mental capacities. An apparently "bemused and angered" Darwin replied that the brain, though undoubtedly built by selection for some complex set of functions, can, as a result of its intricate structure, work in an unlimited number of ways entirely unrelated to the selective pressure that constructed it [19]. Wallace's powerful brain failed to acknowledge that the brain, indeed so powerful, may deviate life choices from afternoon tea social gatherings.

This nineteenth-century academic beef admonishes those still looking for specific causalities and explanations regarding complex entities like humans and nature. Species may exhibit diverse adaptations to the same selective environment due to variations in initial conditions, such as phenotype, genetic diversity, and ancestral constraints, which influence evolutionary trajectories [12]. Chance events, like the sequence of mutations, further contribute to disparate evolutionary outcomes [41]. These historical contingencies, as Gould [18] emphasised, underscore the complexity of evolutionary processes.

Much has evolved since Darwin. Other terms related to evolution emerged: pre-adaptation, non-adaptation, aptation, function, effect, exaptation:

*"We may designate as an adaptation any feature that promotes fitness and was built by selection for its current role (criterion of historical genesis). The operation*

*of an adaptation is its function. […] operation of a useful character not built by selection for its current role [is] an effect. […] But what is the unselected, but useful character itself to be called? […] We suggest that such characters, evolved for other usages (or for no function at all), and later "coopted" for their current role, be called exaptation. They are fit for their current role, hence aptus, but they were not designed for it, and are therefore not ad aptus, or pushed towards fitness. They owe their fitness to features present for other reasons, and are therefore fit (aptus) by reason of (ex) their form, or ex aptus. […]* ***Adaptations have functions; exaptations have effects****. The general, static phenomenon of being fit should be called aptation, not adaptation."* [19].

A classic example of exaptation is birds' feathers, which likely evolved for insulation or courtship displays but later exapted for flight. Exaptation is, thus, "*the missing term in the science of form*" [19], which opposed the orthodox evolutionary theory and reinforced the importance of multiple environmental pressures in biological formations [53].

In Biology, repetitive DNA, or genetic redundancy, plays a crucial role in the evolution of the phenotypic complexes in organisms. It has been known since the seventies that if nature can be called creative, it is only outside of one-to-one correspondence. The Japanese-American geneticist and evolutionary biologist Susumu Ohno [40] said: *"Yet, being an effective policeman, natural selection is extremely conservative by nature."* And later: "Had evolution been entirely dependent upon natural selection, from a bacterium, *only numerous forms of bacteria would have emerged. […] Only the cistron which became redundant was able to escape from the relentless pressure of natural selection, and by escaping, it accumulated formerly forbidden mutations to emerge as a new gene locus."* In other words, genetic redundancy is the door for exaptation to act [19], while "***variety** is a cumulative action of natural selection*" ([8]; emphasis added).

Indeed, variety and chance have always seemed important in biology [57]. It is also known that beneath biological manifestations, from all levels, exists a flux of substances, matter, energy, and information encapsulated in cause-and-effect interactive relationships, which are non-linear and bi-directional to a certain degree. These processes remain obscure and are assessed mainly by their structural manifestations [26]. The claim to transparency between cause and effect is legitimate for real things in nature because we can analyse its material manifestations. And yet, in the realm of biological evolution, new species emerge with traits that distinguish themselves from their predecessors, all without the influence of a guiding creator. Instead, this process unfolds through trial and error [23]. These "self-organising systems" [1] generate novel configurations internally via self-referential feedback loops. That is how seemingly new developments arise, often with minimal involvement from the original initiator [23].

What affords the exaptation? Genetic redundancy [19, 29]. What causes genetic redundancy? That is nature being rogue, unsupervised.

## 3 Exaptation in Architecture

*Finality in the living world thus originates from the idea of organism, because the parts have to produce each other, because they have to associate to form the whole, because, as Kant said, living beings must be ' self-organized '—François Jacob, The logic of life, 1973.*

The forty-year-old term exaptation remains influential [36] and has spread to other fields, such as psychology [30], neuroscience [5], business [6] and, of course, architecture [31–33], urban studies [32], urban resilience [27] and urban informality [34].

In exploring exaptation in architecture, Alessandro Melis [31–33] proposed two main groups–niche construction and redundant structures. The latter is derived from the genetic redundancy mentioned in the previous session, usually observed in architectonics spandrels [4, 33]. Niche construction refers to the idea of a mutual adaptative relationship: the organism adapts to the environment, and the environment adapts to the organism, promoting a co-evolution of trait and constructed niche [39]. This interpretation seems appealing to architects and urbanists because the physical perturbations of the environment also include embedded cultures in environments, such as human agricultural practices [25]; thus emphasising that the development of the human-built environment is not merely a facet of urban evolution but an integral component of human evolution, both historically and presently. Put differently, as [24] asserts, "we co-evolve with the environment we create." Narraway et al. [37] articulated the following niche construction analogy to cities: *"[...] cities are actually ecosystems; our built environments are actually constructed niches; and our built environments have a material effect on our ongoing evolution. Furthermore, these three statements are linked and can be made more visually explicit* via *the biotic perspective."*

Pievani and Serrelli [42] use a complementary approach to distinguish two types of exaptation. First is exaptation as a functional co-optation, meaning a functional shift in a structure that already had a function. A biological example is the feathers of the birds mentioned. In architecture, it can be interpreted as a new functionalisation of existing structures not previously envisaged in the original design. Such interpretation bred a taxonomy: (1) the functionalisation of existing geomorphologies; (2) the re-functionalization of existing designed structures; (3) the integration of new functions to a certain architecture; (4) the change of use [33] and (5) temporary appropriation of space [33]. The second is exaptation as the use of structures without any prior known function. It refers to the redundant portions of a system, like spandrels in architecture.

For Melis and Pievani [32], the re-functionalization of an architectural structure might take place formally, through a conventional design, or informally, as a gauge of the social and communities' resilience [27]. It is essential to distinguish what is being called informal. The authors [33] use the words "deterministic" and "formal" to differentiate what they call "informal", "indeterministic", and "non-deterministic" design or architecture. In their view, transformations or functionalisations delineate a generic idea of informal, which, in turn, they interpret broadly and not exclusively to define slums. For them, informality concerns the uses of space that are not planned in any conventional design. It may happen both spatially—via functional re-purposing and expansion toward unexpected uses—and dynamically, as seen in informal behaviours and temporary appropriations of public and semi-public space worldwide [27].

Melis distinguishes adaptation as the architectural design per se, which assigns a function a priori to the designed structures. At the same time, exaptation is a form of adaptation-non-adaptation because it occurs by functional co-option a posteriori. Melis has a thought-provoking interpretation that instigates reflections. However, a salient observation that becomes apparent is his focus on function. Although the debate around function and form is complex and long-established in Architecture, Dawkings [11] shows that it differs from Biology in at least two seminal points: purpose and agency. In his book "The Blind Watchmaker", Richard Dawkins [11] argued that the "design" by a designer—whom he called the watchmaker—would never excel the "design" perfected by natural selection, evolution, and exaptation in elegance, complexity, and finesse. In Dawkins' [10, 11]) view, biological functions emerge from a long process of evolution, devoid of any a priori purpose or intelligent agent's deliberate design. In contrast, engineers, architects, and designers are conscientious agents who create design stances with specific purposes in mind, even in functional repurposing of spaces a posteriori. Looking through Monod's [35] lenses, in Biology, a function is a contingency, in Architecture, it is a necessity. Although the function of an architectural structure can change over time, it is always driven by human necessity. Ultimately, the very notion of function is human. Back to Darwin:

*"Man can act only on external and visible characters: Nature, if I may be allowed to personify the natural preservation or survival of the fittest, cares nothing for appearances, except in so far as they are useful to any being. She can act on every internal organ, on every shade of constitutional difference, on the whole machinery of life. Man selects only for his own good; Nature only for that of the being which she tends." Darwin,* [8].

Indeed, favelas are places par excellence for constant change, innovation, appropriation, and adaptation. But the question is: which alternative investigations, other than cataloguing and classifying those spatial modes could express exaptation in those complex environments? What are the architecture's internal emergent traits that would favour the 'unsupervised break', the 'exaptation' that occurs internally rather than externally designated function by humans?

## 4 Rendered Parasitism

*I'd like to share a revelation I've had during my time here. It came to me when I tried to classify your species. I realised that you're not actually mammals. Every mammal on this planet instinctively develops a natural equilibrium with their surrounding environment, but you humans do not. You move to another area, and you multiply, and you multiply, until every natural resource is consumed. The only way you can survive is to spread to another area. There is another organism on this planet that follows the same pattern. Do you know what it is? A virus. Human beings are a disease, a cancer of this planet. You are a plague, and we are the cure.*

*—Agent Smith, The Matrix, 1999*

Our case study is an artefact conceived in the context of an academic exhibition called Students as Researchers, a Collateral Event held by the NYIT at the Venice Biennale 2023, with exaptation as the guiding concept. Specific details about the teaching approach and the fabrication methods are detailed in Natividade [38]. In this session, we will cover its philosophical and technical approaches.

From a philosophical point of view, two main works imposed their significance in this project. First is Michel Serres' concept of "parasite" [46]. In the biological realm, a parasite infiltrates an organism's body and absorbs substances intended for the host. The system, be it a city or any other entity, is considered primary, while the parasite, such as favelas, is deemed a disruptive addition that should be eradicated [51]. However, this perspective overlooks the integral role of the parasite within the system. By encountering and assimilating disturbance, the system evolves from simplicity to complexity. Thus, through its capacity to perturb, the parasite ultimately becomes a catalyst for the system's potentiality. In this way, "the parasite attests from within order the primacy of disorder; it produces by way of disorder a more complex order" [46]. Expanding those insights into the interplay of systems and disruptions to a broader context – the Earth scale – we are all parasites [47]. We depend on Earth, not the other way around. We must give up our narcissistic humanism and re-centre: "we must indeed place things in the centre and us at the periphery, or better still, things all around and us within them like parasites" [47].

The second one is Byung-Chul Han's thought-provoking concept of "aesthetics of the smooth", elaborated in the publication "Saving Beauty" [20]. The author criticises the absolute objectivity of modern society that transforms, discretises, and calculates nature as if it were a mirror of humanity: oriented to efficiency, productivity and positivity, to be integrated as merchandise and valued by its utility. Han also inspires an intriguing conflict about the digital by opposing the "natural" and "digital" beauty and, thus, provides a critical guide to contemporary aesthetics. On the one hand, it helps us engage with massive datasets and variables. On the other hand, this same complexity is customarily dissolved into an "average", an "aesthetics of the smooth". Thus, Han denounces eliminating the otherness, the strange, the odd, and the consequent loss of the ability to appreciate beauty in unexpected and informal environments such as favelas.

Regarding the exhibition's theme of exaptation, in the sessions above we demonstrated that, in biology, the door to exaptation is genetic redundancy [19, 29]. In organisms, genetic redundancy is afforded by material fluxes [19, 29] which, in turn, happen in a self-organised manner [56]. Self-organisation [1] describes a process whereby overall order emerges spontaneously from the interactions of individual components within a system without external direction or control. Complex patterns emerge from this phenomenon at the global level of a system through interactions among its lower-level components. These interactions follow local rules and do not rely on global information. In essence, the resulting pattern is not imposed externally but is an emergent property of the system itself. The emergent properties are, thus, characteristics that arise unexpectedly from component interactions, requiring understanding these interactions rather than solely examining individual components' properties [3].

Self-organisation is observed across various scales in the natural world, from the formation of galaxies and weather patterns to the behaviour of social insect colonies and the organisation of cells in living organisms. It frequently plays a significant role in shaping constructed niches, whether among individuals or across environmental components. For instance, social insects construct highly complex nests through self-organised collective action guided by internal cues, communication among nestmates, or environmental factors [37]. Urban researchers have drawn inspiration from this body of knowledge to develop bottom-up approaches for generating urban order [2, 43] tactical urbanism [48] and self-built interventions [37].

Essentially, self-organisation exemplifies the interconnected concept of emergence [7]. Emergence plays a crucial role in any creative process, including generative ones, largely explored in architecture. John S Gero [17], an experienced author and professor on this subject, defines it as follows: *"A property that is only implicit, i.e. not represented explicitly, is said to be an emergent property if it can be made explicit. Emergence is considered to play an important role in the introduction of new schemas and consequently new variables. Emergence is a recognised phenomenon in visual representations of structure. It maps directly onto the concept of changing schemas since a new schema is generally needed to describe the emergent property"*. Emergent properties are not exclusive to structure; they can also arise from behaviour and function [17].

In the end, self-organisation is "an inherent and materially immanent principle of life" anticipated in Ludwig von Bertalanffy's seminal work "Kritische Theorie der Formbildung" [56]. Placing all the pieces of the puzzle together—exaptation, genetic redundancy, self-organisation, and emergency—it becomes evident that material fluxes serve as the fundamental foundation. Thus, the materiality seems to be a convincing way to experiment with exaptation in favelas. Through these lenses, we proposed a piece to express the favelas' material emergent patterns. While favelas' urban fabrics may look purely entropic and unreasonable to an external observer, a closer examination reveals careful calculations, space optimisations and hidden patterns, rich in diversity, variation, and functional redundancies that create those specific niches [38]. So, how could we express the favelas' emergent patterns, and what AI method would convey that?

We elected SOM, an unsupervised learning method that embodies the natural principle of self-organisation to visualise and group complex data by finding patterns and similarities. In 1982, the Finnish computer scientist Teuvo Kohonen introduced a new nonlinearly projecting mapping that he called a self-organising map (SOM). The algorithm plots the overall structure of a given unstructured dataset in one image, grouping similar data closer together while dissimilar ones are placed farther apart. This organisation is achieved through the natural-inspired processes of competitive learning, where neurons compete to become the best match for a given input pattern, and cooperative learning, where neighbouring neurons influence each other's weights to create a topological map that preserves the spatial relationships between input patterns. The resulting grids are not traditional topographic maps but are inspired by neural cells' selective responsiveness to stimuli, forming feature-sensitive assemblies akin to brain maps of real-world events.

To compose the Favelas' SOM, the team filtered 1244 favela-related images in the crawled dataset and cropped them into $224 \times 244$ pixels, originating 25,515 crops. This pixel size fits the VGG16 transfer learning model used in the task. The dataset was split into 12 categories of materials observed from dataset: asphalt, bricks, plaster, plastic, zinc, wires, concrete, humans, sky, vegetation, panoramic and objects. The objective was not to picture favelas' materiality precisely but to have a flavour of their dominant elements.

The final squared SOM with 1,444 squared pictures plotted on a 38 by 38 grid, respecting the classes' proportions. Similar-looking images were placed together, forming clusters while maintaining the integrity of class categories—such a procedure created amusing border zones between types. Then, the algorithm computed the cosine proximity between every image in the grid and the remaining pictures in the dataset. Resembling images were stacked together in the grid with

**Fig. 1** Samples from the dataset. *Source* Author, 2023

the same (x, y) coordinates. Finally, the point positions with the corresponding RGB values were exported to an Excel sheet and translated into geometry in Rhino/Grasshopper. Higher sticks mean more recurrent materials in the dataset. The colours indicate the dominant RGB value of that particular image. The RGB file was extracted using Python code.

As a result, the proposed design reflects a constellation of ideas that are not opposed but rather work as questions or sides of the same unified world that create one shared *metaxu*: natural and artificial, favelas and city, smooth and rough, entropy and negentropy, organic and inorganic, discrete and continuous, transparent and opaque, homogeneous and heterogeneous, negative and positive, agency and inertia, material and immaterial, rational and irrational, formal and informal, form and inform.

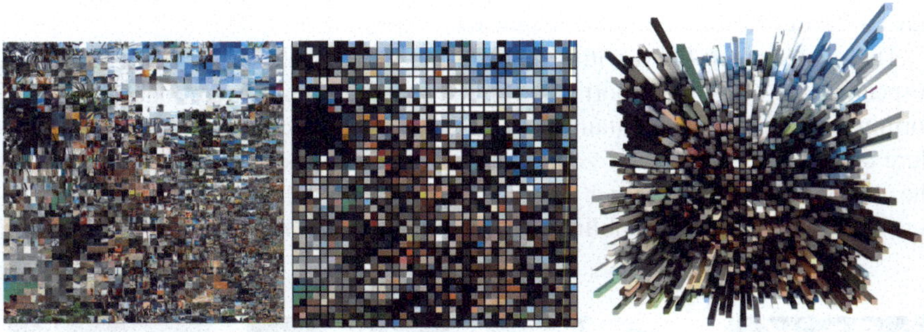

**Fig. 2** Favelas' self-organising map and its translation to RGB and 3D. *Source* Author, 2023

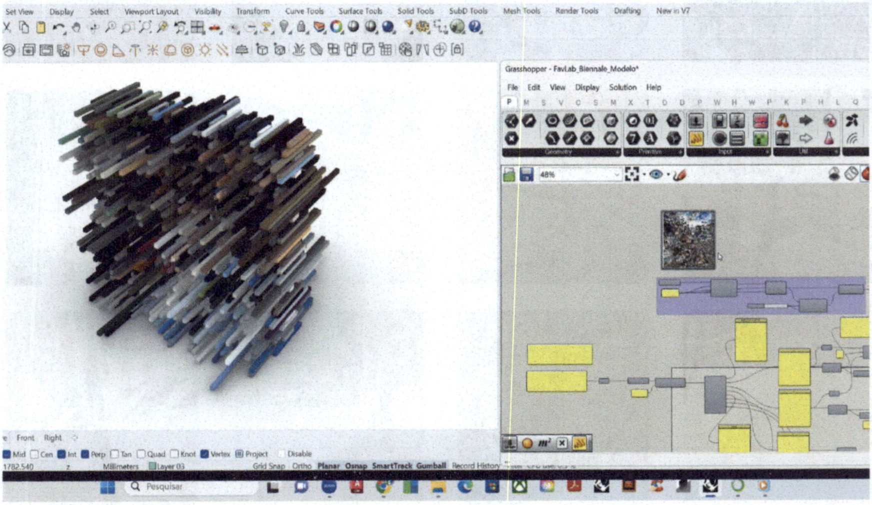


**Fig. 3** Translation from Excel into a geometry in Rhino/Grasshopper. *Source* Author, 2023

**Fig. 4** The object displayed at the Collateral Event "Students as Researchers", Venice Biennale, 2023. CENTRO STUDI E DOCUMENTAZIONE DELLA CULTURA ARMENA. *Source* Author, 2023

## 5 Discussion

The object illustrated in this paper was crafted to express favelas' material emergent patterns within the framework of an academic exhibition guided by the concept of exaptation. While it does not inherently make a political statement, the ethical and social considerations were not overlooked. Specifically, we call attention to three main issues. First, favelas and their variations worldwide are usually defined by their deficiencies or absences. For instance, in a 29-page report by UN-Habitat [54] on adequate housing and slum upgrades, the term 'lack' is used 24 times. From an external perspective, whether in Rocinha or Seelampur, those environments are typically overshadowed by perceptions of impoverishment and dysfunctionality [49]. However, construing them exclusively in terms of what they lack is deemed insufficient [50], as it confines favelas to an entrenched identity associated with 'chaos'.

It leads us to the second aspect, which precisely concerns the prevailing notion of 'order' and its perceived antithesis, 'chaos', particularly in the context of the perception of 'urban chaos' dominating representations, definitions, and interventions in favelas. Souza et al. [50] challenge the inadequate portrayal of favelas as lacking order, arguing that daily coexistence practices have fostered a complex system of codes and social interactions within the absence of state regulation and supervision. Hence, the Self-Organizing Map (SOM) serves not merely as a representation but as a translation of a "displaced rationality" to be juxtaposed with the existing notion of "real rationality" [14].

In this framework, our aim was to present an alternative perspective to Virilio's [55] pessimistic view on the inundation of media, especially images. Instead of allowing internet-derived memories to deteriorate into "junk shops—repositories of diverse and unrefined images, haphazardly accumulated" [55], devoid of narrative or historical context [21], we leveraged the logical architecture of AI to establish causal connections among these images. Through this intentional approach, we intended to discuss the conventional attribution of causality to favelas, resulting in the creation of a novel narrative around those mysterious landscapes that reconcile the structured "order" of the grid with the unstructured "formlessness" of clusters.

Thirdly, the emphasis on transformative evolution rather than fixed identity [21] orchestrates a dynamic interplay between illuminatedreason, facilitated by AI, and the obscurity concealing latent patterns. Recognising each discernible pattern presents an opportunity for subsequent interaction. Viewing exaptation and favelas through the lens of materiality unveils numerous research avenues, such as exploring the modes and materials of urban production and work organisation in their construction as crucial agents for urban transformation [45].

Undoubtedly, inequalities must be overcome in all their senses, including the urban conditions of people's lives and housing. Nevertheless, it is imperative to acknowledge and integrate sociocultural diversities into urban policies while avoiding imposing rigid standards and norms that suppress diverse experiences and communal utilisation of space. This approach does not endorse the contemporary trend of sugarcoating negativity in the name of promoting a positive society, as highlighted by [20], and exemplified by slogans like "favela has won," that replaced the 20th-century notion of "cities without slums" [9]. Instead, the goal is to leverage research to identify more effective strategies for tackling issues and breaking the cycle of pointing fingers without addressing the root causes or providing viable future paths.

## 6 Final Remarks

This chapter endeavoured to develop a distinctive comprehension of architectural exaptation in the context of favelas. Grounded in the ethos of discovery, we delved into the primary sources from biology researchers to engender a critical re-evaluation of exaptation's implications in architectural discourse and advance an interdisciplinary framework accommodating subject complexities. The case study materialised in an artefact was shaped by insights gleaned from philosophical considerations and a sophisticated machine learning technique that unveils hidden materiality patterns within favelas. Inspired by biological mechanisms, our experiment proposes an alternative interpretation to the conventional approach of seeking ever more refined classifications and categorisations. Instead, we advocate for a more abstract understanding of exaptation, seeking its essence in the untapped potential of material performance. Instead of extracting materials like "parasites" do, re-use already extracted resources. Instead of prioritising pure, "smooth" materials, create new composites with exapted performances. Like human bones that structure the body while storing and regulating calcium [16], construction bricks may also sequester carbon, besides their primary function of enclosing space. Like the birds' feathers that thermally isolate and ornate, patterns in building envelopes may enhance environmental parameters, such as daylight, structural strength and thermal and acoustic parameters [28]. The possibilities are immense if we are "enough artists to draw freely upon imagination. Imagination is more important than knowledge. Knowledge is limited. Imagination encircles the world[1]" (Einstein, 1929).

## References

1. Ashby WR (2013) Design for a Brain Springer Netherlands Dordrecht
2. Batty M (1998) Urban evolution on the desktop: Simulation with the use of extended cellular automata. Environ Plan A: Econ Space 30(11):1943–1967
3. Bonabeau E, Theraulaz G, Deneubourg JL (1999) Dominance orders in animal societies: the self-organisation hypothesis revisited. Bull Math Biol 61(4):727–757
4. Buss DM, Haselton MG, Shackelford TK, Bleske AL, Wakefield JC (1998) Adaptations, exaptations, and spandrels. Am Psychol 53(5):533
5. Buzsáki G, Tingley D (2023) Cognition from the body-brain partnership: exaptation of memory annual review of neuroscience, 46(1):191–210, https://doi.org/10.1146/neuro.2023.46.issue-1, https://doi.org/10.1146/annurev-neuro-101222-110632
6. Codini AP, Abbate T, Petruzzelli AM (2023) Business model innovation and exaptation: a new way of innovating in SMEs. Technovation 119:102548
7. Crommelinck M, Feltz B, Goujon P (Eds) (2006) Self-organization and emergence in life sciences (p. 360). Dordrecht, The Netherlands: Springer
8. Darwin C (1859) On the origin of species
9. Davis M (2013) planet of slums new perspectives quarterly 30(4):11–12. https://doi.org/10.1111/npqu.11395
10. Dawkins R (1976) The selfish gene. Oxford University Press
11. Dawkins R (1986) The blind watchmaker. W.W. Norton & Company
12. Donoghue PC, Purnell MA (2005) Genome duplication, extinction and vertebrate evolution. Trends Ecol Evol 20(6):312–319
13. Elhacham E, Ben-Uri L, Grozovski J, Bar-On YM, Milo R (2020) Global human-made mass exceeds all living biomass. Nature 588(7838):442–444. https://doi.org/10.1038/s41586-020-3010-5
14. Fanon F (2008) Black skin, white masks (R. Philcox, Trans). Grove Press
15. Feltz B, Crommelinck M, Goujon P (2006) Self-organization and emergence in life sciences kluwer academic publishers dordrecht
16. Frenkel-Pinter M, Petrov AS, Matange K, Travisano M, Glass JB, Williams LD (2022) Adaptation and exaptation: from small molecules to feathers. J Mol Evol 90(2):166–175
17. Gero JS (1996) Creativity, emergence and evolution in design. Knowl-Based Syst 9(7):435–448
18. Gould SJ (2002) The structure of evolutionary theory. Harvard University Press
19. Gould SJ, Vrba ES (1982) Exaptation—a missing term in the science of form. Paleobiology 8(1):4–15
20. Han BC (2017) Saving beauty. John Wiley & Sons
21. Han BC (2015) The transparency society. Stanford university press
22. Hofkirchner W (2019) Social relations: building on Ludwig von Bertalanffy. Syst Res Behav Sci 36(3):263–273
23. Hovestadt L, Hirschberg U, Fritz O (Eds.) (2020) Atlas of digital architecture: terminology, concepts
24. Kropf K (2017) The handbook of Urban morphology. Wiley, Chichester

[1] "What Life Means to Einstein: An Interview by George Sylvester Viereck" in The Saturday Evening Post (26 October 1929).

25. Laland KN and O'Brien MJ (2010) Niche construction theory and archaeology
26. Lang S, Walz U, Klug H, Blaschke T, Syrbe RU (2008) Landscape metrics–a toolbox for assessing past, present and future landscape structures. In: Geoinformation Technologies for Geo-Cultural Landscapes: European Perspectives. CRC Press, pp 223–250
27. Lara-Hernandez JA and Melis A (2018) Understanding the temporary appropriation in relationship to social sustainability. Sustain Cities Soc, 39:366–374 https://doi.org/10.1016/j.scs.2018.03.004
28. Leschok M, Cheibas I, Piccioni V, Seshadri B, Schlüter A, Gramazio F, Kohler M, Dillenburger B (2023) 3D printing facades: Design, fabrication, and assessment methods. Autom Constr 152:1–18
29. Losos JB (2011) Convergence, adaptation, and constraint. Evolution 65(7):1827–1840
30. Mastrogiorgio A, Felin T, Kauffman S, Mastrogiorgio M (2022) More thumbs than rules: Is rationality an exaptation? Front Psychol 13:805743
31. Melis A, Lara-Hernandez JA, Melis B (2022) Learning from the biology of evolution: exaptation as a design strategy for future cities. Smart Sustain Built Environ 11(2):205–216
32. Melis A, Pievani T (2022) Exaptation as a design strategy for resilient communities. Transdisciplinarity. Springer International Publishing, Cham, pp 307–327
33. Melis A, Lara-Hernandez JA, Foerster B (2020) Learning from the biology of evolution: exaptation as a design strategy for future cities. Paper presented at The 54th International Conference of the Architectural Science Association (ANZAScA). University of Portsmouth, Portsmouth, United Kingdom
34. Melis A, Pievani T, Lara-Hernandez JA (2024) Architectural exaptation: when function follows form. Taylor & Francis
35. Monod J (1974) On chance and necessity. Macmillan Education UK, pp 357–375
36. Moore DS (2023) On the evolution of epigenetics via exaptation: a developmental systems perspective. Ann N Y Acad Sci 1529(1):21–32
37. Narraway C, Davis O, Lowell S, Lythgoe K, Turner S, Marshall S (2019) Biotic analogies for self-organising cities. Retrieved from https://doi.org/10.1177/2399808319882730
38. Natividade V (2023) Metaxu: Reading favelas' material landscapes with self-organised maps (2023). García Amen F, Goni Fitipaldo AL and Armagno Gentile Á (eds), accelerated landscapes – Proceedings of the XXVII international conference of the ibero-american society of digital graphics (SIGraDi 2023), Punta del Este, Maldonado, Uruguay, 1785–1796 https://papers.cumincad.org/cgi-bin/works/Show?sigradi2023_345
39. Odling-Smee F, Laland K, Feldman M (2003) Niche construction: the neglected process in evolution. Princeton University Press, Princeton
40. Ohno S ([1970] 2013) Evolution by gene duplication. Springer Sci Bus Media
41. Ortlund EA, Bridgham JT, Redinbo MR, Thornton JW (2007) Crystal structure of an ancient protein: evolution by conformational epistasis. Science 317(5844):1544–1548
42. Pievan T, Serrelli E (2011) Exaptation in human evolution: how to test adaptive versus exaptive evolutionary hypotheses. J Anthropol Sci 89:9–23
43. Portugali J (1997) Self-organizing cities. Futures 29(4–5):353–380
44. Roy A (2009) The 21st-century metropolis: new geographies of theory. Regionalstudies 43(6):819–830
45. Santos M ([1978] 2013) Pobreza Urbana. São Paulo, Edusp
46. Serres M ([1980] 2013) The parasite. Baltimore, MD: johns hopkins university press
47. Serres M ([1990] 1995) The natural contract, trans. E. MacArthur and W. Paulson. Ann Arbor: university of michigan press
48. Silva P (2016) Tactical urbanism: towards an evolutionary cities approach? Environ Plan B 43(6):1040–1051
49. Simone AM (2019) Improvised lives: rhythms of endurance in an Urban South, after the postcolonial. Hoboken: Wiley
50. Souza J, Barbosa JL, Simão MP (2020) A favela reinventa a cidade. Mórula
51. Stokes CJ (1962) A theory of slums. Land Econ 38(3):187–197

52. Taylor D (2003) The archive and the repertoire: performing cultural memory in the Americas. Duke University Press
53. Thompson D'Arcy (1917) On growth and form
54. UN-Habitat (2020) Monitoring and reporting the SDGs | ADEQUATE HOUSING AND SLUM UPGRADING. Retrieved from: https://unhabitat.org/sites/default/files/2020/06/indicator_11.1.1_training_module_adequate_housing_and_slum_upgrading.pdf
55. Virilio P (2005) The information bomb (Vol. 10) Verso
56. Von Bertalanffy L (1928) Kritische theorie der Formbildung (No. 27). Gebrüder Borntraeger
57. Yates FE (Ed) [1987] (2012) Self-organizing systems: The emergence of order. Springer Sci Bus Media

# Redundancy: From Adaptation to Aptation Design

7

Alessandro Melis, Jose Antonio Lara-Hernandez, and Paola Boarin

**Abstract**

This chapter explores the concept of redundancy within architectural design, challenging traditional deterministic views by introducing evolutionary principles. Historically, design has been perceived as a human-centric, goal-oriented process. However, this chapter posits that redundancy, variability, and diversity are crucial attributes of robust architectural systems, especially in times of global crises. By examining redundancy from an evolutionary biology perspective, the chapter highlights how traits initially evolved for one function can be repurposed, contributing to adaptability and resilience. This approach, termed "aptive design," encompasses both deterministic and non-deterministic strategies, offering a broader understanding of creativity and functionality. Through case studies from Algiers, Naples, and other urban contexts, the chapter demonstrates how informal and vernacular practices, often seen as marginal, play a vital role in urban resilience. By embracing redundancy, architects can innovate in sustainable design, repurposing materials to create adaptable structures. This evolutionary perspective on design challenges the primacy of human intentionality, suggesting that architectural evolution, like biological evolution, thrives on redundancy and adaptability.

A. Melis
New York Institute of Technology, New York, USA
e-mail: amelis@nyit.edu

J. A. Lara-Hernandez (✉)
Auckland University of Technology, Auckland, New Zealand
e-mail: antonio.lara-hernandez@aut.ac.nz

P. Boarin
University of Auckland, Auckland, New Zealand
e-mail: p.boarin@auckland.ac.nz



J. A. Lara-Hernandez et al. (eds.), *The Dusk of Design*, Integrated Science 34,
https://doi.org/10.1007/978-3-031-76867-5_7

**Keywords**

Redundancy • Evolutionary design • Exaptation • Resilience • Informality • Urban resilience • Sustainable design

## 1 Expanding Horizons

In the realm of architecture and design, conventional views often attribute deterministic and demiurgic qualities to the design process. This perspective, deeply entrenched in human-centric ideologies, has dominated the field for centuries. However, as Bernard Rudofsky highlighted in the "Architecture Without Architects" catalogue [10], design can encompass much more than these traditional confines. This book seeks to challenge and expand upon these established views by introducing categories of design that move beyond deterministic and teleological approaches.

These categories represent approaches that do not conform to the linear, goal-oriented processes typically associated with design, especially in architecture. Historical perspectives have often sacralized a human-centric categorization in design, influenced by religious, philosophical, and epistemological frameworks, which have reinforced a view of design as a systematic and progressive activity primarily focused on individual creativity. Consequently, observations of design activities tend to be recorded to conform to established categories and taxonomies, fitting within a framework that does not disrupt traditional categorizations [7].

This perspective overlooks the opportunities presented by design practices that do not fit neatly into conventional categories. For instance, informal and vernacular design, conservation and restoration, as well as organic and empirical approaches to design—recognized more prominently from modernity onwards, which do not involve a systematized, goal-oriented idea of creativity—are well known, but still often seen as collateral rather than central to the discipline.

Even the recent commitments in design discourse, focusing on addressing inequality and non-western-centric perspectives, focus often exclusively on ethical positions on equality and diversity, overlooking their evolutionary potential.

## 2 Redundant Architecture

"The Dusk of Design" has been written as a call for a broader and more inclusive understanding of design, recognizing the significance of diverse approaches that have been historically marginalized.

By moving beyond the confines of traditional, human-centric categorizations, the field of design can embrace a more comprehensive view that values the relationship between resilience, and changing environment solutions.

These expanded perspectives offer an opportunity to extend our understanding of design, recognizing the richness and redundancy inherent in diverse design practices.

The term "redundancy" originates from the Latin "redundantia," meaning "overflow" or "abundance." The Latin root "redundare" combines "re-" (again) with "undare" (to surge), indicating an excess or surplus. Over time, this concept evolved to imply unnecessary repetition or duplication. The Oxford English Dictionary defines it as "the state of being not or no longer needed or useful; superfluity." Merriam-Webster describes it as "exceeding what is necessary or normal; characterized by or containing an excess, repetition, or superfluity." The Cambridge Dictionary defines it as "a situation in which something is unnecessary because it is more than is needed."

Thus, redundancy can be perceived as both an excess and a beneficial surplus, depending on the context.

In literature and architecture, the ciriticism is particularly high as, redundancy refers to the presence of stylistically elaborate elements that are exclusively formal and ornamental. Particularly in architecture, redundancy often carries a negative connotation, suggesting an overabundance of decorative elements that may detract from functional or aesthetic purity.

In technical contexts, redundancy is a crucial characteristic of systems or equipment designed to increase reliability. It involves incorporating multiple elements capable of performing the same function. This way, a failure in one element does not cause the malfunctioning of the system or equipment, as the specific function can be performed by another element. This approach ensures continuous operation and enhances the overall robustness of the system.

In research, redundancy often refers to the repetition of experiments or data points to ensure reliability and accuracy. It acts as a safeguard against errors, providing a means to verify results and enhance the robustness of findings. In quantitative research, redundancy is used to validate statistical results by replicating studies or collecting additional data. In qualitative research, redundancy involves data triangulation, where multiple sources or methods are employed to corroborate findings.

The critical role of redundancy in maintaining the integrity and credibility of scientific research highlights its importance. This principle can also be observed from an evolutionary biology perspective, particularly in the concept of exaptation.

In evolutionary biology, redundancy manifests through the duplication of genes or functions, which can lead to significant evolutionary advantages [3] . Genetic redundancy refers to the presence of multiple genes with similar functions. This redundancy can provide a backup in case one gene mutates or becomes non-functional.

In exaptation a redundant trait may acquire a new function, contributing to an organism's adaptability [11]. For example, feathers in birds initially evolved for insulation but later became crucial for flight. This redundancy in the form of feathers illustrates how surplus traits can be repurposed for new uses, aiding in survival and diversification.

Gould and Vrba [8] also introduced the term "aptation" as a broader category to encompass both adaptations and exaptations (1982). An aptation is any characteristic that provides a function, whether it evolved for that purpose (adaptation) or was co-opted for it (exaptation). Aptive design, therefore, can be defined as the strategy that allows for both deterministic and non-deterministic design.

While the deterministic design, based on the traditional notion of creativity, has been the backbone of architectural literature [2], the possibility of exaptive design, which is pivotal in this book, is only marginally represented in literature, and only through reductionist categories such as conservation, informality, and vernacular design [1]. Adaptive interpretation of design, intended teleologically, is both quantitatively and qualitatively dominant in conventional literature together with the conviction that it belongs to a singular evolutionary line.

If, instead, we accept the notion of exaptive design, which assigns equal importance to both adaptive and exaptive strategies, features like redundancy, variability, and diversity will be crucial in understanding the functional co-optation of creative structures, which is pivotal for design in times of rapid environmental changes. For instance, when a trait or structure in an organism evolves for a new function (exaptation), it doesn't necessarily carry forward the exact purpose that was passed down through generations (ontogenetic inheritance). The present utility of a certain structure might be completely different from its primary function, which was selected for a past utility. This is similar to the case of dinosaur feathers [4], where a fortuitous transition occurred from a primary function (e.g., thermoregulation) to a secondary function (e.g., flight). In architecture, we might say that the characteristics of an archetype are not necessarily transferred to subsequent generations, as in the case of dolmen. While in evolutionary design the term adaptive highlights only its teleological nature, which is complementary to exaptation, paradoxically, in architecture, the term adaptive seems to float still today in a limbo as if even conventional design, implying intentionality, should not, by definition, have always adaptive significance.

We now realize that even the human habitat is not immutable, along with our needs and aspirations, but still we maintain a limited perspective of aptive design, as we exclude exaptation. It is a matter of broadening our perspective, not merely to seek alternatives, but to recognize that the notion of functional perfection—embodied in the concept of the right shape for the right use—is not our only option [9]. The "irrational" placement of columns on the façade of Palazzo Branconio Dell'Aquila, by Raffaello, along with numerous examples from Mannerism, Baroque, and Rococo, illustrates how the possibility of aptive design has sparked extensive and enduring literary debates. These styles deviate, in fact, from classic mimetic theory, which attributes absolute perfection to nature and questions why human designers, like Raffaello, move away from that ideal harmony. Difficult to be understood other than in terms of a self-referential attitude or philosophical concerns, in evolutionary terms, the principle of functional redundancy explains these changes in use.

Every complex characteristic contains a wide margin of functional potential, an extended range of possible uses. Evolution appears as the intertwining of an "anomaly" (in a technical and not just evocative sense, meaning the continuation of a past, hereditary, or selective function in the present) and a reasonable adaptation (adaptation through subsequent selection). The smile of the flamingo [5] is the result of the inversion of the animal's beak, caused by a reversal in feeding posture. The animal developed an anomalous behavior (eating with its head upside down, filtering food), which led to an inversion of the shape and size of the two branches of the beak (hence the inverted smile line). The several examples of organisms living upside down provides the basis for a formulation of exaptation oriented towards changing behavioral rules.

The mannerism of the flamingo's smile, like an inverted tympanum, is therefore the result of a sequence of evolutionary events: some individuals and then populations develop a radical change in lifestyle, which in architecture is reflected in the tastes and interests of artists who redesign the architectural structure of a building in a completely new way, reversing the functions of the parts; the construction plan of the organism is disrupted due to the fixation by natural selection of the new behavior (the adaptive pressure changes the structure as a whole); however, selection cannot arbitrarily change the shape, arrangement, and size of the organs; it reuses the available material, that is, the parts of the previous construction and functional plan, relative to new functions.

A critique of this analogy might be that, in the case of the flamingo, biologists explain a clear practical functionality of the change, while in the case of Baroque redundancies, this functionality is not always clear. However, the hypothesis of evolutionary design also challenges the creationist role of humans, starting from its intentionality as exclusive driver. Humans, as one of the many complex subsystems of evolution, may find it difficult to have a comprehensive perspective. Similarly, while we may understand something about the flamingo's beak, we have probably missed many other organ transformations.

We might even be a flamingo's beak ourselves or realize there is no difference between a human body and a Rococo decoration: what is the utility of humanity? Utility for whom or what? The empirical evidence of adaptationism is, in fact, more simply that redundant, variable, and diverse structures facilitate functional co-optation, which does not need to happen for every structure. Their extreme complexity, in evolutionary terms, might render this function unreadable to humans. For example, the preference for certain forms could be a secondary transformation of other structures or respond to needs whose utilitarian impacts might be reflected in other behaviours that we do not directly recognize. The result of this evolutionary bricolage can be a condition of suboptimal adaptation. The most characteristic form of exaptation is thus illustrated by the imperfections of an adaptation that can rely only on structural resources limited by the previous construction plan. Imperfection is then the authentic sign of history, and the mechanism of exaptation illustrates the nature of natural history better than any other.

Thus, in general statistical terms, the notion of exaptation, capturing the link between morphological potential and the production of functional novelty through a sort of opportunistic assembly, introduces an important principle of redundancy into the conception of history as the "foundation of creativity" [6], p. 131). Evolution is a "process overflowing with redundancy" (ibid.). Functional flexibility is directly proportional to the ability of organisms to respond creatively to changes in environmental rules, and since rule changes are the norm in a contingent process of transformation, flexibility is directly proportional to the possibility of survival: complex organisms exist through imperfection, multiplicity of use, and redundancy [9].

The principle of redundancy as an epistemological extension representing change in complex and evolutionary systems (natural or socio-cultural) is not new. Gould, in particular, was fascinated by the idea that even technological progress is subject to mechanisms of functional exaptation and a general principle of creative flexibility.

In design, exaptation can effectively extend the range of possibilities as it does not exclude adaptation. Both exaptation and adaptation can serve as pathways of change in evolution, depending on which function is active and when in the phylogeny of a group.

An architectural phylogeny may provide an evolutionary history and relationship among various architectural components, much like phylogeny does for species or groups of organisms (Gould, 1977). It can be a method for understanding how different structures, such as domes, tholos, arches, and various types of vaults, evolved over time. These evolutionary paths can be depicted in phylogenetic trees, representing the divergent evolutionary paths taken by different architectural elements and tracing back to common archetypes.

Exaptation highlights the evolutionary significance of polyfunctionality, which is essential for any functional shift, and redundancy, which, despite being a cost to the organism, provides a repertoire of potential exaptations. If highly functional new genes can emerge from what was once considered genetic "junk," redundancy is not necessarily something that natural selection eliminates as useless. Instead, redundancy may be tolerated as a source of evolutionary change (evolvability), particularly during periods of ecological instability.

This concept can also be applied to architectural design, where redundancy can transform the perception of waste from something useless to an opportunity. It opens up ecological ideas regarding the maintenance and recycling of components, even in a non-deterministic way. By viewing redundancy as a resource, architects can innovate in sustainable design, repurposing and recycling materials to create resilient and adaptable structures (Lara-Hernandez and Melis, 2018).

## 3 Redundancy in Architectural Design: A Collection of Case Studies

As described in the previous sections, in the realm of architectural design, redundancy is often perceived through a prism of inefficiency or excess. However, when scrutinized through the evolutionary lens of aptation, redundancy might reveal itself as an essential attribute of robust architectural and urban systems, especially in times of global crises. This section of *Dusk of Design* brings together a series of case studies that highlight innovative approaches to integrating redundancy in design.

Mohamed Yazid Khemri in Chap. 8 opens the discussion with an examination of El Houma in Algiers, presenting a compelling case for the role of redundancy and informality in urban life. Khemri's research elucidates how exaptation—where existing structures acquire new functions—fosters a vibrant and resilient urban fabric. Through fieldwork and comparative analysis, this chapter challenges prevailing prejudices against urban informality, demonstrating its critical contribution to the dynamism and resilience of city life.

Consistently, Maria Simioli and Guido Robazza in Chap. 9 offer a critical perspective on urban informality through their study of Naples. They explore the significance of redundancy through the concept of superfetation—informal growth and adaptation—as a core element of urban resilience. By framing cities as Complex Adaptive Systems, Simioli and Robazza advocate for planning methodologies that embrace the fluidity and adaptability inherent in informal urban processes. Their case studies from Naples exemplify how grassroots creativity and informal practices can significantly enrich urban environments.

From the Mediterranean region of the previous chapters, Alexander Gomez Torres in Chap. 11 shifts the focus to crime in public spaces, where urban design principles are crucial.

Dustin White's exploration of material-based computational design strategies in Chap. 12 moves the discussion of redundancy from the urban scale to the design scale, bridging traditional craftsmanship with modern digital fabrication. Grounded in the theoretical frameworks of David Pye and Bob Sheil, this chapter explores the 'workmanship of risk' and its relevance to contemporary architecture. White's insights particularly describe the symbiotic relationship between computational tools and material innovation, underscoring the role of redundancy in fostering creative resilience.

Antonino Di Raimo in Chap. 13 rounds out the discussion about the relevance of multidisciplinarity as a way to extend architectural traditional silo categorization, with an exploration of trauma-informed design, a field at the intersection of architecture and criminology. His study focuses on the potential of transcribing trauma into architectural spaces through interdisciplinary research methods. By employing a blend of qualitative approaches and visual artifacts, Di Raimo reveals how architectural transcriptions can embody the narratives of trauma. This chapter underscores the impact of trauma on architectural design, advocating for a nuanced understanding that transcends conventional methodologies.

Together, these case studies provide a comprehensive exploration of redundancy in architectural design, highlighting its vital role in fostering resilience, adaptability, and creativity. Each chapter contributes to a deeper understanding of how redundancy, far from being merely superfluous, is integral to the creation of interdisciplinary, sustainable, and inclusive built environments.

## References


1. Melis A (2020b) Informal behaviour as a form of community resilience. In: Di Raimo A, Lehmann S, Melis A (eds) Informality now: informal settlements through the lens of sustainability. Routledge
2. Melis A (2022) The demiurgic position of architecture in times of environmental crises. In: De Finis G, Pecoraro C (eds), Periferi@. Castelvecchi
3. Darwin C ([1859] 1872) On the origin of species by means of natural selection or the preservation of favoured races in the struggle of life, 6th edn. John Murray, London
4. Gauthier J, De Queiroz K (2001) Feathered Dinosaurs, Flying Dinosaurs, Crown Dinosaurs and the Names "Aves". In: New perspectives on the origin and early evolution of birds: proceedings of the international symposium in honor of John H. Ostrom. Peabody Museum of Natural History, Yale University
5. Gould SJ (1985) The Flamingo's smile. W. W. Norton, New York
6. Gould SJ (1993) Eight little piggies. W. W. Norton, New York
7. Gould SJ (1996) Full house. Three Rivers Press, New York
8. Gould SJ, Vrba E (1982) Exaptation, a missing term in the science of form. Paleobiology 8(1):4–15
9. Pievani T (2022) Imperfection: a natural history. MIT Press, Cambridge
10. Rudofsky B (1964) Architecture without architects: a short introduction to non-pedigreed architecture. Museum of Modern Art
11. Gould SJ, (1980) Is a new and general theory of evolution emerging?. Paleobiology 6(1):119-130

# The Interplay Between Exaptation and Urban Life: Redundancy and Informality in El Houma

8

Mohamed Yazid Khemri

**Abstract**

El Houma is a socio-cultural urban concept in Algiers that represents urban life and the space that facilities it, and it is considered in this chapter, as an emblematic case to study architectural exaptation at the urban scale. Exaptation was used in this chapter, to understand the relationship between design elements and social activities, while the concept of redundancy highlights how various structures and elements can promote rich and diverse urban life. In this respect, El Houma as a case study was utilised to illustrate how people use and transform the meaning of urban spaces, in the city of Algiers, in an unexpected, spontaneous, and creative manner. Some of the activities mentioned in this chapter can be considered informal by the global literature, whereas this chapter attempts to address the prejudices that paint urban informality as a state of deregulation and investigates its contribution to the resilience of the urban fabric and the community. The chapter reports on the findings of a fieldwork carried out in three neighbourhoods in Algiers and of a comparative analysis of deterministic and indeterministic designs and their influence on urban life as a measure of Exaptation of the built environment. This research concludes that the intense El Houma is the more enjoyable city life is, and urban fabric is more resilient when and where it is more possible to have a variability of forms, spaces, and redundancy of structures.

**Keywords**

El Houma • Urban life • Exaptation • Redundancy • Deterministic • Indeterministic

M. Y. Khemri (✉)
Lectutrer at Birmingham City University, Birmingham, UK
e-mail: yazid.khemri@bcu.ac.uk



J. A. Lara-Hernandez et al. (eds.), *The Dusk of Design*, Integrated Science 34,
https://doi.org/10.1007/978-3-031-76867-5_8

# 1 Introduction

Informal architecture, understood instead as complementary to the deterministic one, affects a part of the human environment that goes far beyond the slums, or the criticism, often with a social or political background, on the international and globalizing aspiration of the modern movement, in the wake of seminal texts such as '*Architecture without architects: a short introduction to non-pedigreed architecture*' by [39], '*Housing by People: Towards Autonomy in Building Environments*' *by* [45], *and* '*Towards a Critical Regionalism: Six Points for an Architecture of Resistance*' by [11].

The influence / scope of the informal, however, involves many spheres and goes far beyond slums. In fact, in agreement with the Martin School of the University of Oxford, informal cities operate outside the common legal and regulatory framework and yet are intricately intertwined in the functioning of the city. Thus, informality is relevant to the spatial aspects of a city, as well as its economic and political organisation, and represents socio-cultural way of life of people [18, 30].

In this respect, informal in architecture and urban design prevails a wider vision and include urban phenomena like the social use of space, whereas recent studies such as those on the informal use of spaces [26, 30], show, for example, that the informal also concerns the legally admissible use of public and private urban spaces, and of architectural components outside the planning expectations.

However, architectural literature, starting from the pivotal texts of history, is aimed exclusively at the interpretation of architecture as the formal result of a formalized design or creative process, with a precise and usually unique function or final use of a certain place or space [2, 6, 40, 44, 48]. Similar to the early beliefs of Modernist Architects and planners, who underestimated existing urban design and architectural features, and advocated for the concept of radical city as the best approach to design cities ignoring local ways of life.

The lack of interest in the informal, understood as an urban fabric that is the result, in large part, of forms of functional co-option, and of non-design processes, is not, therefore, proportionate with respect to its actual impact. whereas the actual impact of informality in architecture and urban design can be linked to its contribution to urban resilience. For instance, the informal use of urban spaces not only grants people the right to the city and promotes social resilience but also expresses forms of functional co-optation of architectural elements.

Functional co-optation is a term used in biology of evolution to refer to process of exaptation. Exaptation which implies a process by which characters are co-opted to serve new and unexpected functions [15]. This concept will be further explained throughout the text.

This chapter attempts to explore analogies between biology of evolution and architectural design and informal behaviour. This analogy was also confirmed by many biologists who worked on principles of evolution and adaptation [35], which are often interchangeable with those used by architects when discussing architectural design and sustainable projects [33].

Considering the numerous possible analogies between biology of evolution and architecture, the present chapter will focus on the concept of exaptation and its contribution to novelty and resilience. The concept of exaptation will be used to study, analyse and describe an urban phenomenon which is El Houma. The chapter will rely on the seminal work of Stephen Jay Gould [15] on exaptation as the main reference of the study.

Hence, the principle aim of this research is to explore the relationship between the urban design landscape and behavioural patterns in achieving resilience through the lens of exaptation.

This chapter takes Algiers as a case study, and analyses patterns of social use of urban space as an illustration of the concept of El Houma, which is a term used in North Africa to refer to a neighbourhood or urban area that features a great level of social sustainability as a consequence of frequent social use of urban space, be it formal or informal. El Houma is formed of a combination of physical and social constructs, whereas the redundancy of architectural and urban design features is fundamental for maintaining enduring social use of urban space [21].

The choice of Algiers as a context of study is due to its pivotal position as an interface between global north and global south, as well as to the variety of the built environment features the city has accumulated through history, therefore, the investigation will be carried out on three neighbourhoods that represent three influential architecture and urban planning models, which are Medieval, nineteenth century urban planning and modernism.

The chapter will start by clarifying the concept of informality and its relevance and importance to current challenges, next, the chapter explains the concept of Exaptation and explores its analogy with architectural design and informal behaviour. After that, an overview of the concept of El Houma is presented. Finally, case study analysis to report the findings of observations and activity mapping to highlight the intensity of El Houma as forms of exaptation and redundancy of design features.

## 2 Urban Informality, Definitions, and Relevance To Current Challenges

This paragraph intends to demonstrate that the denial, or underestimation of indeterminism in architecture, has also had negative results with respect to the impacts of today's epochal crises. The most obvious example concerns the $CO_2$ emissions that are largely due to deterministic design. The per capita impacts of the designed buildings, especially after the heroic phase of modernism and after the second world war, are the main sources of emissions [3, 41]. Equally, the deterministic design approach employed in many developing countries led to the application of foreign-born models without contextualisation, which resulted in social isolation and degradation of sense of community [23].

If the deterministic approach can be considered the main cause of global crisis, it is worth investigating the potential role of informal behaviour in alleviating its impacts today.

Informality, as an antidote to determinism, offers a variety of scenarios and strategies that potentially can lead to resilience, unlike the invariant presumed rationality of the ordered city.

The complexity/richness of informality has seen the concept referred to in several studies in a more or less indirect way, such as the ecological position of the work of Serge [27] on 'degrowth', Timothy [34] work on the overcoming of the Anthropocene, and more directly, the works of [25] on Temporary appropriation, and Khemri and Melis [20, 22, 30], on the themes of urban life and community resilience.

Informality represents phenomena that are not prescribed by planners, architects or policy makers, and independent from any predefined plans or scenarios for the city and its urban spaces. In fact, informality emerges from people's aspirations and needs, and at the same time showcasing their cultural and social life [36]. Soto et al. [43] states that informality in urban design doesn't refer to individuals as informal, but their activities and way of life. For [38], informality is a complex notion and is difficult to precisely define because it represents a blurry relationship between authorised and unauthorised, legal and illegal, legitimate and illegitimate, for example, possessing and using a land while disregarding the prescribed laws.

Social use of space can be sometimes qualified as informal, when the activities are out of the ordinary or opposite to the intended objectives of the designed space, or what [8] considers unauthorised and unexpected activities. Devlin et al. [7] on the other hand terms them as'off the books' or 'under the radar' activities.

Informal urban practices allow residents to feel part of the community and grant them a right to the city. As coined by [28] the right to the city enables people to adapt the built environment to their way of life; it frees them from the deterministic design of professionals and designers and allow them to show diversity and creativity through unexpected activities in urban spaces [17, 19, 42].

In this respect, Roy [37] argues that urbanity is based on diversity and richness of urban life, and planning approaches should acknowledge and integrate needs, aspirations and peculiarities of people's way of life. [47] added that people whose practices are informal, through behaviour and work, are part and contribute to the overall economy, urban community and space. Additionally, [24] advocates for accepting informal urban practices as creative actions rather than aspects of marginality, whereas their combination with other formal activities develop simultaneous and creative ways of 'making the city'.

On top of that, informal urban practices are fundamental for creating inclusive urban life and are considered drivers for resilience as they promote social cohesion, sense of community, economic growth, and solidarity which encourages residents to collaborate and fight against any disruptions [22].

## 3 Exaptation and Architectural Design: Learning from Biology

The use of the term exaptation in architecture is not ubiquitous and can only be found in few recent papers by [4, 5, 10, 12, 29, 32, 33]. These authors have demonstrated that in design, like in biology, innovation by exaptation can be contrasted to innovation by adaptation, which implies evolution of a trait towards a better function.

Moreover, exaptation indicates the unforeseen relationship between an existing characteristic and a new function, different from the function for which the characteristic/feature was initially designed for [12]. A famous example Faulders [10] used to illustrate exaptation is the wing of a bird, in which the initial function of the feathers was to trap the air and protect against cold weather, which then transformed this function into a new set of possibilities for airborne mobility. Therefore, feathers that were fit for a role of regulating temperature were co-opted to perform a new and unexpected one which is flying.

For [15]: 6 *'characters, evolved for other usages (or for no function at all), and later "coopted" for their current role, be called exaptations'*. According to the authors adaptation refers to a character that promotes fitness and was built by natural selection for its current role, whereas exaptation indicates an evolutionary process expressed by functional change of a character. So, characters that were fit for their current function, were later co-opted for an unexpected one, in other words, *'Adaptations have functions, exaptations have effects'* [15].

The definitions of aptation-adaptation-exaptation [35] reveal a great potential of transferability of this discussion to the field of architecture. However, the two disciplines have developed their own linguistic code based on their independent taxonomies. Therefore, a transdisciplinary approach based on articulated connotations derived from biology could lead to a new way of reading architecture through the meaning of function, away from the rationalism of modern movement [2, 6, 40, 44, 48].

Exaptation operates within an open setting where reactions of the unexpected change the design process, a state in which a combination of (in) dependent characters creates a novel characteristic to respond to a novel function. Thus, exaptation identifies a third channel that inspires the emergence of novelty, through an evolutionary process of a functional shift of an existing character for a function for which it wasn't originally designed [1]. Thus, exaptation does not respond to a pre-defined need, or occur when a new phenomenon is discovered, but the other way around. Exaptation answers the question *"Now that I have come to a result, what problem have I solved?"* [46]: 22).

The process is based on the hypothesis that forms or features have an unpredictable creative potential which is expressed in the emergence of novel functions for which the forms were not originally designed for [1].

Similar to the process described by biology of evolution; in architecture the term spandrel refers to the redundancy of features and relationships that enable nature to adapt to unexpected environmental conditions, whereas exaptation, according to Gould [14], is about the survival and balance of ecosystems.

In a hypothetical new taxonomy of architecture, we should therefore consider the informal as a non-deterministic architectural expression that is, achieved through the redundancy of functionally co-optable structures, or the convergence, even chaotic, of several functions. These architectural components, which operate, in an ecological dimension, beyond the principle of the designed functions, reflect, for example through the post-nationalization of the design, in an architectural key the relationship between associative thinking, creativity and the unconscious [31]. They also confirm the thesis of the geneticist Ewan Birney, according to which all creative structures, in nature, are redundant, without predefined functions, but rich in relationships, like "a jungle populated by strange creatures".

Therefore, thanks to transdisciplinary studies we realise that like in biology of evolution, functional co-optation (exaptation) should be embedded in architecture and urban design in order to achieve resilience. Whereas architectural exaptation in this case is intended as an urban variability and redundancy of forms and relationships that promote the adaptation to constantly changing environments in order to ensure the survival and balance of ecosystems.

[32] discussed how non-deterministic and informal planning as an opportunity to increase the resilience of cities, reflecting on the flexibility and plasticity potential of non-deterministic design of architectural and urban design elements. They argue that in order for architecture to become an opportunity from a problem, it should consist of redundant, diverse and variable design features reflect and enrich functionalism. Moreover, they believe that designers should embed redundancy and flexibility in the early stages of design, instead of pre-defining and imposing a set of functions for the design of cities.

## 4 Algiers as an Emblematic Case Study

The city of Algiers was chosen as a case study to investigate how exaptation of the built environment allows diverse and creative use of space which in turn promotes urban life and community resilience, which in Algeria, is referred to as El Houma.

Algiers, like many other Mediterranean cities, is a cultural melting pot, as it has been the ground for many civilisations and cultures which shaped the city and granted it a variety of built environment features and a diverse mix of cultures and social ideologies. In this respect, the redundancy, variety and richness of forms and architectural features that Algiers possesses have the potential of inspiring forms of exaptation, creativity and innovation. Whereas each part of the city is built and occupied differently, depending on its architectural elements, producing a diverse urban life which sometimes includes informal use of urban space.

In this chapter, the investigation was conducted through activity mapping and observations of practices of El Houma in order to identify the behavioural patterns of people and record spontaneous and unexpected activities, whether formal or informal, that reflect have examples of exaptation of the built environment in

Algiers. The investigation on architectural exaptation is carried out in three neighbourhoods that represent three influential architecture and urban planning models, which are Medieval, nineteenth century urban planning and modernism.

## 5 El Houma as an Expression of Public Realm

El Houma is a term used in Algeria to refer to a neighbourhood or an urban area that is characterised by a strong degree of social sustainability [9, 16, 20], which originates from frequent social use of urban space.

El Houma is defined by tangible and intangible aspects of urban systems. The intangible (social) aspect is enriched by intricate social ties between residents and social use of space, and the tangible (physical design) refers to the built environment that accommodates and facilitates such activities [20].

The intensity of El Houma can be measured by the frequency of social use of space and the variability and redundancy of forms that are functionally co-optable to perform new and better functions. El Houma is a local experience of social life in the city; whereas belonging to El Houma grants residents 'the right to the city' which enables them to appropriate the space and transform architectural and urban space elements to suit their needs while performing spontaneous and various activities [20, 22].

In fact, looking closely at this phenomenon, one can perceive positive forms of exaptation, not foreseen in architectural and urban design processes. Scholars have demonstrated how informal behavioural pattern reflects a high degree of diversity and inclusiveness and increases social sustainability and the resilience of urban spaces. Whereas in their study of informal use of space in El Houma, [22] showed how a wall in neighbourhoods of Algiers can become either an exhibition space, a place to conduct informal commerce, or a place to organise community events such weddings or funerals (see Figs. 1,2, 3).

**Fig. 1** Public spaces of El Houma are co-opted to accommodate various (in)formal activities such as socialising, relaxing, playing or celebration of weddings

**Fig. 2** Street art exhibition by residents of the Casbah in order to improve the image of the neighbourhood and revitalise the space. *Source*: author

**Fig. 3** Groups of people gathering around a deceased's house to show support and solidarity using movable chairs. Left: the Casbah. Right: Bab El Oued. *Source*: author

The concept of El Houma was taken into consideration precisely for these characteristics that define principles and qualities independent from design determinism and linked to the informal vision of the use of space,

## 6 Case Study: Analysis and Discussion

Three neighbourhoods in the city of Algiers were investigated through observations and activity mapping in order to understand the relationship between the built environment and social urban practices as forms of exaptation.

The field work took place in the medieval medina the Casbah, which is characterised by compact and dense urban form with stairs laid out, convoluted narrow streets and courtyard houses of Moorish-style. The second neighbourhood is called

Bab El Oued and represents the inherited French colonial city built according to nineteenth century urban planning with planted avenues and boulevards, public spaces with vegetation and fountains, apartment buildings with arcades and soft edges at the ground floor. The third neighbourhood is Cité Belle Vue, and is a contemporary neighbourhood built according to principles of the modern movement through longitudinal apartment buildings separated by streets, gated communities and functional zoning.

The results of the fieldwork will provide an overview on which urban model among the three studied neighbourhoods reflects a high degree of exaptation and redundancy of design features, thus creativity and innovation of using and co-opting architectural and urban design components to suit people's needs and aspirations, in addition to insights on how informal urban practices contribute to community resilience.

During the field work, several (in)formal activities performed by residents were observed in different parts of the neighbourhoods, as forms of exaptation while contributing to resilience and urban life.

Some parts of the Casbah were in a derelict state due its long existence of and lack of maintenance. However, thanks to strong sense of belonging and solidarity, residents of the Casbah collaborated in beautification projects in order to enhance the image of the place, create a safe environment, and attract visitors to fuel local economy. Belonging to El Houma granted residents a right to occupy and modify the appearance and function of spaces according to their needs, thus, they performed street art and exhibit artwork on housing walls. Therefore, a wall that had a function of an envelope of the dwelling and of separating public (street) from private space (interior of house), was co-opted to be an art gallery (see Fig. 2).

Walls in the Casbah as well as in Bab El Oued, were co-opted to perform other activities like organising community events within the context of a funeral and informal commerce activities. As a sign of solidarity and sense of community, members of El Houma show support to the deceased's family by attending and helping with the funeral and by gathering next to their house as a sign of respect and solidarity. Residents will occupy streets while standing and/or sitting next to walls using urban furniture or entrance steps or folding chairs (see Fig. 3).

While streets are places of mobility, they were attributed different functions in Algiers. Informal street vending was performed in streets where shade is available through trees or the arcades on the ground floor of the apartment buildings, and this was noticed in Bab El Oued which features arcades and planted avenues (see Fig. 4). Shaded streets in residential areas were also used as an arena of football and other games. In Cité Belle Vue Ain Benian, a grass strip that was planted with flowers and used to separate buildings from the pavement, was squatted by children to play marble games due to lack of playgrounds and public space in the neighbourhood (Fig. 6).

Similarly, entrances of apartment buildings in Bab El Oued were appropriated to perform informal commerce activities; these entrances are considered part of the streetscape, and they are wide enough to allow circulation. They are perceived as useful spaces as they provide shade and are located within lively streets. Even

**Fig. 4** Informal commerce activities in El Houma, enabled by the size and design of the street, shade and right to the city. Selling food and various products. *Source*: author

though such activity is unauthorised and thus informal, members of El Houma squat these spaces to conduct informal vending to make ends meet as a reaction to unemployment, while other residents tolerate such activities as a sign of solidarity to their neighbours.

The Casbah is considered a place of peculiar history, architecture and urban planning in Algiers, the streets are convoluted and narrow, and defined by courtyard houses of Moorish style architecture. Some of these houses were co-opted and transformed into museums and restaurants to attract tourists, exhibit and raise awareness of the city's history, architectural heritage and local way of life (see Fig. 2).

Other activities were observed in the neighbourhoods that modified the function of some architectural elements, for instance, urban steps in the Casbah and building steps in Bab El Oued and Cité Belle Vue were originally designed to allow circulation and entering buildings respectively, were co-opted and used as seating areas to rest, chat and relax (see Fig. 5) as well as for commerce activities (see Fig. 4).

In Taleb Abderahmane square in Bab El Oued, a fountain, trees and benches are provided to offer a comfortable microclimate and a resource for seating. The square is inclusive as is attended by different members of the society; however, types of activities have separated the square into two parts. One part is attended by families to chat and relax while their children play different games like marble games, running or cycling. The other part is attended by old and mature men

**Fig. 5** People use urban and building steps as seating areas to play, chat or rest. Left photo: the Casbah. Middle photo: Bab El Oued. Right photo Cité Belle Vue. *Source*: Author

**Fig. 6** Children create their own space of play according to their needs. Left photo: the Casbah. Middle photo: Bab El Oued. Right photo Cité Belle Vue. *Source*: Author

mainly to play domino and cards, and in this part of the square wooden benches are provided. However, through placemaking, people modified function of the benches that were originally designed for seating and adapted them to their needs. Whereas these benches are used by people to play cards and domino which require four players facing each other, thus, sometimes the backrest is removed to free the obstruction and allow the players to face each other. While other benches in the square are made of metal as one element that cannot be modified.

In addition to trees, fountains in the squares of Bab El Oued and bottom part of the Casbah are used to improve the microclimate through evaporative cooling. However, the fountain performed more functions by acting as an aesthetic element and activating the place by attracting children to interact with it. On the other hand, in the Casbah, there is hardly any empty space within the dense compact urban form, therefore, the wall-fountains at the junction of streets were transformed into informal gathering spaces. Whereas they were originally designed to provide water for people to drink, wash, clean the streets or to do the ablutions before entering the mosque, but they were transformed into small public spaces for social interaction.

## 7 Conclusion

This research demonstrated a strong correlation between urban informality and community resilience, highlighting how informal urban practices as forms of exaptation resulted in various resilient strategies of urban neighbourhood residents.

Some of the activities observed in the three neighbourhoods and presented in this chapter could be considered informal because they are out of ordinary, improvised, and/or spontaneous because people improvised and created their own daily life spaces to perform activities in a space that had a different pre-defined function. Nonetheless, these activities contribute heavily to urban life, sustainability and community resilience. For instance, the derelict state of some buildings, lack of safety and unemployment led people to be creative and resort to unexpected and informal urban practices to enhance the image of the city and improve their way of living, like street art exhibition and transformation of a dwelling into a museum, thus architectural and urban design elements were co-opted to perform new functions.

El Houma was used in this chapter as a local model of a resilient community, formed of socio-cultural practices of residents and their interaction with the built environment. El Houma was used to measure urban resilience which was achieved through informal urban practices as forms of exaptation and as an antidote to determinism. It was highlighted in this research that El Houma is not restricted to a specific built environment but is an adaptive concept with the particularity of existing and evolving in various urban settings that have redundant and variable design elements that can offer opportunities for urban life. Nonetheless, the intensity of El Houma varies significantly across the three neighbourhoods, representative of the three urban models.

The observations highlighted that the intense El Houma is, the more enjoyable city life is. Also, urban fabric is more resilient when and where it is more possible to have a variability of forms, spaces, and redundancy of structures, like in the Casbah, because all these elements support the growth of relationships (interconnections) and favour the functional co-optation of the space. These aspects define architectural exaptation.

The informal activities observed in Algiers concern the functional co-option that has been possible in the compact historical city 'the Casbah', or in other urban fabrics, that are mixed use and richer, more redundant and variable in terms of built environment features, and which, unlike the planned city, contribute greatly, thanks to urban resilience, to the mitigation of the effects of environmental crises.

Therefore, most of the activities were observed in the Casbah which is mostly defined by exaptation, since various creative activities took place in addition to the co-optation of forms and functions of buildings and urban spaces to accommodate new uses.

Conversely, the modernist neighbourhood Cité belle Vue Ain Benian does not offer examples of exaptation due to the rigidity of its urban plan and lack of redundant design elements, providing very few opportunities for creativity and innovation to suit the emerging needs of its residents. So, the neighbourhood is not resilient since its deterministic design doesn't allow easily the functional co-optation (exaptation) of its architectural components. On the other hand, the nineteenth century European architecture and urban planning place Bab El Oued rests in the middle. Even though it was designed with a certain deterministic approach, when compared to the traditional medieval town, but its spandrels have allowed more exaptation through the redundancy and richness of its architectural and urban design elements.

It is fundamental to rethink the taxonomy in architectural design, whereas informal behaviour and the co-optation of houses and urban spaces to new functions should not be regarded only through a deterministic or order-oriented approach of planning.

The transdisciplinary study of architecture and biology of evolution seems now more relevant than ever given that global environmental crisis is of anthropic origin and the hope that such combination of studies will form synergies to reduce environmental impacts.

The use of the concept of exaptation in this chapter demonstrated how variability of forms and their potential evolution can inspire creative and innovative ways to face disruptions and adapt to unpredictable conditions.

Through the lens of exaptation, informal use of space could redefine the paradigms of human coexistence with the built environment and should be seen as drivers to the resilience of urban fabric. Therefore, architectural exaptation intended here as an urban redundancy of forms and functions can inspire the survival and balance of ecosystems, by promoting the adaptation to changing environments. This could be considered for further studies, for example, how can informal behaviour contribute (or not) to reduce energy and resource consumption, or to promote human coexistence with other species?

## References


1. Andriani P, Carignani G (2014) Modular exaptation a missing link in the synthesis of artificial form. Res Policy 43(9):1608–1620
2. Benevolo L (1971) History of Modern architecture: the modern movement. Vol 2. MIT Press
3. Carta M, Lino M, Ronsivalle D (2016) Re-cyclical urbanism: Visions, paradigms and projects for the circular metamorphosis. List lab
4. Carta M (2017) The augmented city: a paradigm shift. List lab
5. Carta M (2022) No more masterplan! Resilient communities require incremental, adaptive and generative processes. In: Carta M, Perbellini MR, Lara-Hernandez JA (eds) Resilient communities and the Peccioli charter. Springer, Cham. https://doi.org/10.1007/978-3-030-85847-6_8
6. Collins P (1998) Changing ideals in modern architecture, 1750–1950. McGill-Queen's Press-MQUP
7. Devlin RT (2018) Asking 'third world questions' of First World informality: using Southern theory to parse needs from desires in an analysis of informal urbanism of the global North. Plan Theory 17(4):568–587. https://doi.org/10.1177/1473095217737347
8. Douglas GCC (2014) Do-it-yourself urban design: the social practice of informal "improvement" through unauthorized alteration. City Commun 13(1):5–25. https://doi.org/10.1111/cico.12029
9. Dris N (2005) Formes urbaines, sens et représentations: l'interférence des modèles. *Espaces et Societes*, (122)
10. Faulders T (2014) Radical authenticity: exaptation in architecture. In: ACSA/AIK international conference, Seoul, South Korea
11. Frampton K (1983) Towards a critical regionalism: six points for an architecture of resistance. In: Foster H (ed) The Anti-aesthetic : essays on postmodern culture. Bay Press, Seattle, Wash, pp 16–30
12. Furnari S (2011) Exaptation and innovation in architecture: the case of Chicago's millennium park
13. Gould SJ, Lewontin RC (1979) The spandrels of San Marco and the Panglossian paradigm: a critique of the adaptationist programme. Proceed R Soc Lond Series B Biol Sci 205(1161):581–598
14. Gould SJ (1996) Full house: the spread of excellence from Plato to Darwin. Cambridge, MA: Harvard University Press

48 Gould SJ, Vrba ES (1982) Exaptation-a missing term in the science of form. Paleobiology, pp 4–15

16. Grangaud I (2013) La hawma: les processus de disqualification d'une institution ottomane. (Algiers 1830). Insaniyat
17. Harvey D (2003) Debates and development, the right to the city. Int J Urban Reg Res 27(4):9–44
18. Hernández F, Kellett P, Allen LK (eds) (2010) Rethinking the informal city: critical perspectives from Latin America. Berghahn books, New York
19. Huchzermeyer M (2011) Cities with slums: From informal settlement eradication to a right to the city in Africa. UCT Press, Cape Town
20. Khemri MY, Melis A (2022) Algiers as a paradigmatic mediterranean city: El Houma as a measure of Urban life. Disp Plann Rev 58(2):16–39. https://doi.org/10.1080/02513625.2022.2123162
21. Khemri MY, Melis A, Caputo S (2020) Sustaining the liveliness of public spaces in El Houma through placemaking. J Publ Space 5(1):129–152. https://doi.org/10.32891/jps.v5i1.1254
22. Khemri MY, Melis A (2021) Achieving community resilience through informal urban practices: the case of El Houma in Algiers. In: Di Raimo A, Lehmann S, Melis A (eds), Informality now: informal settlements through the lens of sustainability routledge

23. Khemri MY, Caputo S, Melis A (2021) The drawbacks of a global concept of sustainable neighbourhood in developing countries. In: Mohareb N, Versaci A, Mahgoub Y, Maruthaveeran S, Alberti F, (ed). Cities' vocabularies: the influences and formations. Springer Nature Switzerland AG
24. Landry C (2006) The art of city making. Earthscan, London
25. Lara-Hernandez JA, Coulter CM, Melis A (2020) Temporary appropriation and informal activity: exploring the subtle distinction. Cities 99(102626):1–10. https://doi.org/10.1016/j.cities.2020.102626
26. Lara-Hernandez JA, Melis A (2018) Understanding the temporary appropriation in relationship to social sustainability. Sustain Cities Soc 39:366–374
27. Latouche S (2009) Farwell to growth. Polity Press
28. Lefebvre H (1968) Le Droit a la Ville (The right to the city). Antropos, Paris
29. Melis A, Lara-Hernandez JA, Melis B (2022) Learning from the biology of evolution: exaptation as a design strategy for future cities. Smart Sustain Built Environ 11(2):205–216. https://doi.org/10.1108/SASBE-08-2021-0141
30. Melis A (2020) Informal behaviour as a form of community resilience. In: Di Raimo A, Lehmann S, Melis A (eds), Informality now: informal settlements through the lens of sustainability Routledge
31. Melis A (2019) The introduction of nature in the Austrian radicals practice. In: Planning cities with nature. Springer, Cham pp 45–63
32. Melis A, Pievani T (2022) Exaptation as a design strategy for resilient communities. In: Rezaei N (eds) Transdisciplinarity. Integrated Science, Springer, Cham. vol 5. https://doi.org/10.1007/978-3-030-94651-7_15
33. Melis A, Pievani T, Lara-Hernandez AJ (2024) Architectural exaptationwhen function follows form. Routledge
34. Morton T (2016) Dark ecology: for a logic of future coexistence. Columbia University Press
35. Pievani T (2008) Exaptation. Il bricolage dell'evoluzione. Bollati Boringhieri
36. Revell K (2010) Working with informality : increasing resilience in cities of the Global South. Isocarp, 1–13
37. Roy A (2005) Urban informality: toward an epistemology of planning. J Am Plann Assoc 71(2):147–158. https://doi.org/10.1080/01944360508976689
38. Roy A (2009) Why India cannot plan its cities: informality, insurgence and the idiom of urbanization. Plan Theory 8(1):76–87. https://doi.org/10.1177/1473095208099299
39. Rudofsky B (1964) Architecture without architects: a short introduction to non-pedigreed architecture. UNM Press
40. Scully VJ (2003) Modern architecture and other essays. Princeton University Press
41. Sharifi A (2016) From Garden City to Eco-urbanism: The quest for sustainable neighborhood development. Sustain Cities Soc 20:1–16. https://doi.org/10.1016/j.scs.2015.09.002
42. Simone A (2005) The right to the city interventions. Int J Postcolonial Stud 7(2005):321–325. https://doi.org/10.1080/13698010500268189
43. De Soto H (1989) The other path. Harper and Row, London
44. Tafuri M, Verrecchia G (1980) Theories and History of Architecture. Granada, London
45. Turner JFC (1976) Housing by people: towards autonomy in building environments. 2nd edition. Marion Boyars
46. Wiener N (1993) Invention: the care and feeding ofideas. MIT Press, Cambridge, MA
47. Yiftachel O (2009) Theoretical notes on "gray cities": the coming of urban apartheid? Plan Theory 8(1):88–100. https://doi.org/10.1177/1473095208099300
48. Zevi B (1978) The modern language of architecture. Australian National University Press

# Reimagining Informality: Lessons from Naples on Growth, Adaptability, and the Value of Incompleteness in Urban Planning

9

Maria Simioli and Guido Robazza

**Abstract**

This paper explores the interplay between structured urban planning and the unpredictability of informality in urban development. At its core, it examines how cities, exemplified by Naples, evolve through informal, superfetative processes that stem from a grassroots level, contributing to their adaptability, resilience, and vibrancy. By challenging conventional planning paradigms, superfetation embodies a collective urban creativity that redefines urban landscapes in response to changing community needs and environmental conditions. The discourse is framed within the theoretical context of systems theory and biological paradigms, advocating for a planning methodology that can embrace flexibility and adaptability. This approach reflects the principles of autopoiesis in biology, where systems are self-creating and continuously evolving. The paper argues that cities can be interpreted as Complex Adaptive Systems (CAS), where informality plays a crucial role in fostering a symbiotic relationship between urban environments and their inhabitants' individual experiences, needs, and creative solutions. The paper presents case studies from Naples, specifically focusing on the practices of superfetation within this complex and rich urban settings. These examples demonstrate how informal architectural practices, while often at odds with legal frameworks, provide essential contributions to the city's architectural and social fabric. The study suggests that such practices should not only be tolerated but actively integrated into the planning paradigm to enhance urban resilience and adaptability. In conclusion, the paper calls for a re-evaluation of urban governance and planning regulations to

M. Simioli
Dipartimento di Architettura, Università Degli Studi di Napoli Federico II, Napoli, Italy
e-mail: maria.simioli@unina.it

G. Robazza (✉)
School of Architecture, Art and Design, University of Portsmouth, Portsmouth, England
e-mail: guido.robazza@port.ac.uk



J. A. Lara-Hernandez et al. (eds.), *The Dusk of Design*, Integrated Science 34,
https://doi.org/10.1007/978-3-031-76867-5_9

acknowledge and incorporate the value of informal processes. It proposes a shift towards more inclusive planning frameworks that accommodate the fluidity of urban life and recognize the contributions of all citizens to the urban design process. This approach would not only enhance the vitality of cities but also ensure they are better equipped to respond to future challenges. The insights gained underscore the need for further research into the integration of informal practices within formal planning to maintain their essence and vitality in urban development.

**Keywords**

Superfetation • Adaptive architecture • Informal urbanism • Naples

## 1 Introduction

The potential for innovation that informality can bring to contemporary design, to make it more adaptive, plural, and open to the complexity of living, is fundamental to reconstituting planning as a resilient process in an increasingly crisis driven global reality. The phenomenon of urban informality is at the heart of a broad international debate that today seeks to recognize informality as the value of a new paradigm of the contemporary city [1] and the foundation for constructing a new epistemology of planning [34].

Informality has, in fact, long been an essential part of the organic design and self-construction of urban spaces, which can constitute community created shanty towns that elicit pity or anxiety from traditional planners [45] but equally generate 'architecture without architects', structures with often more general and enduring appeal than the divisive modernism of Le Corbusier [35]. In truth, as Ward points out, much of global urban growth in the twentieth century came in the form of 'popular', 'unofficial' growth, existing often partly or mostly outside the formal recognized economy of the cities being grown by this process [47].

The key change in a post-Quito Agenda phase is to adopt an attitude of optimism towards the idea of melding planning and informality, or rather planning for informality [23]. In this perspective, informality need not be seen as a separate and opposed process to planning, but rather planning can be used to generate space and platforms for informal innovation, the space-making to create settings for bottom up and DIY social uses which Pablo Sendra defines as 'infrastructures for disorder' that are "a point of departure for a continuous and open process" [39]. Such 'minor projects' could be seen as cracks in the 'constellation of thoughts' around the purpose of space allowing new and different ideas to emerge in a fertile environment [8]. The reference to this last term, "Minor Project" by Camillo Boano, reestablishes the concept of the 'minor' not as referring to a minority but to a difference "The project, when it clashes with becoming, with the possible, becomes imprecise, indeterminate, latent; possible because it interprets the possible not as inventions and novelties—simply not thought of—but as differences" and to the possibility of being a de-constituting action of the project "at the centre is the generative process and not the final image".

Through superfetation, the 'parasitic architecture' [28] that sees new purposes and new structures built onto existing buildings, we may see cracks in the otherwise closed 'constellation of thoughts' of urban planning, and a possibility, if this informality is tolerated, to generate bottom-up processes of urban transformation that can see a closed city become more open.

## 2 Theoretical Framework: Evolution in the Design Paradigm

The attitude of city development has been shaped for decades by the perspective of The Charter of Athens [9], in spite of the creation of many competing perspectives and critiques during that period but there has been increasingly a shift in the top down, rigid approach in favour of flexibility and informality as laid out in the Habitat III Quito Agenda and models that mirror biological processes have been found to act as useful perspectives on the value of this approach.

There has long been a syncretic relationship between thought relating to the built environment and the openness and variability as witnessed in biological systems. As early as The Doorn Manifesto by Team X from 1954 [43] began to question whether housing should not be conceived as a closed system but needs to weave relationships with the surrounding environment, a few years later, in 1959, the "Open Form" manifesto by Oskar Hansen [37] encouraged participation in artistic works, considering art as a process capable of promoting flexibility and variability, deliberately seeking to limit the architect's action to the creation of a "perceptual background" where users can manage and adapt their architectures. Hansen's manifesto is influenced by Geddesian theories, which already in 1915 argued that cities must evolve together with the environment, as an open and autonomous organism [19], with the term "biopolis" indeed it supports the need for a city and a form of planning as a living and open system. The "Mobile Urbanism" manifesto by Friedman [17] also experiments with a construction system that allows occupants to determine the design of their constructions incorporated into large supporting structures, spatial frames, with low impact on the ground. In the '60s and '70s, Jorn Utzon's reflections, according to the Additive Architecture model [46], show how the concept of assembly allows at the same time to create unified organisms while being incremental, flexible, and economical. The concept underlying these considerations is that it is possible to generate an infinite variety of configurations with a modest number of elements. Finally, the provocative manifesto-experiment "Non-Plan: An Experiment in Freedom" by Paul Barker, Peter Hall, Reyner Banham, and Cedric Price [3], while recognizing the need for economic planning, opts for greater flexibility in the production of space, identifying areas, limited zones where regulations would have been removed.

These incremental approaches, building on the use and reuse of limited basic elements, lend themselves to a biological interpretation of the city, whose form can evolve, and change based on pressures and circumstances. Fundamental to this perspective is the concept of exaptation, as recently highlighted by Melis & Pievani, implying that elements with one purpose or no purpose, put to work to solve new or different problems, can develop unexpected uses that can expand the life cycle of the existing structures, vital for increasing resilience and reducing harmful environmental impacts [31]. In biology such unexpected uses are described as 'spandrels', a term appropriately borrowed from architecture [22] but while a spandrel is only an aesthetic imposition on a redundant space created by the necessity of form, architecture is capable creating functional forms with enough openness to allow its users to define fundamentals of how they are used or lived in, as with Alejandro Aravena's Iquique housing project [32].

As Sennett laid out in The Open City [40], reflecting on Jane Jacobs' emphasis about the importance of the 'unexpected encounter' and 'chance discovery', cities must be spaces where uses and users overlap and mix to be creative and adaptive enough to be resilient. Of particular importance is the concept of 'incomplete form', a condition that leaves the full uses and interpretations of the built environment open, both for reinterpretation and for future dialogue with the adjacent development which will succeed it [40]. Any design is essentially a single creative act in a persistent and ongoing process which must acknowledge the unknown needs of the future to be negotiated with and imposed upon it, in opposition to creating something that must be destroyed and replaced once its design purposes are outlived. Rahul Mehrotra perceives the characteristics of elasticity—in relation to the unimaginable uses that are inscribed in space through practices of spontaneous appropriation and re-appropriation—incrementalism, as capable with small moves, micro-actions, gradually, of improving over the course of time cycles both the image and the physical form of spaces, which are "consumed, reinterpreted, and recycled" [30]. This principle is of ancient formation, if one considers the Milanese Courtyards of the early twentieth century, spontaneous agglomerations built very often with debris recovered and leftovers from construction sites where many of the workers worked [44].

## 3 Superfetation as Variability

A key concept that architecture and urbanism can adopt from biology, and that has been adapted into a range of disciplines, is autopoiesis. Atopoiesis is the capacity of a system to self-generate and self-maintain. This term, created by biologist Humberto Maturana [29], has become a key influence on systems thinking. Within urban theories, the term lends itself naturally to a more organic way of conceiving of the city as a self-generating but also continuously evolving organism [31]. Evolution is not a process of design from without, where forms are created with a specific purpose in mind to adapt the organism to a purpose, rather change must come endogenously and existing forms exapted to new purposes. Conceiving of a city in this way allows us to see not just the value of informal, bottom-up changes in self-generating change to suit the urban environment, but also to recognise the vital importance, when planning or designing, of creating forms that allow for reinterpretation, reuse, or even absence of immediate purpose, for informal use to shape. Thus, the embrace of variability emerges as a fundamental design principle, vital for producing dynamic, adaptable environments adapted to the changing human needs and contexts. The capacity to integrate change and adaptability into the built environment is essential for the evolutionary vitality of cities and their ability to respond to the citizen's needs.

From a systems theory perspective, both biological organisms and cities can be seen as Complex Adaptive Systems (CAS) [24]. A CAS consists of a dynamic network where the interactions among its parts create complexity, and its overall behaviour cannot easily be deduced from the actions of its individual components. Such a system is adaptive, meaning, the behaviour of both the individual parts and the system evolves and self-organizes in response to specific events or a series of events that introduce change. This system is essentially a large-scale collection of smaller, somewhat similar, and interconnected structures that come together to adjust to environmental changes, aiming to enhance their collective ability to endure and thrive [2, 27]. Seen and understood under this perspective, a parallel between the evolutionary processes in biological organisms and the development and adaptation processes in cities can be drawn. Recent studies on urban morphogenesis have compared them to CAS [36, 48]. By viewing urban evolution through this lens, it's possible to appreciate the nuanced and multifaceted ways in which cities adapt and transform, driven by the collective intelligence and actions of their inhabitants.

The phenomenon of superfetation plays a key role in this discourse. Superfetation, in architecture, identifies a volume extraneous to the original building, added as a makeshift element in a subsequent phase, often to respond to a functional need for which no other space was available [10]. A new built element is added to an existing building allowing new and different unplanned purposes to become part of the building, showcasing a creative adaptation to changing needs, space constraints, or aesthetic desires. More than a mere architectural extravagant addition, superfetation embodies the collective intelligence of city dwellers, their innate creativity, and their unmediated contributions to the urban fabric. It represents a grassroots form of urban evolution, where the atomization of decision-making through individual acts of design and adaptation reveals a rich, complex layer of urban intelligence.

Cities, seen through the lens of a System Theory perspective, are not static entities shaped solely by formal planning and design but are living, breathing organisms that grow and evolve, through the contributions of every inhabitant. This evolutionary process creates a rich palimpsest, a layered tapestry of historical and contemporary urban forms that embody the city's myriad phases of development and adaptation.

The informal creations by citizens, acting as ad-hoc urban designers, brings a diverse array of functionalities and forms into the urban landscape. This decentralized approach to city-making harnesses the collective wisdom and creativity of the population, contributing to environments that are more flexible, interactive, and inclusive. Such spaces offer a wider range of activities and respond to the citizens' needs but also embody the capacity to evolve, mirroring the dynamic nature of human societies.

By emphasizing variability and adaptability in design, we challenge the traditional notions of architecture and urban planning as creators of fixed, rigid spaces. Instead, we advocate for a paradigm where spaces are conceived to be as alive and evolving as the communities they serve. Naples is an incredible city that exemplifies the power of architectural and informal practices to leverage variability, creating built environments that are not only responsive to their users but also contributors to the vibrant, ever-changing fabric of urban life.

City planners, with good reason, often oppose this phenomenon, as it is frequently illegal and carried out in disregard of planning rules and building regulations. Indeed, there are profound ethical considerations; these illegal activities cannot be condoned when they harm health and safety or infringe upon the property of others. However, to focus solely on their illegality, without nuanced consideration, would be to fail to recognise the essential and historic nature to such incremental and informal repurposing that has contributed to create the fabric of cities in the global north and south.

Superfetation, as a manifestation of the collective intelligence of urban inhabitants, offers an alternative narrative on the biological evolution of cities. It highlights the significance of variability in the evolutionary process, presenting a city's ability to adapt and transform as a direct reflection of its inhabitants' capacity for innovation and creativity. Through this process, cities evolve not as monolithic entities but as vibrant ecosystems of human ingenuity and adaptability.

## 4 Case Study: Naples as an Urban Palimpsest

Discussing informality in Naples, the city par excellence of the "art of arrangiarsi".[1] requires a necessary clarification. "Arrangiarsi," derived from the French "arranger," meaning "to adjust, combine, organize," assumes a significant meaning and implies an attitude that oscillates between legality and illegality, translating into the creative capacity, flexibility, and adaptability of a community whose sole imperative is survival. Exaptation [21, 22], Naples becomes "arrangement" as the optimization of the potential of available resources through an adaptive and creative process to new developed conditions. This characteristic not only roots itself in the lifestyles of a community but also becomes an expression of the culture, history, and form of the city stratified over centuries. It builds palimpsest [11]

[1] The title of the 1954 film by Luigi Zampa.

a long process wherein not everything found on the surface corresponds to the most recent layer, and vice versa [38]. This represents that mode of arrangement, capable of compensating for the deficiencies of basic services according to a model of "informal welfare" [12]. The metropolitan area of Naples is configured as a set of forms built between the planned and the unplanned, which refers to an informal growth of and cities seeking solutions that go beyond traditional modes and organizational systems: from practices of political and social squatting, including processes of occupation of buildings and public spaces for forms of community self-organization [13] to the privatistic one of building abuse [14]. This issue highlights the different dimensions of informal and abusive phenomena, one as a creative mode capable of generating commodities, the other as a process of undue appropriation of resources aimed at increasing one's profit.

The superfetation, synonymous with makeshift spontaneous architecture, loses its negative connotation and becomes a vehicle for new design values [20]. Superfetation is a sort of spontaneous evolution of the building, uncontrolled in terms of design and unforeseeable at the time of construction. If we trace a sort of history of Neapolitan architecture, alternative to the official one, less ostentatious and closer to the narrative of the real use of spaces by people, additions often fulfilled a specific need: the addition of a sanitary service, a small storage room, or a kitchenette. In the plans for the recovery and enhancement of the city's architectural heritage, particularly in the historic center, superfetations end up conditioning the writing of subsequent chapters of the project and determining the city's external image [4]. Often, superfetation, viewed from a purely regulatory perspective, is an element to be eliminated, but there are exceptions that can create an opportunity for rethinking and awareness of different valorisation possibilities. Emerging as an example of this form of attractiveness and local assertion are the "bassi" of the city, now protagonists of a new vital dimension thanks to the tourism boom, thanks to spontaneous actions by groups of people that have led to a regeneration of the "bassi", conferring on them a growing and innovative importance identified as "laboratories of change" [6]. Resuming Benjamin's words that define Naples, "porous as this stone is architecture. […] Everywhere, vital space capable of accommodating new, unforeseen constellations is preserved. The definitive, the characterized are rejected" [5]. Superfetation, therefore, serves as a useful device to express differences, understood as the functional needs of the communities that inhabit them, as an architectural element made of makeshift, recycled materials, constantly changing, adaptable, and modifiable, which places the citizen as an active user of transformation. These realizations, based on individual initiative, cannot be considered simple abusive works to be eliminated, as they have responded to a latent demand for basic functions and services that the original projects had failed to intercept or that have matured in recent years and represent modes of spontaneous adaptation. In contexts of illegal construction, on the other hand, superfetation has often become a useful element for subsequent transformations over time, becoming, in its redundancy, a local characteristic of the places. The theoretical background of this new forms of informal planning [7, 26], or DIY urbanism, which delineate the geography of the infinite design variations of bottom-up practices. The real success of these initiatives lies in their unique potential to intertwine with longer-term renewal processes and a greater sensitivity to the reading of local specificities: cultural, social, and ecological. They promote new forms of participation and the interaction of new social and ecological competences and awareness [33], today ecology itself represents the field through which many of the mentioned approaches are trying to overcome the conflicts and dichotomies that characterize the design of the existing city, placing landscape design at the centre. The influences between the world of DIY urbanism and Ecological urbanism [25] are evident appropriation, finds its maximum expression in common goods [18], the former Asilo Filangieri and Scugnizzo Liberato; but also practices of reappropriation of open space, such as of the urban gardens in Ponticelli, as a form of reuse of a large green space in the heart of the public city. In this essay, we examine the role of superfetation as an architectural device capable of producing diversity and accommodating the unexpected, in the contexts of public residential construction and in the Municipality of Lacco Ameno on the island of Ischia.

The theoretical premises of this work lay the foundation for conducting an exploratory analysis on the role of superfetation as an informal device in the city of Naples, specifically aimed at the creation of new and diversified forms of territorial engagement. In particular, informality is here assumed as a value field to reorient the project, capable of enabling harmonious relationships and more accessible and adaptable transformation processes, while related to the economic feasibility of private actors. This approach, besides being more sustainable, ensures the necessary openness towards incremental, organic, and grassroots development. Specifically, the solutions identified stem from the observation of informal practices, highlighting their collective value, and seeking to understand the methods, processes, and interventions through which the phenomenon can be brought into the legal sphere of use. This work was conducted through an inductive methodological approach, which, starting from sensory experience, was able to recognize informal uses and define their collective value, aiming to understand the methods, processes, and interventions through which the phenomenon can be brought into the legal sphere. The choice of the study area, the complexes of public housing and the unauthorized buildings on the island of Ischia, was supported by the intention to record and interpret superfetation as an expression of informality in both planned contexts, the first case study, which evidently failed in its observance and compliance, the result of a top-down project, and in an unplanned context, the unauthorized one of the second case study.

## 5 Superfetation in Public Housing: Domitian Coast and Island of Ischia

### 5.1 Public Housing on the Domitian Coast

Public housing in Italy, Edilizia Residenziale Pubblica (ERP) has often been subject to unauthorized alterations and spontaneous modifications of private and collective spaces, known as abusive superposition, which have often been considered synonymous with urban decay and functional obsolescence, as they are constructed with poor materials, quickly and by unqualified individuals within the family network. However, if superposition is assumed as an expression of individuality, as a possibility for personalizing the living environment, and as a mode of local construction and cultural memory of the territory, then these elements become design tools in the construction of knowledge. Working from the observation of places and listening to communities re-centralizes the figure of the end user as a "user in the meantime", allowing us to define an open project "that welcomes conflicts and dissonances" [41], because while it is true that this type of spontaneous work forms a strong bond between inhabitants and site, a process that we could define as "appropriation", on the other hand, it entails the "privatization" of common spaces, ending up exacerbating differences and asymmetries in access to certain types of services connected to residence. The case study presented here does not represent the expression of how superposition is an architectural device that allows variability, but how it is a tool for reading differences, allowing the regeneration project to explicitly express plurality, flexibility, and the indeterminacy of solutions in legal forms. Below are the results of the experimentation conducted within the PINQuA[2] project on the ERP complexes of four municipalities on the Domitian Coast: Castel Volturno, Cellole, Mondragone, and Sessa Aurunca [16]. We will show the work on the Duca degli Abruzzi neighborhood is located on the historic center of Mondragone (CE) [15]. The external space of pertinence of the buildings is devoid of green areas and is made of impermeable materials used for parking, the few existing permeable spaces located behind the buildings are in a state of neglect and abandonment. The ground floors of the buildings originally built on stilts have been illegally occupied over the years by the construction of tuff masonry walls between the pillars of the ground floor, to create private garage boxes, while the balconies of the facades, enclosed with aluminium verandas (Fig. 1). Based on these factual conditions, improper uses and

[2] PINQuA "Innovative Housing Quality Program" conducted by the Department of Architecture of (DiARC), University of Naples Federico II, financed with PNRR funds and currently undergoing the executive design phase. Scientific director: Filippo De Rossi, Michelangelo Russo; scientific coordinator: Enrico Formato, Alessandro Sgobbo; Project coordinator: Giovanni Multari. Project team: Eduardo Bassolino, Gilda Berruti, Claudia Colosimo, Salvatore Della Corte, Cinzia Didonna, Nicola Fierro, Giovanni Laino, Cristina Mattiucci, Eugenio Muccio, Giuliano Poli, Maria Simioli, Federica Vingelli.

**Fig. 1** Identification of informal uses of public and private space

unauthorized superpositions have been identified: Verandas, covered and uncovered car boxes, garages, gardens, spaces for worship, dog areas, collective spaces, shops.

The location of these practices suggests their collective value, whether they occur in public or private places, enclosed or open spaces, collective or individual use, whether they generate profit or not, as criteria for understanding which of these elements can be recomposed in a systemic overview of the project.

Based on these considerations are identifies some design actions according to two macro-categories of intervention:

(a) Buildings: It is involves not only the redevelopment of the properties and energy efficiency improvements but also the construction of a structural framework, in addition to the original building, as a space that offers freedom of use and generates possibilities for evolution; on the roof, as well as on the ground floors, volumes will be identified in which it will be possible to establish a variety of temporary uses, subject to change according to needs (Fig. 2).
(b) Open spaces: Interventions for the regeneration of open spaces involve the integration of some functions and uses of collective space, particularly the redesign of public space through the rationalization of pedestrian and vehicular paths. Furthermore, in spaces identified as "Playgrounds", these are imagined

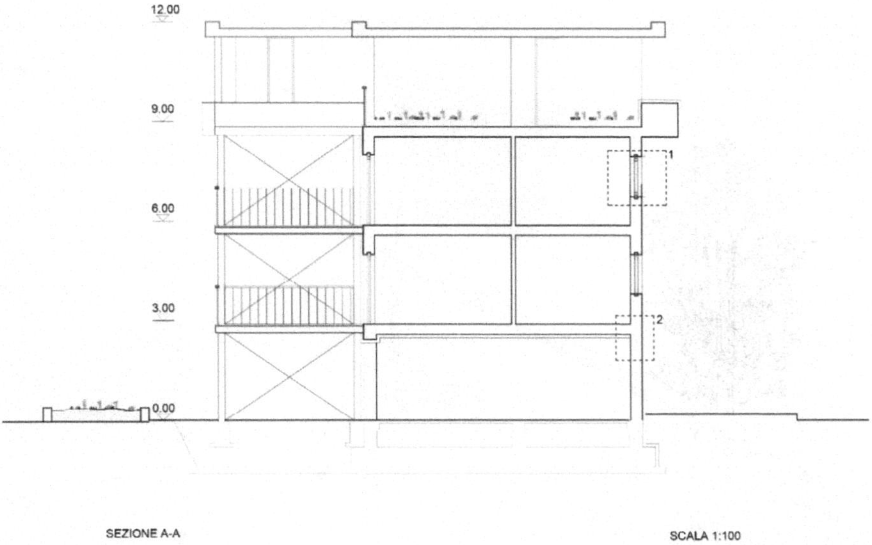


**Fig. 2** Addition of volume on the facade for flexible and adaptive uses

as "blank squares", i.e., places for which no specific use, spatial configuration, or user typology is defined, but rather unplanned places left to the indeterminacy of future actions, capable of accommodating "perturbances" (Fig. 3).

## 6 Island of Ischia

The case study of the island of Ischia examines how spontaneous practices of territorial modification have translated into adaptive architectures concerning the morphological characteristics of the local landscape, incrementally expanding in response to the capacity for enlargement, and flexible in accommodating new uses with variations in family and market dynamics. The study of superfetation, as an element of local architecture, was conducted as part of the studies supporting the Reconstruction Plan of the island of Ischia (PdRI) following the 2017 earthquake, by the DIARC, Univeristy of Naples Federico II [15]. The study aims to establish guidelines for an "informally designed" approach through the definition of minimal rules and the identification of informal architectural characteristics.

The case study of Ischia provides a fascinating insight into how spontaneous architectural practices can adapt and respond to the evolving needs of a community, particularly in the aftermath of significant events like earthquakes [42]. Let's break down some key points from the study:

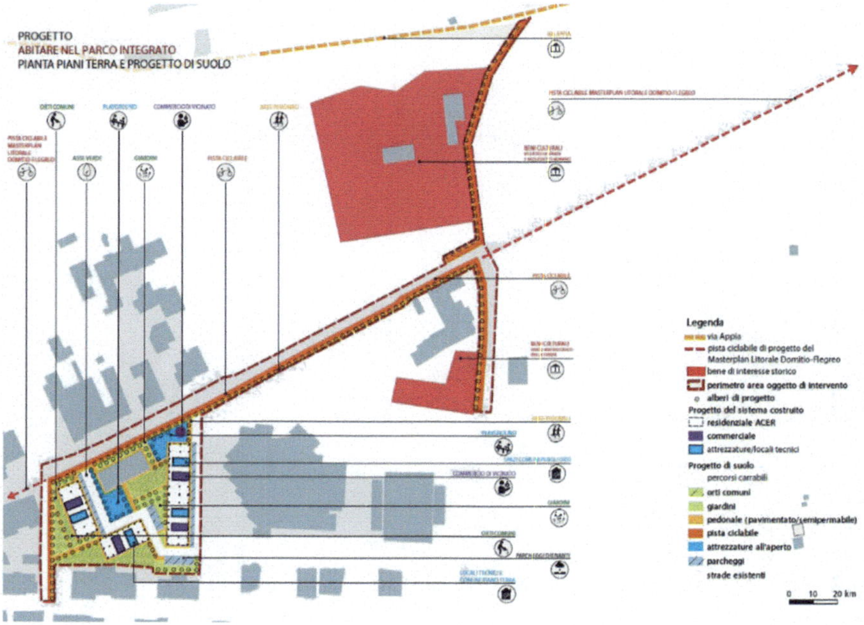


**Fig. 3** Plan of the interventions in the open spaces

- *Adaptive architectures:* the architecture on the island of Ischia has evolved in response to the local landscape and changing socio-economic dynamics. Structures incrementally expand to accommodate new uses and adapt to the morphological characteristics of the terrain.
- *Superfetation in Architecture:* superfetation, as observed in Ischian architecture, refers to the incremental addition of built volumes in successive epochs. This often results in the construction of visible vertical ascent systems, such as the "a collo d'oca" stairs, which integrate with the natural topography of the area.
- *Territorial modification and landscape integration:* the architecture of Ischia reflects a close relationship between built environment and landscape. Elements like lodges, and terraces are not just functional but also blend harmoniously with the surroundings, sometimes integrating with agricultural terraces.
- *Informal design approach:* the Reconstruction plan of Ischia emphasizes an "informally designed" approach, characterized by minimal rules and the recognition of informal architectural characteristics. This approach aims to guide reconstruction efforts while preserving the unique vernacular architecture.

The plan, therefore, identifies a design strategy characterized by principles of incrementality, elasticity, weakness, interaction, flexibility, circularity, which are manifested through the establishment of a system of "minimal rules": selection and spatial arrangement of plots, surface ratios, sustainability principles, and permitted uses. The design strategy outlined here seems to prioritize adaptability and sustainability, aiming to create a resilient framework for development while respecting existing natural and built environments. By emphasizing principles like incrementality, elasticity, and circularity, the plan seeks to foster organic growth and minimize environmental impact.

The concept of "minimal rules" suggests a light-touch approach to regulation, allowing for flexibility and innovation within set parameters. This approach seems well-suited to accommodate diverse needs and contexts, particularly in areas where existing infrastructure may need to be repurposed or where community-led initiatives are encouraged. The emphasis on preserving permeable areas aligns with contemporary priorities around green infrastructure and biodiversity conservation. By prioritizing the reuse of existing developed areas and minimizing soil consumption, the plan promotes resource efficiency and helps mitigate the urban heat island effect.

The allocation of space for self-construction within planned developments reflects an understanding of the value of vernacular architecture and community involvement in shaping the built environment. This approach not only empowers residents but also fosters a sense of ownership and continuity with local traditions. Furthermore, the focus on public and collective functions in areas earmarked for relocation underscores a commitment to fostering social cohesion and enhancing quality of life. By prioritizing amenities like parks, gardens, and communal spaces, the plan promotes inclusive urban development and addresses the needs of diverse communities.

The experience of the Island of Ischia demonstrates the intention, on a territorial scale of vast area, to delineate a planning tool capable of accommodating "perturbances" and guiding transformation.

## 7 Conclusions and Implications for Urban Governance

The paper explored the interplay between the structured world of urban planning and the chaotic beauty of informality in urban design. By delving into examples such as Naples' public housing, it explored the value of informal, superfetative processes from bottom-up practices, which are indispensable for city adaptability, resilience, and liveliness, and identified potential design and regulatory frameworks to accommodate them. Superfetation processes, often operating on the fringes of legality, embody a collective urban creativity that challenges conventional planning creativity, and being able to include them in the ecology of the city can propose a paradigm where cities evolve organically, mirroring the complexity and adaptability of natural ecosystems.

This study thus identifies informality to play the major key role in fostering a deep, symbiotic relationship between the city and its people, one that is active every day, where urban fabric finds weaving threads of personal experiences, needs, and creative solutions. This perspective enriches the urban milieu and at the same time suggests one more layer of flexibility and resilience—something that is most critical for a city under continuous change and battered by all kinds of unexpected circumstances.

Urban planning could embrace an open project methodology, capable of accommodating conflicts, diversity, and indeterminacy. This requires a shift towards frameworks that prioritize flexibility, adaptability, and the capacity to evolve over time, reflecting the organic nature of urban growth and the variety of human needs. Planning systems can comprise flexibility, allowing the city to stretch and change without losing its identity or functionality. This implies that urban planners will have to encourage the design of infrastructures and public places that can adapt for more than one kind of purpose or activity and answer the community's changing needs. There is a pressing need to redefine urban regulations in a manner by which informal and bottom-up processes should be recognized and absorbed in official planning paradigms. This involves developing new forms of regulatory inclusion that recognize the value of superfetation and other informal practices, transforming them from liabilities into assets for urban development.

The insights gained from this research have profound implications for urban policy, practice, and theoretical frameworks. Policymakers and urban planners are, therefore, invited to revisit the question of informality in urban development, probably through lenses that would bring forward its potential as a source of urban innovation, adaptiveness, and resilience. The result would support a call for a review of current planning regulations and practices, including a more inclusive approach that acknowledges the contribution of all city dwellers to the urban design process.

This is an area where future research must delve deeper: how these practices can be integrated with efficacy into formal planning without losing their essence. The works also establish the potential role of technology and new design methodologies for a more participatory and inclusive urban design process. Comparative studies across the cultural and geographical contexts may, therefore, shed more light on the universal and unique characteristics of informality and superfetation in urban development.

**Authors contribution** All parts of this article have been jointly reviewed and approved by both authors. However, the primary contributions to specific sections are as follows: Guido Robazza

**Table 1** Practices and Locations

| Practices | Public/Private | Open/Closed | Collective/individual | Profit |
|---|---|---|---|---|
| Veranda | Private | Closed | Individual | No |
| Parking | Public | Open | Collective | No/Yes |
| Garage | Public | Open | Collective | No/Yes |
| Shop | Public | Closed | Collective | Yes |
| Gardens | Public | Open | Collective | No |
| Playground | Public | Open | Collective | No |

contributed to the *Introduction, Theoretical Framework: Evolution in the Design Paradigm, Superfetation as Variability*, and *Conclusion*, while Maria Simioli contributed to the *Case Study: Naples as an Urban Palimpsest* and *Superfetation in Public Housing: Domitian Coast and Island of Ischia.*

## References


1. AlSayyad N, Roy A (2003) Urban informality: transnational perspectives from the Middle East, Latin America, and South Asia. Lexington Books
2. Anish S, Gupta A, (2010) Insights from complexity theory: Understanding organizations better. Indian Institute of Management
3. Barker P, Price C, Hall P, Banham R (1969) Non-plan: an experiment in freedom. New Soc 338
4. Becchi A (2001) Napoli sostenibile? Logiche di sopravvivenza e modelli di sviluppo, Napoli Sostenibile
5. Benjamin W (1978) Reflections: essays, aphorisms, autobiographical writings
6. Berritto A, Mazza R, Gabriella P (2018) Tradizione, informalità, innovazione: i bassi e il turismo a Napoli tra dimensione esperienziale e nuova forma di rivendicazione del locale. In: Fasolino, I. (ed), Arte, Artigianato, Prodotti Locali, Turismo, Creatività. pp 6–12
7. Bishop P, Williams L (2012) The temporary city. Routledge, London
8. Boano C (2021) Progetto Minore: Alla ricerca Della minoritá nel progetto urbanistico (ed) architettonico. LetteraVentidue Edizioni
9. Burdett R (2018) Contemporary Urbanism. The Quito papers and the New Urban Agenda. Routledge, New York, pp 140–149
10. Coppola A (2017) Roma: la metropolizzazione parassitaria e i suoi modi informali. In: Cellamare, C. (ed), Fuori Raccordo. Abitare l'altra Roma. Donzelli, Roma-Bari
11. Corboz A (1985) Il Territorio come Palinsesto. Casabella 516:22–27
12. De Nicola A (2022) L'emergere del "Welfare informale", tra possibilità e contraddizioni. In: Gissara M, Percoco M, Rosmini E (eds), Città Immaginate. Riuso e Nuove Forme Dell'abitare. Manifestolibri
13. Esposito E (2020) L'occupazione informale di uno spazio di edilizia popolare come tappa di passaggio: la dimensione materiale Del percorso abitativo Di Un occupante abusivo nella città di Napoli. In: Atti Della XXII Conferenza Nazionale SIU| L'urbanistica Italiana Di Fronte All'Agenda. p 202
14. Formato E (2020) Learning from do-it-yourself Naples. Imparare dalla Napoli fai-da-te. Proyecto Y Ciudad 11:69–80
15. Formato E, Di Iorio F, Gioffrè V, Russo M (2023a) Un Cretto abitato. Ricostruire sottraendo. In: Galderisi A, Mininni M, Presta IG (eds), La Declinazione Della Sostenibilità Ambientale Nella Disciplina Urbanistica. Atti Della XXIV Conferenza Nazionale SIU: Dare Valore Ai Valori in Urbanistica. Planum Publisher, Roma—Milano
16. Formato E, Simioli M, Fierro N, Vingelli F (2023b) Abitare il territorio periurbano. Una proposta metodologica di rigenerazione per l'edilizia residenziale pubblica. In: Perrone C, Marchigiani E, Savoldi P, Tosi MC (eds), Forme Di Welfare e Dotazioni Di Servizi, Un'eredità in Continua Evoluzione. Atti Della XXIV Conferenza Nazionale SIU: Dare Valore Ai Valori in Urbanistica. Planum Publisher, Roma—Milano
17. Friedman Y (2022) Yona Friedman/pro Domo. Actar D, Inc
18. Gargiulo E, Cirulli A (2016) Gli spazi occupati a Napoli: informalità, trasformazioni urbane e discorsi Sui "beni comuni." Punziano G (a cura di), Società, Economia e Spazio a Napoli Esplorazioni e riflessioni 83–94
19. Geddes P (1915) Cities in evolution: an introduction to the town planning movement and to the study of civics. Williams, London

20. Giovanni Di A, Ranzini A (2022) I valori dell'informalità abitativa. In: Atti Della XXIV Conferenza Nazionale SIU Dare Valore Ai Valori in Urbanistica
21. Gould SJ, Vrba ES (1982) Exaptation a missing term in the science of form. Paleobiology 8:4–15
22. Gould SJ, Vrba ES (2008) Exaptation. Il bricolage dell'evoluzione, Paleobiology. Bollati Boringhieri
23. Haas T, Mehaffy MW (2019) Introduction: the future of public space. Urban Design International 24
24. Holland JH (2006) Studying complex adaptive systems. J Syst Sci Complex 19:1–8
25. Lentini M (2020) Il Progetto di Paesaggio come modalità di lavoro nei contesti urbanizzati contemporanei. Ricomposizioni di antinomie ostinate. Ri-Vista. Ricerche per la Progettazione del Paesaggio 18:38–61
26. Lydon M, Bartman D, Garcia T, Preston R, Woudstra R (2014) Tactical urbanism. Street Plans, New York, US
27. MacLennan B (2007) Evolutionary psychology, complex systems, and social theory. Soundings Interdisc J 90(3):169–189
28. Marini S (2015) Architettura parassita Strategie di riciclaggio per la città. Quodlibet, Macerata
29. Maturana HR, Varela FJ (1991) Autopoiesis and cognition: The realization of the living. Springer Sci Bus Med
30. Mehrotra R (2010) Simultaneous modernity. Ruins of modernity pp 244–250
31. Melis A, Pievani T (2022) Exaptation as a design strategy for resilient communities. In: Transdisciplinarity. pp 307–327
32. O'Brien D, Carrasco S, Dovey K (2020) Incremental housing: harnessing informality at Villa Verde. Archnet-IJAR Int J Architect Res 14:345–358
33. Robazza G (2020) Build art, build resilience. Co-creation of public art as a tactic to improve community resilience. J Publ Space 5
34. Roy A (2005) Urban informality: toward an epistemology of planning. J Am Plann Assoc 71:147–158
35. Rudofsky B (1964) Architecture without architects. An introduction to non-pedigreed architecture. The Museum of Modern Art, New York
36. Rybski D, González MC (2022) Cities as complex systems—Collection overview. PLoS ONE 17:e0262964
37. Scott FD (2014) Future Adaptability. In: Wieder A, Zeyfang F (eds) Open form space, interaction, and the tradition of Oskar Hansen. Sternberg Press Ltd., London, pp 34–46
38. Secchi B (2000) Prima lezione di urbanistica. Gius. Laterza & Figli Spa
39. Sendra P (2016) Infrastructures for disorder. Applying Sennett's notion of disorder to the public space of social housing neighbourhoods. J Urban Des (Abingdon) 21:335–352
40. Sennett R (2018) The open city. In: The Quito papers and the New Urban Agenda. pp 90–95
41. Sennett R, Sendra P (2020) Designing disorder: experiments and disruptions in the city. Verso Books
42. Simioli M (2024) L'informalità come struttura d'azione nel progetto dei paesaggi dell'abusivismo. In: Bricocoli M, Pezzagno M (eds), Servizi, Dotazioni Territoriali, Welfare e Cambiamenti Sociodemografici. Atti Della XXV Conferenza Nazionale SIU: Transizioni, Giustizia Spaziale e Progetto Di Territorio. Planum Publisher, Roma—Milano
43. Smithson A (1982) The emergence of team 10 out of C.I.A.M. Architectural Association
44. Staid A, Aime M (2017) Abitare illegale: etnografia Del vivere ai margini in occidente. Milieu
45. Turner J (1976) Housing by people. Towards autonomy in building environments. Pantheon Books, New York
46. Utzon J (2009) Additive architecture. Mogens Prip-Buss
47. Ward C (1996) Anarchy in action, Second Edition. (ed). Freedom Press
48. Wohl S (2018) Complex adaptive systems and urban morphogenesis: analysing and designing urban fabric informed by CAS dynamics. A+ BE| Architecture and the Built Environment pp 1–238

# 10 Enabling Flexibility and Residents' Agency in Shaping Social Housing: Case Studies from South America

Priscila Besen

**Abstract**

Architects, engineers, planners and other built environment professionals often plan spaces with an assumption of permanence, commonly neglecting flexibility for adaptation. Buildings are usually designed as if they will be standing forever in the same place fulfilling the same function, and subject to the same climatic conditions. This has significant implications for the people who inhabit the spaces—due to the lack of possibilities to make changes to suit their needs—and for the environment, leading to issues such as demolition waste due to an inability to adapt structures or reutilise materials. Throughout history, social housing projects have demonstrated some of the most critical examples of these inflexible approaches in the built environment. By standardising ways of living, without any possibility of change, adaptation, or expansion, these rigid models of housing have led to various issues. Due to cost-effectiveness considerations, there are usually many limitations to the ways that buildings can be customised for residents to suit their preferences and needs, resulting in mass production and standardisation. The result is that buildings become obsolete almost instantly after being built and that residents end up taking a "do-it-yourself" approach to make alterations and additions. However, alternative approaches are emerging to make social housing more adaptable and flexible and to give residents more agency. Certain projects reverse this cycle by thinking of the possibilities of alterations and additions from the very beginning. This chapter looks at innovative ways of thinking of adaptation and creation of a sense of belonging from the very beginning of a project. Case studies from South America, including incremental housing by Elemental in Chile and progressive housing

P. Besen (✉)
Te Wānanga Aronui o Tāmaki Makau Rau—Auckland University of Technology, Auckland, New Zealand
e-mail: priscila.besen@aut.ac.nz



J. A. Lara-Hernandez et al. (eds.), *The Dusk of Design*, Integrated Science 34,
https://doi.org/10.1007/978-3-031-76867-5_10

propositions in Brazil, demonstrate these possibilities. These case studies challenge architects in their aim to "control" how buildings look like, allowing the creativity of residents to flourish and take centre stage in these neighbourhoods over time.

**Keywords**

Incremental housing • Social housing • Building adaptation • Self-construction • Progressive housing

## 1 Introduction

More than 2500 years ago, the Philosopher Heraclitus acknowledged that "there is nothing permanent except change" [11]. Yet, architects, engineers, planners and other built environment professionals worldwide still plan spaces with an assumption of permanence, often neglecting flexibility for adaptation. Buildings are designed as if they will be standing forever in the same place fulfilling the same function, and subject to the same climatic conditions. This has significant implications for the people that inhabit the spaces—due to the lack of possibilities to make changes to suit their needs—and for the environment, leading to issues such as demolition waste due to an inability to adapt structures or reutilise materials.

It is common practice to consider the completion of building construction as the final stage of a project. Designers have little opportunity to implement ideas beyond this stage, and assessments of projects after completion, such as post-occupancy evaluation, are limited in practice. However, considering our pressing environmental challenges, thinking about the entire life-cycle of a project is gaining importance. Considerations such as maintenance, retrofit, expansion, end-of-life options and deconstruction are starting to attract increasing attention in the architecture profession. By understanding the importance of embodied carbon in architecture, architects are starting to think more carefully about the entire life-cycle of buildings and working on extending the life of existing structures [4]. One prominent voice in this context is Lacaton & Vassal, a practice that applies creative approaches to extend the life of existing buildings, and provides a radical vision: "Never demolish. Never subtract, remove, or replace. Always add, transform, and utilize, with and for the inhabitants" [1].

There are also emerging ideas on the evolution of architecture and how to adapt the built environment that already exists. The concept of exaptation refers to "the functional shift of a structure that previously served a different 'purpose' (that is, a trait related to a selective pressure) or no function at all (for example, as a side effect of other processes)" [7, p. 1]. In the realm of evolutionary biology, Darwin's basic concept is that "natural selection and other evolutionary factors do not work from zero, but continuously tinker with the already existing material, re-functionalizing it and transforming constraints into opportunities" [7, p. 1]. There are multiple applications of this concept in the context of architecture, where this functional shift can apply to tectonics, changes in the use of architectural spaces, and integration of decorative elements into architectural components.

Considering the social and ecological challenges of our time, adaptability to change and life-cycle thinking are important considerations for the future of architectural design. This chapter focuses on the challenges of designing social housing considering the full life-cycle of projects, and it discusses new possibilities for adaptation, expansion and self-construction through case studies from South America.

## 2 The Challenges of Inflexible Housing

Throughout history, many building typologies faced issues due to lack of flexibility, of expansion, and of adaptation, with some of the most critical examples being social housing projects. By attempting to standardise residents' ways of living, without any possibility of change or expansion, these rigid models of housing have led to various social and environmental issues. Due to cost considerations, there are usually many limitations to the ways that social housing units can be customised for and by residents to suit their preferences and needs, resulting in mass production and standardisation. There are several examples of the negative impacts of standard social housing worldwide, with one of the most well-known cases being the Pruitt-Igoe housing project, located in the USA, which fell into such a state of despair that led to its demolition less than two decades after its completion [6].

This issue is not limited to social housing only. Charles Montgomery, in his book Happy City, writes about the irony of lack of freedom in shaping the environment of American cities:

> a nation that celebrates freedom and weaves liberty into its national myth rarely gives regular people the chance to shape their own communities. Municipal governments, often with the counsel and assistance of land developers, lay down community plans complete with restrictive zoning long before residents arrive on the scene. Residents have no say about what their streets and parks and gathering spaces will look like. And once they move in, it is illegal for them to tinker with the shape of the public spaces they share, or (...) to use their homes for anything beyond the dictates of strict zoning bylaws. Most people in wealthy cities do not make their own homes or neighbourhoods. They simply decorate and inhabit the dwellings they are offered [8, p. 306].

The same can be said about many masterplanned neighbourhoods and housing projects around the world. By standardizing living spaces without room for change or expansion, these rigid models often fail to meet residents' evolving needs. The result is that buildings become obsolete almost instantly after being built. Depending on the local context and policies, some residents end up taking a "do-it-yourself" approach to make alterations and additions. This is more commonly observed in the Global South, where there is less control of building alterations by municipalities.

Recent research and practice encourage the involvement of residents in the design process and in the construction of their homes and public spaces. Bernheimer states the following about the connection between residents' agency in shaping the built environment and well-being:

> Communities which design their own buildings are more likely to be happy and healthy. (...) To create something better suited to human needs, you need to give people the tools to co-create their own homes, streets and workspaces. When people are involved in creating and nurturing their own environments, they also feel a greater sense of agency, community and pride – a quality known as "collective efficacy". (...) Where there is collective efficacy and the ordered complexity it creates, we see vibrant, healthy places. To build a resilient future, we must take an active role in the shaping of our own environments – the shaping of us" [3].

While in modernist projects, architects tried to control all aspects and details of buildings, new perspectives emerge on the role of the designer as a facilitator and enabler of communities to shape their own spaces. New pedagogies and practices challenge the notion of the architect as a singular designer model in favour of an approach where "the architect is a facilitator of culturally meaningful and co-designed dialogue and visioning" [13]. New housing models are emerging to respond to these new paradigms.

## 3 New Approaches: Incremental and Progressive Social Housing

Alternative approaches have emerged in the last few decades to make social housing more adaptable and flexible and to give residents more agency in shaping the places where they live. Many designers now question the cycle of redundancy, inflexibility, and inadequacy, by thinking of the possibilities of alterations and additions from the very inception of projects. The case studies presented in this chapter demonstrate innovative ways of thinking about adaptation and creation of a sense of belonging from the beginning of a project. Case studies from South America, including incremental housing by Elemental in Chile and progressive housing propositions in Brazil, are utilised to discuss these possibilities. These case studies challenge architects in their aim to "control" how buildings look like, allowing the creativity of residents to flourish and take centre stage in these neighbourhoods over time.

## 4 Incremental Housing by Elemental in Chile

In the early 2000s, Elemental's 'half-good house' strategy in Chile gained attention internationally due to its flexibility and residents' co-production principles [5, 14]. This concept emerged in 2003, when the Elemental team was appointed to develop the social housing project at Quinta Monroy in Iquique. The project redeveloped, on the same site, an informal settlement community of one hundred families that had been occupying public lands for over 30 years. Following the positive outcomes from this project, in 2009 they were commissioned by leading timber company Arauco to design 'Villa Verde', an estate planned to house timber workers and families in the city of Constitución under the Housing Solidarity Fund program.

Both housing projects, and others implemented later, applied the idea of "incremental housing" or "progressive housing", also described by Aravena as "half a house" [2]. Each typology is a $40m^2$ house, similar in size to other social housing schemes for informal settlements. However, considering that the traditional middle-income house is about 80 $m^2$, the design provided a structural layout that allowed future expansion, hence "half a house". This system overlays a self-build framework over an industrialized framework—informal overlaying the formal [5]. This incremental housing approach has been described as a touchstone in the housing discourses in low and middle-income countries and has been adapted to fit other sites and climatic conditions across Latin America [9]. Figure 1 shows both projects in two stages—after completion of the "half-houses", and after alterations and expansions were implemented by residents.

As these projects were completed years ago, there have been evaluations to understand their performance in practice. At Quinta Monroy, a post-up-grading survey—administered to all 100 families 18 months after the intervention—revealed that 35% of households took loans for home extensions. In this project, materials dismantled from the clearance of the original informal settlement were reclaimed and reused for the expansion of the new homes [9].

A post-occupancy evaluation conducted at Villa Verde four years after completion showed that almost all the houses in the settlement had been extended to complete their missing 'half', revealing residents' endorsement of the incremental process. The study revealed that overall satisfaction levels had improved in the four years after occupation. The most significant rise in residents' satisfaction related to the capacity to improve their houses—residents were supportive of the strategy to place responsibility in their own hands to self-manage improvements according to financial capacities and individual aspirations. The variety and extent of informal additions, demonstrated that the incremental process is likely to continue and take on more complex forms over time [5].

**Fig. 1** Incremental housing projects in Chile: Villa Verde (left) and Quinta Monroy (right), shown before and after expansions (Image credit: Elemental)

## 5 Incremental Housing Propositions in Brazil

In Brazil, there have been propositions to implement incremental housing models in various cities. A book by Toledo et al. [12] explored innovative social housing models based on experiences with interventions at the community of Rocinha, Rio de Janeiro, Brazil. Composed of eleven housing models with different characteristics, the projects range from the rearrangement of apartment plans, the introduction of the practice of expansion popularly known as "puxadinho", to the introduction of commercial spaces in social housing buildings. One of the examples is shown in Fig. 2.

The proposed social housing models were based on the requalification of urban areas and the creation of new public spaces, with the intention of addressing well-being of residents and mobility needs. Furthermore, preliminary studies aim to meet the diverse needs of families, with the variations in area and number of rooms. The project includes "evolving" social housing, which allow for the relocation of layouts and the controlled expansion of the units, enabling residents to build their own spaces according to their needs [12].

In the South of Brazil, an international Student Design on Sustainability and Social Housing—(CHIS 2014), organized by the PUCRS University, asked for proposals to address social housing and sustainability at the neighbourhood of Azenha in Porto Alegre. The winning proposition, developed by Piovesan et al. [10], explored the potential of incremental housing in higher-density models. Each

**Fig. 2** Example of an innovative social housing model in Brazil proposed by Toledo et al. [12]. (Image credit: Toledo et al.)

apartment in the proposed complex had an area that could be left either as balcony and open space, or be converted by residents into an additional room according to their needs (Fig. 3).

Both of these projects in Brazil were conceptual propositions, and were not implemented in the communities. However, these projects brought new conversations to the possibilities of social housing adaptation and expansion to engage residents in the construction of their living environments. Influenced by the ideas of Elemental in Chile, these propositions explored ways to implement incremental housing in apartment typologies, addressing the challenges of higher-density and limited land availability for housing.

## 6 Reflections and Conclusions

Incremental and progressive housing brings new possibilities to architectural and urban design, demonstrating that it is possible to achieve positive outcomes by empowering communities to participate in shaping their built environments. Now that the buildings developed by Elemental in Chile have been occupied for a few years, it is possible to assess the real benefits of these projects, with post-occupancy evaluation demonstrating residents' increasing satisfaction with their built environments over time. While these projects demonstrate innovation and new ways of designing, the next steps might be to rethink the base home provided, and think how those spaces could also be designed and built in more flexible ways. With new construction technologies, simple considerations such as moveable partitions could help residents change their living environments according

**Fig. 3** Winning proposal for Student Design Competition "Sustainability and Social Housing" organised by PUCRS, Porto Alegre, Brazil, 2014. Project by Eduardo Piovesan, Priscila Besen and Sofia Arrias Bittencourt

to different needs and life stages. Another challenge for this approach is how to apply it in higher-density environments, as enabling the expansion and adaptation of apartments brings additional technical challenges. This idea has been explored by various conceptual propositions in Brazil, but it has not been implemented in practice so far.

Considering the global ecological crises, the role of the architectural profession is shifting. While in the Global North, the concept of exaptation might be better implemented through repurposing existing buildings, in the Global South there are still high demands for new construction, especially when it comes to housing. By rethinking the scope of architectural design and enabling change over time, the role of the architect might shift to co-designing the basic infrastructures to empower communities to create their own spaces. By re-functionalising existing materials and spaces and transforming constraints into opportunities, architects can embrace flexibility. This shift away from rigid permanence toward adaptable and community-oriented architecture celebrates creativity, challenges conventions, and underscores the importance of considering long-term impact and the evolution of the built environment over time.

## References


1. Anne L (2022) "Never demolish. Always transform, with and for the inhabitants": Anne Lacaton delivers inaugural Jaqueline Tyrwhitt Urban Design Lecture. Harvard University Graduate School of Design. https://www.gsd.harvard.edu/2022/04/never-demolish-always-transform-with-and-for-the-inhabitants-anne-lacaton-on-urban-design-and-architecture/
2. Aravena A (2011) Elemental: a do tank. Architectural Design 81(3). https://doi.org/10.1002/ad.1235
3. Bernheimer L (2017) The shaping of us: how everyday spaces structure our lives, behavior, and well-being. Trinity University Press
4. Besen P, Boarin P (2023) Integrating energy retrofit with seismic upgrades to future-proof built heritage: case studies of unreinforced masonry buildings in Aotearoa New Zealand. https://doi.org/10.1016/j.buildenv.2023.110512
5. Carrasco S, O'Brien D (2022) Re-thinking elemental's incremental housing: residential satisfaction and resident-driven adaptations in Villa Verde, Chile. Urbe, 14. https://doi.org/10.1590/2175-3369.014.e20210171
6. Comerio MC (1981) Pruitt Igoe and other stories. JAE 34(4), 26–31. https://www.jstor.org/stable/1424663
7. Melis A, Pievani T, Lara-Hernandez JA (2024) Architectural exaptation: when function follows form. In architectural exaptation. Routledge
8. Montgomery C (2013) Happy city: transforming our lives through urban design. Farrar, Straus and Giroux
9. Núñez Collado JR, Wang HH (2020) Slum upgrading and climate change adaptation and mitigation: lessons from Latin America. Cities 104. https://doi.org/10.1016/j.cities.2020.102791
10. Piovesan E, Besen P, Bittencourt SA (2014) Concurso de Estudantes—Sustentabilidade e Habitação de Interesse Social—CHIS 2014. Concursos de Projeto. https://concursosdeprojeto.org/2014/06/09/concurso-de-estudantes-chis-2014/
11. Sinitsky I, Ilany BS (2019) The concept of invariance and change: theoretical background. In change and invariance. Brill. https://doi.org/10.1007/978-94-6300-699-6
12. Toledo LC, Natividade V, Vrcibradic P (2014) Repensando as Habitações de Interesse Social
13. Yates A, Pedersen Zari M, Bloomfield S, Burgess A, Walker C, Waghorn K, Besen P, Sargent N, Palmer F (2023) A transformative architectural pedagogy and tool for a time of converging crises. Urban Science 7(1). https://doi.org/10.3390/urbansci7010001
14. Zilliacus A (2016) Half a house builds a whole community: elemental's controversial social housing. ArchDaily. https://www.archdaily.com/797779/half-a-house-builds-a-whole-community-elementals-controversial-social-housing

# 11 Public Space and Safety: Access Control a Means for Territorial Reinforcement

Alexander Gomez-Torres

**Abstract**

For several decades, the potential to reduce crime through modifications to the built environment has been widely debated and implemented within the field of Environmental Crime Prevention. Universal design principles such as Natural Surveillance, Target Hardening or Access Control have all been found to lead to important drops in criminal activity rates; though, the efficacy of these principles is still the subject of continued research. Safety is one of those human dimensions the remains within the thresholds of form and function. Functional places are not always safe, and the same is valid for aesthetically pleasing ones. One of the most notorious changes to the built environment is gained through the implementation of vertical barriers under the umbrella of Access Control. These measures are widely recognized as an effective means to regulate people's access, egress, and movement. Vertical barriers may be also implemented via exaptation, by existing built environment features originally created for a different purpose. Nonetheless, implementing Access Control measures for safety is useless without understanding its influence on Territorial behaviours. Territoriality recurrently highlights the patterns of behaviour associated with the defence or protection of a defined place. A better understanding of Access Control thresholds was gained thanks to using a Graphic Mapping Overlay methodology. To understand vertical barriers in a public space, it is important to quantify them in terms of their impact on human scale. How Access Control gets implemented must pay close attention - such controls must be wary to avoid creating unwelcoming, undersurveilled spaces. Safer and liveable urban places need a multidisciplinary approach from urban planning, environmental studies, and architecture under design-crime prevention frameworks.

A. Gomez-Torres (✉)
University of Auckland, Auckland, New Zealand
e-mail: jgom498@aucklanduni.ac.nz



J. A. Lara-Hernandez et al. (eds.), *The Dusk of Design*, Integrated Science 34,
https://doi.org/10.1007/978-3-031-76867-5_11

**Keywords**

Architecture • Urban design • Public space • Territorial reinforcement • Access control • Vertical barriers • CPTED

## 1 Introduction

The Design of Redundancy argues in favour of reimagining conventional binarism that separates function from form when considering architecture and design. Instead, this chapter points out that fixated design approaches never constrain the human imagination, and when left to one's natural capabilities, creativity can be as functional as any design. Design is anchored in learning processes; it is a part of trial-and-error, where humans have an innate tendency to experiment with all things design. Therefore, redundancy as an active living learning implies a combination of layers and interconnected modes for both architecture and built environment transformation. Urban spaces inherently offer design function through a building or a public space, yet these spaces are shaped by layers and lines tangled with a specific purpose. In other words, architectural and designed structures imply redundancy in flexibility, adaptive reuse, and strength.

Safety is one of those human dimensions the remains within the thresholds of form and function. Functional places are not necessarily safe; the same applies to aesthetically pleasing ones. Safety is a fundamental element of human environments, and form and function are bound to each other inextricably (Melis and Lara-Hernandez, 2022). Functional settings can be designed to satisfy specific practical requirements. However, they are not necessarily safe; aesthetic appeal or visually engaging places are not guaranteed risk-free. Although it is generally accepted that pleasant environments may also be secure, that may not be the case as not all types of crime (factual or perceptual) relate to design features. Any attempt to create spaces that support human life through form and function must be interpreted through safety principles. This multidimensional approach is necessary for creating spaces that not only fulfil human needs and aesthetic standards but also ensure the security of all their users.

Another dimension of safe places in the context of urban development, urban renovation or regeneration is the re-imagination of existing facilities or built structures for alternative purpose is the concept of "exaptation". As a term, borrowed from evolution and biology, exaptation is a treat, feature or structure that takes on a function that differs from its original purpose (Merriam-webster, n.d.). In other words, exaptation, refers to a trait that evolves for one purpose but is co-opted for another. In the context of urban design and Environmental Crime Prevention, exaptation could imply the use of existing architectural or engineering features in ways that were not originally intended to enhance safety and functionality. For instance, a retaining wall initially designed for structural purposes might later be recognized as for it potential to improve boundary delimitation to increase safety through Access Control, thus serving a new function without having been originally built for this purpose. "Adaptations have functions; exaptations have effects" [19], p. 6).

Enhancing safe, secure, and liveable urban spaces requires a thorough comprehension of the impact of design on human behaviour. As such, architecture, urban planning, and design can yield more opportunities when one goes beyond traditional separations of function and form. This paper focuses on understanding the environmental principle of access control on the role it plays on territorial reinforcement. This exploration will seek to identify the aspects of the physical characteristics of the built environment that align with crime prevention and safe urban environments. Access control is a central concept in several disciplines include architecture, planning and urban design and as it helps to regulate and manage people's access through specific place [11]. Placing regulation on who can access and leave a location limits the ability to commit crime. Territorial reinforcement supported by access control promotes the sense of ownership and enhance the means to increase place attachment. Nevertheless, the understanding of territoriality within safety and built environment is still under research (Ekblom, 2011a-b, [18]). The notion behind relies on people are more prompt to safeguard and care of an area they feel attached to. An understanding of the works of access control and territorial reinforcement is essential as we advance safes and well-designed urban places.

## 1.1 The Theory Underpinning Access Control and Place

Thus, the Environmental Crime Prevention Theory [1, 6] and its subsequent development since the 1970s has been a major impact on how urban spaces and architecture are understood and designed concerning crime prevention mechanisms. The basic idea behind this theory and other underpinning sub-theories such as defensible space (Crime Prevention Through Urban Design _ CPTUD) [27–29], and Crime Prevention Through Environmental Design (CPTED) [14], is that the physical environment may also be designed to impact the behaviour and, in this way, prevent criminal activity. It is based on the idea of how much the design influences the individual's surrounding place and how this design may be used to create a safer environment [3, 4, 10, 18]. CPTUD and CPTED approaches agreed on a series of design principles that can influence human behaviour such as Natural Surveillance, Access Control, Activity Support, Target Hardening, or Image Management/maintenance.

One of the most common misinterpreted design principles is the role of physical and perceptual barriers in the creation of safe spaces; this is a measure under Access Control which relies on the conscious and planned aspect to influence traffic movement throughout a place [3]. It can be achieved by various engineering structures, such as physical barriers for example, fences or walls, psychological (perceptual) barriers by using indicator species—there are elements of landscape design, which indicate boundary zones, and more complex systems, such as surveillance cameras and security checkpoints. Thus, the ultimate goal is to exclude unauthorized access and to influence people's behaviour patterns, promoting it for the safety and security of a place [5, 9].

The creation, use or manipulation of physical or perceptual barriers in terms of safety is useless without understanding its influence in Territorial behaviours. Territoriality is defined as the "persistent attachment to a specific territory" or "the pattern of behaviour associated with the defense of a territory" (Merriam-webster, n.d.). As to design outcomes, the application of Access Control measures may be limited to mainly enhance Territorial Reinforcement [11]. In terms of the built environment, ownership behaviour is related mainly to the concept of "Territorial Control," where access is authorized by the group in control [20]. Through enhancing territoriality, access control acquires meaning in finally allowing a sense of ownership to the inhabitants or users of a certain space. Owning the place, people are likely to look after this area and monitor for any unwelcome behaviours or activities taking place [22]. The design upholds Territoriality by making it visible through the design attributes to which certain people presume everyone else recognises them as inherent values [2, 7]. This is achieved through such means as personalized landscaping, signs, maintenance, and visual enhancement of property. Founding Territoriality, the design techniques make the boundary more visible transform respect to the physical owners or managers of the space.

The relationship between access control and territorial reinforcement is an important aspect of environmental design in relation to crime prevention. Access control concerns defining of who have a right of access while territorial reinforcement concerns inviting those authorized users who assume responsibility for the space to be there [8, 10, 12]. The two work together to create psychological and physical atmospheres that are not only physical safer but feel that way to those who acceptably use the space, which is critical in fostering a successful urban environment. However, as powerful as these principles are as architectural definition, they may create results leading to the space perceived as unfriendly and underly surveyed. Thus, implementation of this principle of access control and territorial reinforcement should be critical, avoiding measures leading to the creation of places with low levels of appropriation. It should also be non-discriminatory to avoid creating unnecessary exclusion [24]. Understanding of access control and its link with territorial reinforcement in turn implies undertaking a multidisciplinary approach which combines urban planning & design, environmental psychology, and criminology, particularly within the frame works of Crime Prevention and Environmental Design. This principle is important in creating safe and secure places in the urban environment.

# 2 Assessing the Presence of Access Control

## 2.1 A Graphic Mapping Overlapping Methodology

Understanding the physical attributes of the built environment that relate to safety is the quest for the presence or absence of a specific quality. The search for that answer may end in a problematic and often dangerous binary oversimplification of yes/no or 1 / 0. Classifying complex phenomena into two distinct and opposing groups—can oversimplify the multifaceted nature of safety in the built environment. These considerations open the door to whether there are any limits while implementing a particular attribute. A question for this case, "Is there a threshold for physical or perceptual Access Control?" is critical. It stresses the challenge of quantification in creating safe environments.

## 2.2 Finding an Access Control threshold

Interpretating a threshold unveils key understanding for reaching optimal limits. A threshold may be defined as a point or level at which something begins or changes; it is a critical transition point. However, what constitutes a threshold is highly related to the presence or absence of a specific attribute. The proposed methodology aims to understand the Access Control presence through a mixed research—indirect and non-participant observation through a case study. The place (public space) is investigated through a graphic mapping overlapping technique [28].

**Case study**: Third Millenium Park, one of the most significant and controversial public spaces in Bogota, D.C. Colombia, was selected for this study. The place is centrally located within the foundational area a few blocks from the Presidential Capitol, the Congress, and the City Council's main headquarters. It is also the result of demolishing the previous Santa Ines neighbourhood by the end of the 1990s, after being declared as one of the most dangerous places in Latin America [26], Llorente et al. 2005). The environmental conditions assessed in this paper correspond to the park that opened to the public from 2000 to 2019, when it used to be a public park always opened to the public. The park has been renovated and shifted to a more recreational place [21], but it is enclosed with peripheral controlled and restricted access (Fig. 1).

## 2.3 Graphic Mapping Overlay Methodology

A plausible way to assess the presence of a particular attribute relies on the **Graphic Mapping Overlay Methodology,** a fundamental approach within landscape architecture and ecological planning that is especially useful in multi-layered spatial analysis [28], Fehler n.d.). This is a mapping technique that enable designers and planners to visualise and assess physical attributes by assembling thematic

**Fig. 1** Access control: Gated place (Left-2024) versus open space (Right-2013) –The Third Millennium Park

maps in layers. Each layer represents the different presence of a specific attribute or features of the area under study. After overlapping these layers, it is possible to analyse interactions between different capture data, improving the understanding of how specific built environment features interact with others within the same space.

The construction of a composite map involves at least the three following milestones:

**Base Map**: This map contains key elements of the place (case study). It identifies the boundaries of the design intervention and unveils the design features present in the place. In large-scale analysis, it may be labelled as the "Urban Design Inventory", and in smaller scales, e.g., block, it may be named the "Architectural Design Inventory" (Fig. 2).

1. **Attributes Maps**: An attribute denotes a defined property that can be used to identify, classify, or comprehensively describe an entity. It can be a design quality, or a built environment features subject to quantification. By overlaying these attribute maps, architects, urban designers, and planners can visualize complex data comprehensibly, making informed decisions that balance design qualities, function, and safety features. Attribute maps unveil critical information and

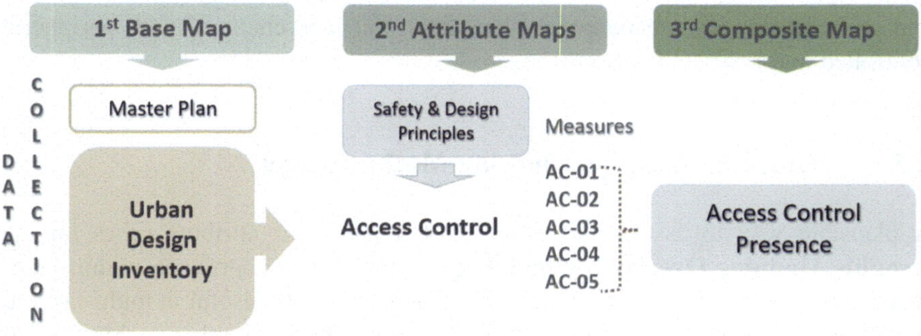


**Fig. 2** Data collection scheme for access control by author

**Table 1** Access control measures by author

| CPTED principle | Code | Attributes | Measures |
|---|---|---|---|
| ACCESS control | AC-01 | Delimitation | The area aims to be controlled must be cleared limited |
| | AC-02 | Vertical barriers | Landscape barriers may include changes in topography (pavement, surface levels, Terrain slopes). Though, non-exit areas and entrapped spots with lack of visibility must be eliminated (including cul-de-sacs) |
| | AC-03 | Horizontal barries | Landscaping, waterways, textures can create human scale non-obstructive barriers |
| | AC-04 | Sidewalks width (minimum) | The minimal width for pedestrian routes must be no less than 1.80 M |
| | AC-05 | Sidewalks length (maximum) | Pathways must be designed straight and shorter as possible. No more than 70.00 M length |

understanding that helps to ensure that urban environments are pleasant, safe, and responsive to community welfare (Table 1).

2. **Composite Map**: Attribute maps are overlaid over the Base Map to be assessed and measured. The collected overlapped data renders a final image that unveils the presence/absence of a Design principle that can be labelled as a Composite Map.
3. **Vertical Barriers (AC-02)**: even though Access Control can be implemented through multiple measures (Figs. 2 and 3); this paper focuses on understanding the opportunities and limitations of using Vertical Barriers.

### 2.3.1 The Case of Access Control Through Vertical Barriers

A vertical barrier serves as an Access Control measure, a physical element within the built environment that regulates (enables, restricts, and manages) human movement. In the case of study, vertical barriers included landscape-designed features (green walls, terrain slopes) shaped the public space. Additionally, for the Third Millenium Park existing buildings, especially gated or enclosed facilities above human scale, played a similar role (Fig. 3).

Like most built environment areas, the Third Millennium Park hosted perceptual barriers. Urban features provide strong or even rigid delimitation to clusters or subareas within places and usually function as Edges, as defined by Kevin Lynch (1964). Such perceptual elements are usually human hazards that constrain human activity within a specific place, such as the case of Highways, arterial roads, railway corridors, and interstate power lines.

The presence of vertical barriers on the case study can be interpreted in at least three relevant ways. First, in relation to the entire park area; second, in relation

**Fig. 3** Access control—vertical barries on the third Millennium park. captured from Google street view, 2013

and impact to the pedestrian footpaths; and third, in relation to the human scale. (Fig. 4).

The Third Millennium Park, like similar major public spaces, enables urban environments that provide communal spaces for human well-being, such as exercising, relaxing, and interacting. Nonetheless, the physical response of design attributes can certainly impact their accessibility and safety. The excessive use of Access Control responses through vertical barriers within a park is a feature that can influence the physical and perceptual experience of visitors. At first, the number of vertical barriers in the case study might appear surprisingly low, accounting for less than 25% (33,928 M2 within 161,574 M2) of the entire area. Yet, this minor percentage can impact the park's accessibility and natural surveillance (Fig. 5).

Another aspect unveiled after the measurement is the distribution of vertical barriers in the Third Millennium Park; it becomes evident that they are located along the public footpaths. Almost 50% (13,295 M2) of these obstacles are placed alongside pathways, increasing concealment and impeding movement in critical public areas. Pedestrian links facilitate movement within the public space, which is why the concentration of barriers alongside footpaths acts against enabling human accessibility. Public spaces allow people to explore various places and enjoy the landscape features. Furthermore, for the case study, the presence of vertical barriers above human scale obstructs public footpaths, limiting access but also interrupting people's movement, increasing a sense of limitation. (Fig. 6).

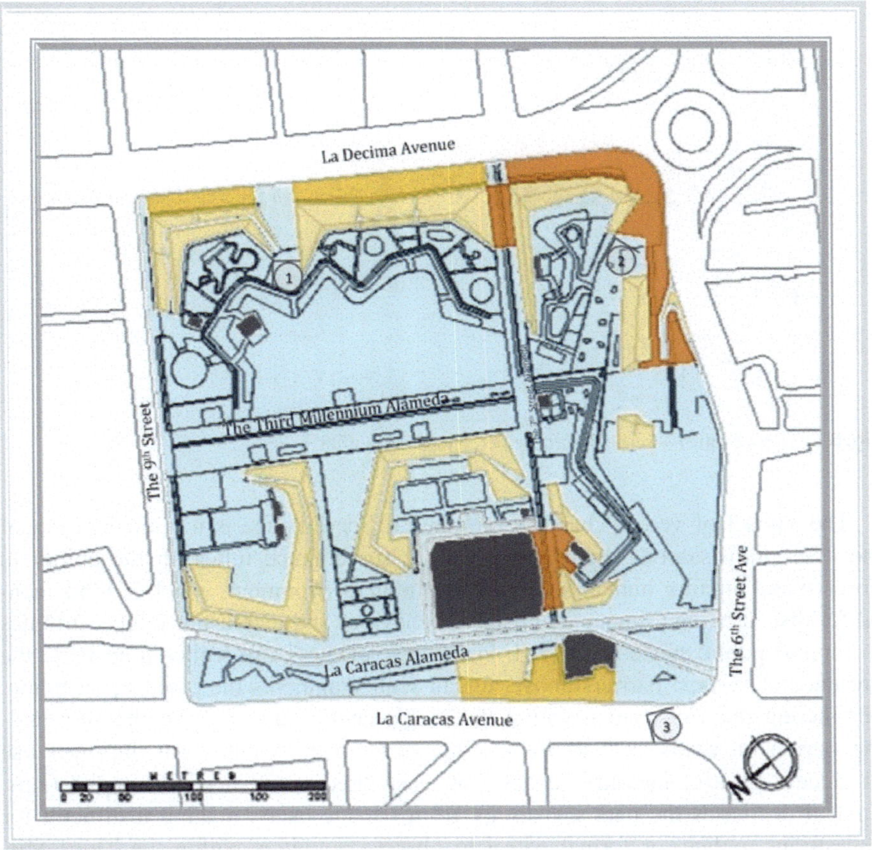


**Fig. 4** Access control—vertical barriers map (AC-02) by author

| Third Millennium Park Features | Code | M2 | % |
|---|---|---|---|
| Vertical Barrier (Non-enclosed) | | 20.633,16 | 12,77 |
| Sidewalks with Barrier Limitations (Arterial Rds) | | 6.553,69 | 4,06 |
| Sidewalks within Entrapment Areas | | 6.741,01 | 4,17 |
| Buildings (UD-09) | | 6.931,34 | 4,29 |
| Non-Vertical Barriers | | 120.714,98 | 74,71 |
| | | **161.574,17** | **100,00** |

**Fig. 5** Access control—vertical barriers map table (AC-02) by author

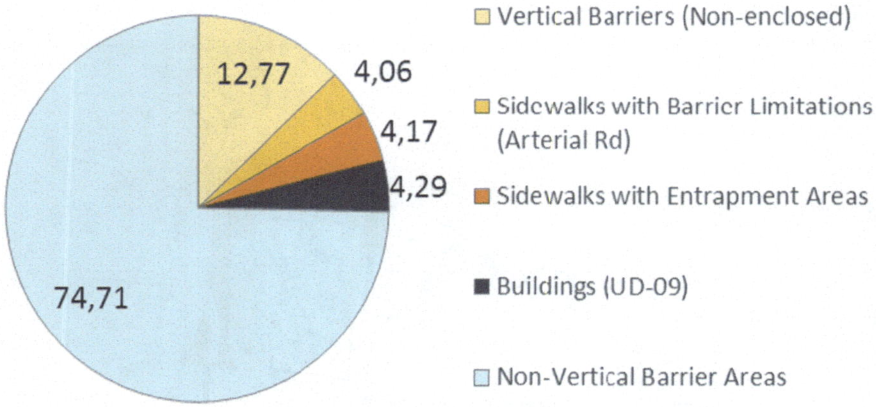


**Fig. 6** Access control—vertical barriers percentage by author

The impact of vertical barriers on Natural Surveillance is a critical aspect of the case study safety's constraints. Natural surveillance relies on the ability to observe and monitor human activities within an environment, which is fundamental for discouraging crime and antisocial behaviour and increasing safety attributes to support public space users [35]. In the case of the Third Millennium Park, the presence of vertical barriers above human scale fragments the place's perception, obstructing the view and isolating the park's central spaces from the surrounding peripheral roads such as La Decima or Caracas Avenue. An improper use of Access Control measures enables at least three major negative implications. First, it decreases the park's visibility from outside users, making it difficult for pedestrians, police, or formal patrol to observe activities within the Third Millennium. A low presence of Natural Surveillance translates into a lack of open sightlines increasing the sense of insecurity among park visitors; these constrains discouraging them from using the place and weakening the park's role as a communal gathering area. Furthermore, obstructing visibility can create concealment within the park, especially besides existing buildings, e.g. National Institute of Legal Medicine and Forensic Sciences, with a high levels of target hardening (see Image 04) or along highway structures and bridges such the intersection between the 6th Av and la 10th Av, enabling anti-social behaviours that negatively impact on the sense of place and further compromise the park's safety attributes.

The partitioning caused by vertical barriers in the Third Millennium Park presents not only a significant deterrent to visitors but also a potential alienating effect. Thus, the experience of certain portions of the park becomes altogether different, if not unavailable, to the public and especially to those with disabilities or young children. Instead of a public space serving as a means to attract people together, it is reduced to a collection of small-scale parks through which socialization with the local community remains unattainable.

## 3 Discussion

Vertical barriers are primarily an Access Control measure that can be enforced to public spaces to enhance their safety. Vertical barriers are non-negotiable; the intention is to eliminate unauthorized access and control how people will be allowed to use the place. Although these barriers succeed in securing the space and in some case reduce crime the impact on the space appropriation and social engagement may limit the public nature of the place.

In the case of the Third Millennium Park the percentage of barriers taller than the typical heights of ordinary humans, appear to impose fragmentation and ostracism among the people that utilize the space in Bogota. In addition to the physical separation, there also occurs a social landscape that indicates diminishing human interactions within the space. The design of tall walls and barriers turns some areas of this park into a fortress rather than an open public area which only emphasizes the control that overrides accessibility in favour of openness. This prioritization of control tends to fracture the experience of the space, preventing the natural flow of human interactions.

Furthermore, the physical segregation that emanates from these barriers, mainly along La Decima Av and the 6th Av, deter people from assembling due to limited interaction bringing little public value that public spaces should create [17, 18]. Proper public areas' design encourages the clustering of people rather than scattering them away, the overuse of tall barriers creates spaces more daunting and prohibitive which contrasts to the principles of good public space design that prioritize openness and inclusiveness. After experiencing from years of low levels of human activity and being considered by 2015 as one of the most dangerous areas in Bogota, DC, the original awarded winning landscape Third Millenium Park was partially demolished and renovated [6]. With less than two decades of service, the open metropolitan park was turned into a gated facility as opposed to open public space.

With the environmental safety perspective, even though the rationalization for these barriers is typically to prevent crime, they can oddly make spaces unsafe. This is because barriers do restrict access points and provide build-in pedestrian management that makes crime hard to accomplish in an unrestrained manner, hence reducing disrupting behaviours. Unfortunately, they can also create literal entrapment zones. It can increase physical highly enclosed environments by providing sealed end corners and reveal spots that will target people consequently making more vulnerable to antisocial behaviour. It can also strongly constraint pedestrian appropriation by restricting human presence in the space that requires free movement and interaction that supports good sense of place.

## 4 Conclusion

To create urban environments that are more functional, creative, and less dangerous, it is necessary to rethink traditional binary frameworks (form vs function) in urban design and architecture. Human adapting learning is an evolutionary trait that shapes places through testing and examination. Redundancy is a crucial functional quality in creating adaptable, resilient urban places that go further than predefined uses and can evolve to meet the challenges of the built environment. Like safety itself which is a fundamental part of any urban system. Functional areas or pleasing designed spaces are not necessarily safe without a comprehensive understanding of access control and territorial human activity. The same applies to the repurposing of existing structures for new functions, exaptation, that also plays a significant role in shaping-built environments. Original public space features are subject to be transformed and enhance safety and functionality beyond their original built purposed. Urban design and Architecture practice must consider the interaction between access control (human movement regulation) and territorial reinforcement (sense of ownership and care). These principles enable places that are factual and perceptually safer.

Public spaces and pedestrian walking ways provide a means for dimensionality of accessibility whereby people will move easily and safely to access either adjacent structures or day to day public space activity. However, the overuse of the vertical barriers can create obstacles and interruption that can make the path be less flow. Also, an excessive presence of vertical barriers acts against safe perception which means bringing fear instead of cohesion. On the other hand, instead of relying mainly on barriers, the balance of Access Control and Natural Surveillance's brings better qualities for gathering and participation. That is why one of the central features of access control is the impact it has on territorial behaviours. Territoriality, refers to a constant attachment to a certain territory or the behaviours linked to protecting that territory, is critical to determining how access control mechanisms operate. Aided by the concept of territorial reinforcement, access control measures can promote territoriality by fostering a feeling of ownership among an area's residents or users. Through the right level of access control, public space users experience better mechanism to oversee specific areas, feeling more responsible for maintaining it and more willing to monitor people's access and actions within that space in which some party is not supposed to act or be present.

**Acknowledgements** The author acknowledges Emeritus Professor Errol Haarhoff and Dr Emilio Garcia, supervisors of the PhD research, on where some of these findings are based. Furthermore, this investigation was possible thanks to the Doctoral Scholarship granted by COLCIENCIAS (Administrative Department of Science, Technology, and Innovation), Colombian Government.

## References


1. Andresen MA (2014) Environmental criminology: Evolution, theory, and practice. Routledge, Abingdon, Oxon
2. Armitage R (2017) Crime prevention through environmental design. In: Wortley R, Townsley M (eds) Environmental criminology and crime analysis, 2nd edn. Routledge, New York, NY, pp 259–285
3. Armitage R (2014) Crime prevention through environmental design. Encyclopedia of criminology and criminal justice. Springer, New York, pp 720–731
4. Armitage R (2013). Crime prevention through housing design: policy and practice. England. Palgrave Macmillan.
5. Atlas RI (2013) 21st century security and CPTED designing for critical infrastructure protection and crime prevention 2nd edn. Taylor and Francis Group, CRC Press, Boca Raton
6. Brantingham PJ, Brantingham PL (eds) (1981) Environmental criminology. Sage Publications, Beverly Hills, CA, pp 61–70
7. CanalCapital (2018). Alcalde Peñalosa inició la renovación del Parque Tercer Milenio. Canal Capital, Alcaldía Mayor de Bogotá. https://www.canalcapital.gov.co/top10/alcalde-penalosa-dio-inicio-la-renovacion-del-parque-tercer-milenio
8. Cozens P, Love T (2017). Crime prevention through environmental design (CPTED). Oxford research Encyclopedia of Criminology. Retrieved from 27 Nov. 2017. From http://criminology.oxfordre.com/view/10.1093/acrefore/9780190264079.001.0001/acrefore-9780190264079-e-2.
9. Cozens P, Love T (2015) A review and current status of crime prevention through environmental design (CPTED). J Plan Lit 30(4):393–412
10. Cozens PM (2014) Guidelines for design against crime think crime! : using evidence, theory and crime prevention through environmental design (CPTED) for planning safer cities. Praxis education Quinns Rocks, Western Australia
11. Cozens PM, Saville G, Hillier D (2005) Crime prevention through environmental design (CPTED): a review and modern bibliography. Prop Manag 23(5):328–356. https://doi.org/10.1108/02637470510631483
12. Crowe TD (2013). In: Fennelly LJ, ebrary I. (eds) Crime prevention through environmental design 3rd edn. Elsevier, Amsterdam
13. Crowe TD (2000) Crime prevention through environmental design: applications of architectural design and space management concepts, 2nd edn. Butterworth-Heinemann, Oxford
14. Crowe TD, National Crime Prevention Institute (University of Louisville) (1991) Crime prevention through environmental design: applications of architectural design and space management concepts. Butterworth-Heinemann, Boston
15. Ekblom P (2011) Guest editor's introduction. Eur J Crim Policy Res 17(1):1–5
16. Ekblom P (2011) Deconstructing CPTED … and reconstructing it for practice, knowledge management and research. Eur J Crim Policy Res 17(1):7–28
17. Fehler D (n.d.). The graphic overlay method. Retrieved from http://www.ou.edu/class/webstudy/fehler/E3/go/index.html
18. Gehl J (2013). In: Svarre B (ed) How to study public life. Island Press, Washington, p 2013
19. Gehl J (2010) In EBSCOhost (ed.) Cities for people. Island Press Washington, DC
20. Gibson V, Johnson D (2016) CPTED, but not as we know it: Investigating the conflict of frameworks and terminology in crime prevention through environmental design. Secur J 29(2):256–275
21. Gould SJ, Vrba ES (1982). Exaptation-a missing term in the science of form. Paleobiology, 8(1):4–15. http://www.jstor.org/stable/2400563
22. Habraken NJ (1987) The control of complexity. Places, 4(2), 3. Retrieved from http://escholarship.org/uc/item/9c52v1mw

23. IDRD (2018). Inician Obras de Reconstrucción del Parque Tercer Milenio. Alcaldía Mayor de Bogotá. https://www.idrd.gov.co/noticias/inician-obras-de-reconstruccion-del-parque-tercer-milenio
24. Kim ML (2004) Space, time, and crime. Carolina Academic Press, Durham, NC
25. Kramer R (2012) Political elites, "broken windows", and the commodification of urban space. Crit Criminol 20(3):229–248. https://doi.org/10.1007/s10612-011-9137-9
26. Lynch K (1960) The image of the city. Technology Press, Cambridge MA
27. Llorente MV, Rivas Á (2005) Estudio de caso La caída del crimen en Bogotá: una década de políticas de seguridad ciudadana. [Case Study the fall of Crime in Bogota: a decade of policies in citizen safety]. World Bank, Washington, D.C.
28. McHarg IL (1969) Design with nature. Natural History Press, Garden City, NJ, USA
29. Melis A, Lara-Hernandez JA, Melis B (2022) Learning from the biology of evolution: exaptation as a design strategy for future cities. Smart and Sustainable Built Environment 11(2):205–216
30. Merriam-Webster. (n.d.). Territoriality. In: Merriam-Webster.com dictionary. Retrieved From 20 May 2024. https://www.merriam-webster.com/dictionary/territoriality
31. Morris I (2011) En un lugar llamado El Cartucho. Crónica. [In a placed called "El Cartucho": chronicle]. Alcaldía Mayor de Bogotá, Cultura, Recreación y Deporte. Instituto Distrital de Patrimonio Cultural. [Internet]. [Consultado 2014 jul 09]. Disponible en: http://idpc.gov.co/publicaciones/?product=en-un-lugar-llamado-el-cartucho
32. Newman O (1972) Defensible space: People and design in the violent city. Architectural Press, London
33. Newman O, Franck KA (1982) The effects of building size on personal crime and fear of crime. Popul Environ 5(4):203–220
34. Newman O, Institute N, of Law Enforcement and Criminal Justice. (1973) Architectural design for crime prevention. National Institute of Law Enforcement and Criminal Justice, Washington, DC
35. New Zealand Ministry of Justice (NZMJ). (2005). National guidelines for crime prevention through environmental design in New Zealand, Part 1: seven qualities of safer places. Retrieved from 01 Aug 2017 https://www.justice.govt.nz/assets/Documents/Publications/cpted-part-1.pdf

# Material-Based Computational Design Strategies

# 12

Dustin White

**Abstract**

Exploring the concept of "*workmanship of risk*," in which a craftsman is constantly exposed to some level of risk and uncertainty, they interact with materials, responding to their properties and adapting their approach as the work progresses. In contrast, industrial production seeks uniformity and predictability. Incorporating *workmanship of risk* principles into material-based computational design strategies introduces elements of adaptability, creativity, and a deeper understanding of materials. Material-Based Computational Design Strategies are defined as the use of computational methods in conjunction with a thorough understanding of materials and their properties to analyze, simulate, and optimize materials during the design process. To investigate novel materials and their interactions with environmental factors, structures, and design parameters in order to produce more efficient, responsive, and sensorial architectural designs. The adaptive design approach, influenced by the *workmanship of risk*, incorporates a dynamic component into computational strategies. This allows for real-time adjustments based on material simulations and prototypes. Designers are encouraged to adopt algorithms that mimic craftsmen's problem-solving strategies, fostering an iterative and learning-driven approach to optimizing designs. The opportunity to work toward human–machine collaborations in which algorithms handle complex calculations and simulations and craftsmen contribute their knowledge and intuitive understanding of materials. This collaboration can produce designs that combine computational precision with the creativity and adaptability of human craftsmanship. Contrasting the two ideas, computational design frequently lacks the qualities of human touch; however, embracing imperfections in which the *workmanship of risk* challenges the absolute perfection and uniformity of material allows us to celebrate the

D. White (✉)
Florida Atlantic University, Boca Raton, USA
e-mail: whited@fau.edu



J. A. Lara-Hernandez et al. (eds.), *The Dusk of Design*, Integrated Science 34,
https://doi.org/10.1007/978-3-031-76867-5_12

inherent qualities of materials in a nuanced way. By encouraging designers to understand and appreciate the material aspects of their creations, we can promote the development of skills beyond coding and algorithmic proficiency. This comprehensive approach encourages a more well-rounded, empathetic, and contextually responsive design process that capitalizes on materials' unique properties while accepting the uncertainties inherent in the creative process.

**Keywords**

Material-based computational design • Workmanship of Risk • Adaptive design • Digital fabrication • Human–machine collaboration

## 1 Introduction

Drawing upon insights from David Pye's seminal work "The Nature and Art of Workmanship," we can evaluate contemporary approaches to material manipulation and production in architecture through the lens of craftsmanship. Pye's philosophical exploration of craftsmanship, particularly the concept of the "workmanship of risk," sheds light on the interaction between craftsmen and materials. In this paradigm, craftsmen navigate uncertainty and adapt their methods in response to the properties of materials, fostering adaptability and creativity. This stands in contrast to industrial production, which prioritizes uniformity and predictability.

Incorporating principles of the workmanship of risk into material-based computational design strategies introduces a dynamic element to the design process. Material-Based Computational Design Strategies involve the use of computational methods alongside a deep understanding of materials and their properties to analyze, simulate, and optimize materials during design. These strategies aim to explore novel materials and their interactions with environmental factors, structures, and design parameters. By doing so, architects can create designs that are not only more efficient and responsive but also engage the senses, enriching the architectural experience.

Pye's distinction between the "workmanship of certainty" and the "workmanship of risk" becomes a poignant framework for analyzing the evolving relationship between architects, materials, and manufacturing technologies. In his exploration, Pye contrasts the predictable outcomes of mechanized, industrial processes with the unpredictable and nuanced results of skilled craftsmanship. This conceptual duality underscores the essence of architectural production where the balance between precision and imperfection, control, and spontaneity, plays a crucial role in defining material effects.

Bob Sheil's "Manufacturing the Bespoke" delves into the realm of digital fabrication and prototyping, highlighting how technological advancements offer architects unprecedented opportunities to realize complex and bespoke designs.

However, Sheil's work also raises questions about the preservation of craftsmanship and tactile engagement in an increasingly digital age—a concern that resonates with Pye's philosophical inquiries into the nature of workmanship.

Both Pye's and Sheil's works underscore the significance of material experimentation and innovation in contemporary architectural practice. The concepts of "material performance" and "material ecology," which are pertinent to the work I will delineate, align with Pye's emphasis on the inherent qualities and limitations of materials that influence the approach of the craftsperson. This discourse prompts a critical examination of how architects navigate the balance between the certainties and uncertainties of material manipulation in their quest for meaningful and adaptive design solutions. Moreover, Sheil's investigation into prototyping and iterative design echoes Pye's view of craftsmanship as an ongoing process of refinement and adaptation. In this process, the craftsperson's engagement with materials leads to iterative insights and discoveries, driving the evolution of the design.

## 2 Materialization

The projects discussed below leverage materials and computational design workflows to attain specific spatial, sensorial, optical, and monolithic objectives. Each project employs tailored computational workflows to analyze, simulate, and optimize materials in the design process, investigating their interactions with environmental factors, structures, and design parameters. These innovations aim to push the boundaries of digital technology and fabrication, enhancing the relationship between material and machine, as well as material and design. By developing and utilizing novel software techniques, our goal is to facilitate the seamless translation from the virtual world to physical media, merging craftsmanship with technology to hybridize design and making.

Instead of fixating on binary conditions of True/False or right or wrong, my interest lies in processes with a range of possibilities that can be exceeded to produce variety. I'm particularly fascinated by the manufacturing process and how we can engage or intervene to create more efficient and responsive architectural designs.

In the initial stages, projects featured lightweight systems enhanced by sensors, lighting, iPads, and simulation tools to accentuate their material effects. Progressively, they evolved into stereotomic monocoque and additive stereotomic designs, incorporating stone and upcycled stone waste. While my research and practice don't always prioritize inherent material qualities, I recognize materials as a medium for manifesting specific behaviors or effects. This reciprocal relationship suggests that materials serve as a platform through which certain behaviors or effects are expressed. For instance, wood's properties allow it to be shaped into objects exhibiting strength and flexibility. Conversely, behaviors and effects can also influence materials, such as repeated stress leading to deformation, highlighting the dynamic interaction between materials and the behaviors/effects they exhibit or experience.

Throughout the design process, both physical materials and the observable phenomena or effects they produce are indispensable components, inherently intertwined and inseparable from each other. As tangible materials are brought into existence, they simultaneously enrich the possibilities of intangible components, including emotions, experiences, aesthetics, and conceptual aspects of the project. This integrated approach underscores the intricate relationship between materials and the overall project outcomes. The aim is to create work that is intelligible for people to engage with yet layered with complexity for deeper understanding.

The projects showcased here embody a material practice that seeks new material performativity through precise deployments of materials, forces, and effects. Infinity Helmet is a wearable prototype equipped with sensors, LED lighting, and an iPad, creating an interactive event kiosk. Stereoma is an immersive and spatial exhibition design constructed from a thin, veil-like material with interactive LED lighting. Coral Table is a minimalist dining table design constructed from a single block of brown marble. An artifact, Gothic Reimagined, is a bas-relief column and the first and only A.I.-assisted 3D printed stone project in the world.

In each instance, the progression from initial concept to final realization is viewed as a blueprint for research and development. It entails a series of trials, prototypes, mock-ups, and setbacks to ultimately achieve the desired design outcome. Furthermore, each project is seen as a segment within a larger trajectory of design exploration, undergoing testing within specific contexts to engage diverse audiences, subjects, and users. The interaction between the artifacts and the behaviors they elicit contributes to the ongoing, evolving nature of the work. A project reaches its culmination only when it integrates the interactions and spontaneous reactions from users, viewers, or the public.

The collaboration initially sparked by an invitation led to the creation of a revolutionary information point or kiosk for The Paseo Project, an annual public art event in Taos, New Mexico. This event intertwines performance, projection, and installation, showcasing globally invited artists throughout the vibrant streets of Taos. However, the organizers outlined one recurring challenge faced by visitors was navigating the labyrinthine streets to discover and engage with the artists' works during the evening event.

In response, our design looked to transcend the normative event kiosk or information point and we took this opportunity to envision an interactive information object that redefined the idea of public vs. private interactions when a user accesses and engages in social activities both online and in person. The object entitled "Infinity Helmet" (Fig. 1) is a transformative creation intended to interact with users and users interact with it and play off the idea of a virtual reality headset in the sense that it is an immersive virtual and physical reality experience. Where it differs is that its intentions are not to simply cover the eyes of a user but instead to envelope a portion of the body. It looked to blur the lines of a two-dimensional matrix of standard screens, into a spatial immersive environment through its material and formal qualities. Its intentions are also more diverse and include other embedded experiences that your typical headset does not such as infinity mirrors, iPad with connected video and event application, connection to a natural environment, virtual and physical social interaction, and audio-influenced artificial lighting.

To realize our vision, we began our research into polyhedral (Archimedean solids), crystalline, and minimal surface geometry, leveraging parametric and computational tools to empower architects with design input from the initial idea through the fabrication process. Archimedean geometry forms the foundation of our exploration, referring to a branch of geometry focused on the properties and relationships of figures and shapes within Euclidean space, particularly regular polygons and polyhedra.[3] Within this framework, emphasis is placed on figures

**Fig. 1** Infinity Helmet

possessing specific symmetrical properties, constructed by combining regular polygons. These figures, known as Archimedean solids, feature regular polygonal faces and identical vertices, distinguishing them from the more renowned Platonic solids.

For the helmet, we utilized a truncated tetrahedron, resulting from the truncation of the vertices of a regular tetrahedron, which yields hexagonal and triangular faces. A regular tetrahedron, comprising four equilateral triangular faces and four vertices, serves as the starting point. Through a computational modeling workflow, we crafted a faceted geometry using the triangular edges as "anchor edges." Subsequently, we decoded and digitally scripted the process using Kangaroo for Rhino/Grasshopper, dynamically relaxing the new faceted geometry into a minimal surface condition. The resulting minimal surface condition's spatial beauty presented numerous opportunities for the geometry to mold and contour the body seamlessly. This massing logic facilitated a streamlined process: once a polyhedral shell was designed, the minimal surface reacted directly to the exterior shell. With the two geometries working together, we applied programmatic elements to negotiate the open sections of the surfaces.

Once we established the geometry, we initiated a tessellation process for the minimal surface, aiming to determine the most effective approach for handling its double-curved form. Our goal was to apply tessellation strategies that would optimize both visual and experiential outcomes when the surface interacted with light, all while addressing structural concerns. We explored various tessellation strategies, including quad, Delaunay, and hexagon tessellations, at different resolutions. Ultimately, we found that a projected hexagonal mesh netting provided the best solution, dividing the minimal surface into four hundred discrete panels. It was crucial to ensure that each cell was planar or developable to facilitate the unrolling of each cell as a 2D shape. To enhance the spatial quality further, we refined our Grasshopper script to subdivide each hex cell into twelve triangular facets, creating a captivating kaleidoscopic effect within the volume's interior.

In the final computational stage, our focus shifted to developing batch and automated processes to streamline and optimize fabrication procedures. These processes encompassed tabbing, labeling systems, mass unrolling, setting up layers for mountain and valley folds, accommodating tab scale and geometry, and managing two-way folding perforation geometries based on material and thickness. Additionally, we tackled the challenge of mass part nesting. These tools were instrumental in establishing an efficient workflow with optimized layers, ensuring the smooth execution of the fabrication process.

One challenge we encountered was the unrolling of morphed or irregular hexagons, which tend to self-overlap. To overcome this issue, we refined our scripts to split each of the four hundred panels into two parts and built-in additional tab connections, resulting in a seamless fabrication process.

Prototyping was an essential and critical step in the process not only to test the accuracy and resolution of our computational tools but also to find the optimal material that would not only achieve the form and hold shape but also create a dynamic interior spatial condition. Where the equilibrium between precision and

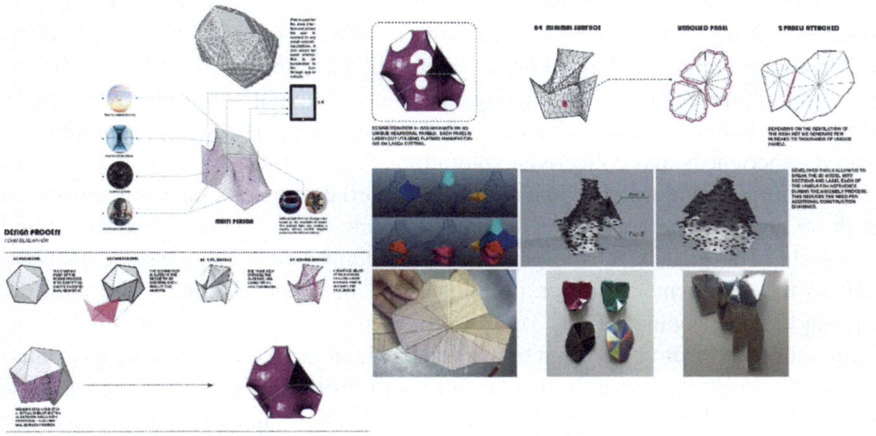

**Fig. 2** Diagrams outlining the various processes of form finding, penalization and material strategies, and embedded experiences

imperfection, control, and spontaneity, plays a pivotal role in defining material effect or the apparatus.

We conducted extensive prototyping using a variety of sheet materials, including aluminum-clad paper, anodized gold aluminum, wood veneer, PETG plastics, and various paper-based products. Small mock-ups consisting of 20–25 panels were produced to test different connection methods such as pop rivets, zip ties, lacing, and gluing. Each material responded differently to these connection types, with reflective paper proving most effective when glued. Ultimately, we evaluated the entire process—from modeling and laser cutting to assembly and on-site installation—ensuring that the helmet assembly process could be accurately executed with the aid of panel assembly diagrams referencing the digital model (Fig. 2).

The unique design of this wearable device resonated seamlessly with both the physical dimensions of the body and the intricacies of the mind. Structured into four distinct layers, the helmet offers a multi-dimensional sensory experience. The first layer employs layered parallel planes to create an immersive infinity mirror effect, while the second layer seamlessly integrates an iPad, enhancing user interaction with personalized content and dynamic lighting effects. This layer also provides users with the flexibility to access various apps, videos, and capture their experiences through photography or video.

Incorporating natural daylight conditions, the third layer allows users to experience the influence of the environment, establishing visual connections with the sky or other individuals. Finally, the fourth layer introduces dynamic artificial lighting that changes color based on sound amplitude, creating a visually captivating display on the helmet's exterior. Each layer produces unexpected ephemeral material effects influenced by the tessellation pattern and reflective material color, amplifying the kaleidoscopic effect, and generating geometric color overlays on the exterior shell.

Beyond its sensory experience, the Infinity Helmet possesses the potential to foster social interaction at events. Imagine a network of interconnected helmets dispersed throughout a large public gathering, enabling users to share experiences and influence each other's navigation and interaction within the event space. This networking capability enhances social engagement, creating opportunities for shared experiences and collective exploration.

This project explores the interaction of material geometry and the utilization of these architectural techniques to create an immersive spatial experience filled with light. The goal is for the geometry to become latent, allowing the focus to shift to the experience of space illuminated by light and its interaction with the surrounding environment (Fig. 3). The project's layers, ranging from shadows to illumination and from projection to the projected, are derived from this experiential approach. An interactive light display serves to further the project's aim of questioning the possibilities for spatial occupation. This lighting system responds to the presence of users through sensor-controlled regions, enhancing their engagement with the space.

This project is rooted in the exploration of materials. Rather than simply conceiving a form and then applying a material skin, our approach seeks to integrate material and form as a cohesive system. We begin by understanding the fundamental capabilities of the materials involved, and engaging in a meticulous process to generate components. By employing various media and representations, we aim to confront, challenge, and experiment with the formal capabilities of these materials. Ultimately, the form, experience, and effect of the project emerge from the collaborative interplay of these systems, driven by an intrinsic dialogue with the act of making.

As a venture focused on form-finding, this project draws heavily from the pioneering work of Frei Otto. Otto's exploration of tensile surface structures led to the

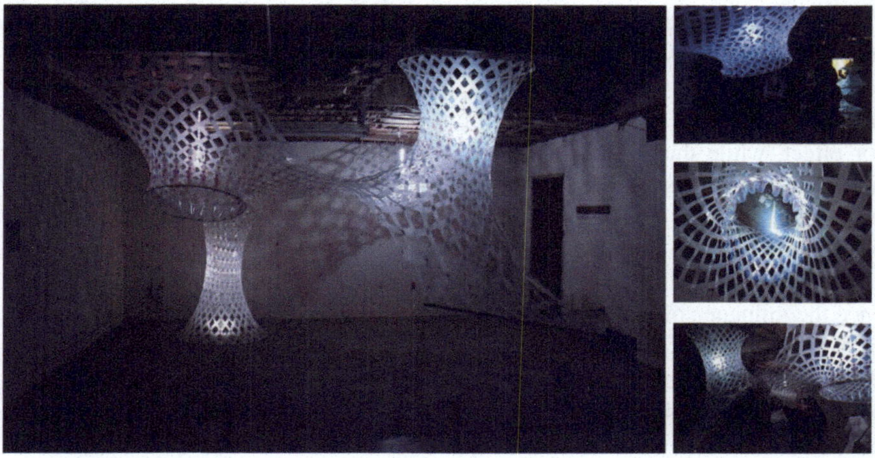

**Fig. 3** Spectroline

development of optimization methods for determining their geometry. The form-finding process aims to achieve equilibrium between the surface's stresses and geometric boundary conditions. In the 1950s, Otto pioneered engineering principles for constructing membrane structures, employing physical models made from fabrics, nets, and soap bubbles to experiment with equilibrium. His innovative use of soap bubbles to create minimal surface geometries like the Costa, catenoid, and helicoid not only presented formal and structural opportunities but also influenced the visual and representational aspects of design. The dichroic effect of soap films established a material precedent, inspiring the search for materials capable of exhibiting similar behavior.

Otto's exploration of soap films from a mathematical perspective led to the discovery of minimal surfaces. In the 1970s, Otto observed that when given a set of fixed points, soap film naturally spreads between them to create the smallest possible surface area, embodying the principle of "doing more with less." Depending on the support provided, the soap film naturally assumes the shape of the minimum surface, resulting in form-finding experiments of exceptional precision.

In our efforts to refine the production processes of the Infinity Helmet, we turned to the computational design tool *Kangaroo* for *Rhino/Grasshopper* and its capacity to integrate geometry, structure, and material performance. Our objective was to achieve a cohesive outcome by balancing considerations of material, surface, environment, space, geometry, and form. While structure played a significant role, it did not dictate the result. The Catenoid minimal surface geometry that created a vaulted effect was adapted and modulated to suit new plans and material configurations. For example, one of these configurations planned to add an Enneper geometry to the end of the Catenoid (Fig. 4), but this proved to be impossible given the final material systems that were determined. These adjustments were made possible by computational methods and custom-scripted tools that we refined over a series of projects and prototypes capable of modeling, structurally analyzing, and organizing large quantities of non-uniform elements.

The project faced a significant challenge due to an extremely tight budget of only six hundred dollars, limiting our options to lightweight, flexible, and robust plastic-based materials. Despite this constraint, our exploration of materials played a pivotal role in shaping the architectural concept, which was realized through the

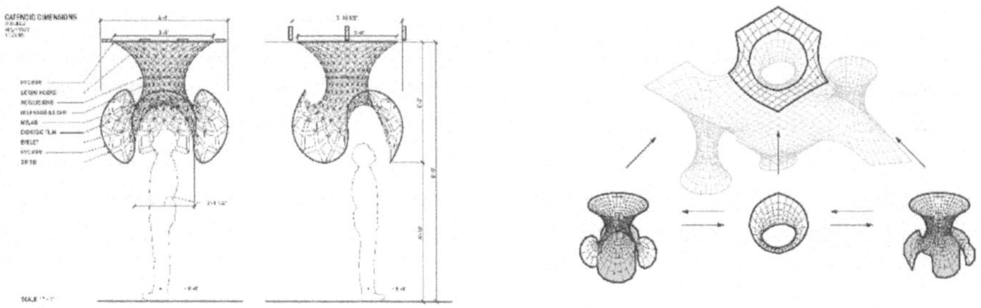

**Fig. 4** Design of Enneper geometry and diagram of installation parts

use of light, paper-thin surface elements. The dynamic visual effect, transitioning from reflective paper sheets to delicate luminous surfaces, was central to the atmospheric intentions of the project.

After thorough consideration, Mylar sheets were chosen for their availability, tensile strength when tested with a pop rivet connection, translucent properties, and suitability for flatbed manufacturing processes. We sought a material that straddled the line between solidity and void, ultimately transforming Mylar from a drawing material into a building material, forming an interactive hypersurface. Despite its paper-thin nature, we embraced the concept of adding depth to the material through LED lighting effects.

The material's systematization is achieved through its penalization and its effects on the surface. Developing the digital model solely through quad-based tessellation proved to be a dynamic workflow, allowing for control over surface resolution and panel density through quad-remeshing processes in Rhino/Grasshopper. The surface was divided into eight parts, consisting of four quadrants and four catenoids of varying sizes and diameters, totaling 1200 discrete quad panels. Similar to the Infinity Helmet design, we employed automated batch processing for mass labeling and unrolling the panels. However, in this case, each quad face was unrolled in 2D.

We optimized a process to determine the desired "X" panel shape, incorporating parameters to control the surface aperture size, pop rivet diameter, and panel outer edge offset. This ensured that when four panels met at a corner, there was sufficient material to accommodate the pop rivet. The developed scripts not only facilitated the production of material mockups but also reduced the need for labor-intensive drawings. These were not drawings or shop drawings per se, instead, they were drawings that contained all the thresholds and range of that element designed for that piece. A panel or smart panel that has all the parameters to become any part of the design (Fig. 5).

Once all parts were laser-cut and organized, the 3D model part tags were referenced, and the surface was assembled. An additional layer of panels was cut from dichroic multicolor film, duplicating panels layered into the catenoid shapes.

The project's final layer integrated four custom LED light groupings strategically positioned within each catenoid opening. Each LED package featured a Wi-Fi-enabled Arduino board (Fig. 5), an open-source electronic prototyping platform, attached to a custom laser-cut and 3D-printed frame directing the light projection. Additionally, each board included an integrated motion sensor and a series of lighting simulations programmed to respond to speed, occupancy, density, and sound. When the space was unoccupied, the lights dimmed and emitted a slow pulsing effect. However, upon occupancy, the lighting transformed: colors changed, brightness increased, and the pulsing ceased. By assembling discrete components, we created a diverse array of spatial conditions, fostering a part-to-whole relationship where individual elements collectively produced emergent phenomena.

**Fig. 5** LED wiring and assmembly. Laser Cut "X" panel and panel assembly

Spectroline's material presence is minimal and translucent, sometimes appearing to almost disappear. These immaterial effects, combined with seamless activation, construct a sensorial spatial experience.

The previous two installations can be viewed as prototypes or preparatory steps toward achieving permanence in the work. Coral Table represents a departure from the intricate indexing and aggregation of numerous part-to-whole relationships, adopting instead a "monolithic" approach (Fig. 6). A monolith is something perceived as a singular, unified entity, devoid of joinery or individual parts. Unlike previous works that yielded complex material systems requiring calibration to reconcile tensions between smooth, uniform, and faceted surfaces, this strategy imposes both formal constraints and a singular surface effect. The process of crafting this monolith involves extracting and subtracting from a single block of brown marble (Palissandro Bronzetto) native to the Tuscan region of Italy.

The table was created for the "Natural Things" exhibition at the Marmomac Stone Exhibition in Verona, Italy, the largest and most significant trade show for stone and stone fabrication. Measuring three meters in length and one meter in width, the table's design adopts a format suitable for dining. Inspired by naturally occurring formal systems, such as landscape erosion patterns revealing geological characteristics and coral reef growth patterns exhibiting branching and bifurcation as they grow vertically, the design strategy integrates these elements. The result of this analysis is a variable porous support system comprising non-repeating, asymmetrical, natural branching support members that attach to a standard rectangular tabletop geometry. Computationally, the design employs similar scripted tools and topological mesh modeling as previous work, albeit in this case, the minimal surface geometry forms a compression structure rather than one under tension. The

**Fig. 6** Coral Table

modeling process aims to uncover unexpected or unknown shapes characteristic of coral growth.

The sculpting of natural materials like stone inherently entails challenges. The very characteristics that enrich the material during its formation also present obstacles to those working with the material. This process aligns directly with David Pye's concept of the "workmanship of risk." Stone, by its nature, resists carving, chiseling, and sanding. The skill of individuals working with the material directly correlates with their ability to navigate this resistance. Even with the use of digital design and computational processes, which enable the simulation of various production techniques, there remains an inherent risk when working with a material that has formed naturally over time. While the software allows for optimization and control of shape, density, and scale, the placement of complex vaults and arches may appear random. However, in reality, the legs were strategically placed in four locations, branching into nine connection points to support the top mass.

In digital architecture, there's often a predominant focus on thin, tectonic surfaces, with mass, thickness, and depth sometimes overlooked. Stereotomy, defined as the art and science of cutting three-dimensional solids, particularly within the context of digital architecture and fabrication, underscores the importance of interlocking parts, rationalization, and the role of mathematics and geometry. The project's development explored the subtractive qualities of stereotomy, aiming to enhance the aesthetics of the material's striation through digital eroding and carving.

The utilization of a robotic arm in this project stands out from other digital fabrication platforms due to its impartiality towards any specific fabrication method. The capabilities of the robot are determined by the end effector attached to it. In this instance, two subtractive methods were employed: a 1.5 m diamond-coated articulating saw for cutting large sections from the block, and a CNC milling spindle capable of operating on 7 axes. Robotic milling offers advantages in terms of carving options, including undercutting. However, industrial robots, despite their motion flexibility, pose challenges due to their inherent lack of stiffness compared to gantry-style machines. Robotic stone milling is a complex process, involving a slow progression of the bit engaging incrementally into the surface and requiring a continuous stream of water to clear the milled material. Achieving an exhibition-quality surface finish necessitated twelve hours of milling production over seven days, followed by several weeks of craftsman sanding and polishing.

Ultimately, risks can arise regardless of the tools and techniques chosen for a project. Coral Table serves as an example highlighting the importance of collaborating with individuals and companies who possess a legacy of working with the selected materials, even when leveraging the latest technological advancements in design and fabrication.

As sustainability concerns grow, there has been a renewed interest in locally sourced materials such as soil, earth, and stone, offering opportunities for more environmentally friendly construction practices. In the context of a previous project, which utilized stereotomic construction methods involving material subtraction through milling, questions arise about the management of waste streams and the potential for reimagining construction approaches through additive stereotomy.

Gothic Reimagined looks at a novel method of design by collaborating with creative text to image AI platforms such as MidJourny and Stable Diffusion. The future of material practice and manufacturing has made it possible to achieve aesthetics of complexity with relative simplicity, offering a unique chance to delve into the history of architecture and provoke new possibilities for form and ornament (Fig. 7). Through this process of exploration, there was an intrigue into thinking through concepts of bits, particles, aggregates, and amalgamation of concepts that explored a range of architectural styles and concepts and understanding the value of integrating the AI image-making process into the design process to produce a physical series of 3D printed stone artifacts.

The goal of MidJourney's AI diffusion models is to replicate and comprehend how AI technologies spread throughout complex systems like social networks and organizational networks. These models consider many variables that influence or hinder the diffusion of AI, including network structure, technology type, and organizational or individual characteristics. By analyzing the interactions and spread of AI technologies, these models can thereby preserve the memory of technological innovations and their impact. To create new images during the training of AI diffusion models, massive image databases are sampled. These images can be used in a variety of applications, such as art, design, and architecture. AI diffusion models can be applied to architecture to produce distinctive visions and complex forms that were previously unimaginable. This is because AI diffusion models have this unique ability through text prompting and image referencing to recompose, create, interpret experiences, and embed historical or cultural references to elevate

**Fig. 7** Gothic Reimagined

our architectural design process. I developed a fascination with reimagining historical movements like Gothic, Art Nouveau, Classical, and Baroque architecture throughout my early studies with the tool.

In order to test and challenge these new AI-assisted design tools a series of images were created using MidJourney and outlined through a rigorous rule-based expansive design practice created to analyze Gothic elements. The process of collaborating with AI in order to distort, split, and rearrange elements to fit into a given space, but it's important to note the essential body is apparent and to some

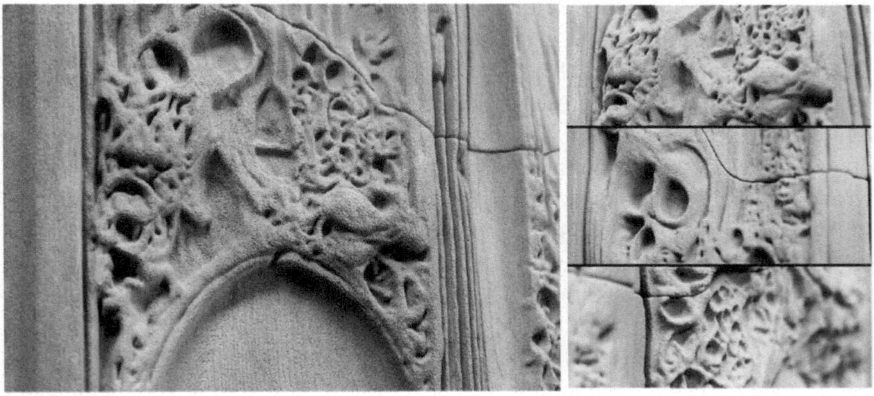

**Fig. 8** Printed stone gothic images. Prints create a new architectural aesthetic through AI

extent, their relationship with one another is maintained. Specifically, I was analyzing the key elements in Gothic architecture such as the Pointed Arch, which led to the Rib Vault, Flying buttresses, and elaborate tracery. The following are examples of rules that were created to establish the prompt structure: (1) stylizing as opposed to realistic representation, (2) schematic characterization by accentuating certain features (3) dislocating split details. (4) an illogical transformation of details into new Gothic representations. When developing image iterations of these historic Gothic elements, the ambiguity of language was a crucial component of the creative process. The final image interpretations were a dynamic flow or reorganization of materials and elements that appeared to have been taken from earlier construction projects but reinvented and preserved in a new novel structure (Fig. 8).

Participating in the "Marmomac Meets Academies" exhibition at Verona, Italy's Marmomac Stone Theater provided an opportunity for collaboration with Conc3de, a Dutch 3D-printing company. Conc3de specializes in a bio-binder-based printing process, enabling the creation of intricate parts with unique material properties using upcycled waste material. For our collaboration, we utilized upcycled limestone sourced from local industries, marking a significant step in reducing the extraction of new virgin materials in the stone industry. This approach taps into vast quantities of unused stone powder found worldwide, offering a promising avenue for mass construction through additive stereotomy.

In the case of Gothic Reimagined, we faced the challenge of a relatively short six-week timeframe and the stylized, medium-resolution output of new AI tools. To address this, we developed a procedural modeling approach to generate the base surface output. This approach allowed us to refine surface features through manual editing and sculpting, facilitating the transformation of 2D photos into high-resolution 3D relief models. The final prints consisted of stone fragments and particles arranged to create two stone artifacts, each featuring four distinct images. These artifacts comprised multiple self-supporting parts, with dimensions of 150 cm x 50 cm. "Gothic Reimagined" marked a milestone as the first successful conversion of entirely AI-generated imagery into physical stone forms using 3D printing technology (Fig. 8).

"Gothic Reimagined" represents an initial foray into the exploration of a novel architectural aesthetic using A.I. text-to-image technology. However, the true authorship lies in the development of a workflow that translates images into models and subsequently into stone aggregates for production. As one delves into the details, a distinct understanding emerges, contrasting with mere images. This presents a unique opportunity to blend traditional material craftsmanship with digital technology and creative AI, paving the way for innovative architectural expressions.

While we are still in the early stages of fully grasping the potential of these new AI tools in design and architecture, it's evident that they offer a tremendous opportunity to enhance our architectural design process. This enhancement can take various forms, including technological innovations, accessing architectural history to inspire future explorations of form and ornament, or utilizing novel additive manufacturing processes. Through these avenues, we can delve deeper into the creative possibilities and push the boundaries of architectural expression.

**Acknowledgements** I would like to thank the many students, colleagues, and collaborators whom I had the privilege of working with.

*Infinity Helmet:* Karla Murillo, Christopher Varrette

*Spectroline*: Karla Murillo, Wesley Thomas, Fabiola Vazquez, & Patrick Walker

*Coral Table*: Generelli SA., Gruppo Tosco Marmi, Maurizio Barberio, Giuseppe Fallacara

*Gothic Reimagined*: Concr3de 3D Printing & Christopher Pope

## References

1. Pye David (1995) The nature and art of workmanship. Revised edition, Herbert Press
2. Sheil B. (Ed) (2012). Manufacturing the Bespoke : making and prototyping architecture edited by Bob Sheil. John Wiley and Sons Ltd
3. Borden Gail Peter, Michael Meredith (2012) Matter : Material processes in architectural production. Routledge
4. Andrusko PA (2014) Stereotomy: Stone Architecture and New Research by Giuseppe Fallacara. Nexus Netw J 16(2):501–504
5. Fallacara G, Gadaleta R (2019) Stereotomy: Architecture and Mathematics. In: Sriraman B (ed) Handbook of the Mathematics of the Arts and Sciences. Springer International Publishing, pp 1–20
6. Asperl Andreas et al (2007). Architectural geometry. Bentley Institute Press
7. Indrawan Stephanus Evert (2017) Digital fabrication, Architectural and material techniques Iwamoto Lisa. IJCAS (Int J Creative Arts Stud 3(1):87–91 https://doi.org/10.24821/ijcas.v3i1.2074
8. Skavara Marilena et al (2017) Fabricate. UCL Press

# The Possibility of Transcribing Trauma into Architecture: Exploring Potential Interdisciplinary Research Methods Across Architecture and Criminology

13

Antonino Di Raimo

**Abstract**

This research article is part of a broader project focused on Trauma-Informed Design (TID). It explores the relationships between trauma experiences and architectural spaces using qualitative approaches. Initiated in 2020, this study involved dialogues between an architect and two criminologists, using visual artefacts to explore trauma in architecture. Exhibited at the Italian Pavilion "Resilient Communities" during the Architecture Venice Biennale 2021, the study employed architectural transcriptions of a semi-structured interview to create an imaginary urban setting embodying trauma narratives. In this work, I explore the possibility of new research methods that can unfold embodied experiences of trauma in architecture. I will firstly recall the notion of trauma and how it has recently impacted fields like criminology and architecture. Using a meander approach, I will briefly refer to the process of "informal conversations" as a research method to identify the research problem, which is beneficial when research is conducted across different disciplines. Then, drawing on other scholars' works, I will argue for trauma as an assumption capable of holding together notions like architecture, memory, and the body. Based on the stories that traumatised individuals tell, I will use a transcribed interview conducted by a criminologist with an individual as an explorative point to investigate deeper the relation between trauma and architecture. Assuming a practice-based research framework and aligning with Candy (2006) definitions, I will use digital drawings to explore these relations. I will briefly recall the importance of ethnographic drawings and consider transcriptions in architecture and their intrinsic value through the work of other architects and theorists. Assuming an

A. Di Raimo (✉)
Portsmouth School of Architecture, Art & Design, Portsmouth, UK
e-mail: antonino.diraimo@port.ac.uk



J. A. Lara-Hernandez et al. (eds.), *The Dusk of Design*, Integrated Science 34,
https://doi.org/10.1007/978-3-031-76867-5_13

embodied cognition approach, I will argue for the human body as the main vehicle in making transcriptions between individual narratives externalised in words and transcribed into architecture through the medium of drawing. According to practice-based research, as new knowledge is gained through creative outcomes, I would argue that representing the human body in these genres of transcriptions can constitute a "relational affordance," a medium capable of operating in the transcription process while unfolding aspects often subsumed by written words. Through the practice of transcribing contents into architecture and externalising them in terms of architectural and urban drawings, I sought to unveil the possible meanings of representing trauma within architecture before looking for any trauma-informed "solutions."

**Keywords**

Trauma-informed design (TID) • Interdisciplinary research • Architecture and criminology • Embodied cognition • Ethnographic drawings

# 1 Introduction

## 1.1 Approaching the Notion of Trauma, a Gap in Architecture

> As a child I felt different to anyone in the family. I had to separate myself from what was happening. I was bullied at school. Abused as a child. It took me away. I felt different… I attempted suicide at 15. My dad was being bullying. I was sticking up for my mum. It was hard. Horrible. I wasn't allowed to be a child. I feel like a little girl when I talk about it…I never thought of myself in terms of me being a woman. What is a woman? I've the body of a woman but inside I feel like a child.[1] (L. Jones, personal communication, 2015)

Trauma-informed approaches, a growing field of research and practice, focus on the awareness of organisations and service providers regarding the importance of understanding trauma's effects on survivors. The American Psychological Association states that "Trauma is an emotional response to a terrible event like an accident, crime, natural disaster, physical or emotional abuse, neglect, experiencing or witnessing violence, the death of a loved one, war, and more. Immediately after the event, shock and denial are typical. Longer-term reactions include unpredictable emotions, flashbacks, strained relationships, and even physical symptoms

[1] For ethical reasons, we decided to cite in text only three quotations from the interview conducted by Dr. Madeline Petrillo. The interview is about the story of Lynne Jones, which is a pseudonym. As Dr. Petrillo claimed, "Lynne was sentenced to more than two years in prison for an offence of wounding her former partner. She is assessed as 'high risk.' She is an alcoholic, has mental health problems, and a history of acquisitive offending. This is the story the criminal justice system tells about Lynne."

like headaches or nausea. While these feelings are normal, some people have difficulty moving on with their lives. Psychologists can help them find constructive ways of managing their emotions" [1, 2]. Often, traumatic events develop into Post-Traumatic Stress Disorder (PTSD). Yehuda et al. [3] provide a useful definition for PTSD: it "is a condition that can develop following exposure to extremely traumatic events such as interpersonal violence, combat, life-threatening accidents, or natural disasters. Symptoms of PTSD include distressing and intrusive memories and nightmares of the trauma, irritability, hypervigilance (an enhanced state of threat sensitivity or preoccupation with the potential for danger), difficulty sleeping, poor concentration, and emotional withdrawal." From a global perspective, the World Health Organization, based on a recent survey, estimates that 3.9% of the world population has had PTSD at some stage in their lives. Most of us will experience events in our lives that can be considered traumatic. However, most people, although exposed to traumatic events, do not necessarily develop PTSD [4]. While trauma-informed approaches are increasingly being adopted by a significant range of service providers, there are still research gaps in understanding how these approaches might be relevant to the design quality of the places where affected people live. Very relevant to this study is that it has been acknowledged that "Individuals with PTSD frequently avoid places, activities, or things that could remind them of the trauma" [3]. Therefore, as with many other disorders, the spaces where people live, including architectural spaces, can play a significant role in either fostering distress or mitigating it. Integrating trauma-informed principles into physical spaces through architecture and interior design is defined as Trauma-Informed Design (TID). While there is no commonly adopted definition of TID, its relevance has been fully acknowledged, especially in the design of architectural settings where the emotional and physical recovery of individuals is a pivotal reason for the existence of those spaces. Between 2021 and 2022, we conducted an exploratory thematic literature review about TID [5]. While acknowledging that TID still seemed to be a pioneering research stream, we also observed its increasing adoption in the design of prisons, confinement structures, and healthcare centres. We also highlighted qualitative research conducted by scholars from the field of criminology, such as the work of Jewkes et al. [6], and from architecture. These experiences emphasise principles that can be relevant during the design stages. For example, the work of the Practice Research Centre at the University of Colorado remarked on the importance of TID as part of the broader field of urban and architectural design and human health and well-being [7]. However, from that preliminary and indicative review, which focused on the notion of trauma and its extension to architecture, we concluded that: (a) There is a need to establish a common ground across different disciplines regarding how the notion of trauma is relevant to them and to what extent. (b) When focusing on architectural design, recommendations seem to be very general and aligned with diverse health and well-being approaches within the built environment.

The work of [8], for instance, identifies the impact of the built environment on both physical health and emotional well-being, linking environmental attributes like light, air quality, and spatial design to health outcomes. In their scoping review,

[9] explore the varied definitions and dimensions of well-being in buildings. Their study highlights the conceptual ambiguity of well-being and the need for standardised definitions and measures to enhance the applicability of research findings in real-world building design and policy. Many of the findings of these works might well align with the recommendations suggested by the Trauma-Informed Design Process [10]; "incorporating trauma-informed principles in the design of housing can significantly improve the mental health and well-being of residents" as they put it.

Apparently, looking for trauma in architecture might easily dissolve into health and well-being principles. Therefore, from the point of view of architecture, we felt a significant gap at least because trauma as a notion, although rooted in psychology, seemed to imply the need for an interdisciplinary perspective. Additionally, despite the vast amount of studies focused on Trauma-Informed Care, we found a lack of evidence in understanding how the notion of trauma could be related to architecture as a discipline, extending the focus from the observation of traumatic symptoms and the healing of them to the broader lived experience of traumatised individuals within the space.

## 2 Attempting to Set an Interdisciplinary Perspective: Underpinning the Chain of Relationships Across Mind, Body, and Architecture

In terms of trauma studies, we found the work developed by Bessel van der Kolk, culminating in his book "The Body Keeps the Score" [11], very relevant to this research. Van der Kolk argues for the pivotal role of the body in not only storing the trauma experience but also in the healing processes. Based on his long-term research and experience in defining and dealing with Post-Traumatic Stress Disorder, Van der Kolk advocates for a comprehensive approach that positions the body (and not only the mind) at the centre of understanding trauma. Van der Kolk's work insists on the hypothesis that trauma is a state of the body, targeting the soma rather than implying that trauma is exclusively a physical or cognitive condition. However, it is worth noting that despite the global popularity of his work (the book has been translated into 36 languages), Van der Kolk's work has also faced criticism concerning: (a) The healing proposals underpinned by the research. (b) The political dimension implied by trauma becoming perhaps too popularised and therefore generic. Regarding the healing proposal, [12] state, "Advocates of research-supported treatments for PTSD are losing in lots of ways." The criticism here is: (a) The holistic approach might lack the linearity of research-supported treatments. (b) The notion of trauma developed by Van der Kolk and further developed by his followers is subtly criticised, for example, by Carr [13], regarding its popularity, which seems suspicious. Carr suggests that the way Van der Kolk has popularised trauma results in everyone making sense of individuals' lives, particularly in the US. This might reflect the contradictions and justifications of modern societies, particularly the American one. While we are not in a position to assess

Van der Kolk's work in terms of trauma treatments, we consider it very relevant to our study. In our view as an architect and two criminologists, we assumed that the hypothesis and evidence provided by Van der Kolk, especially those described in his long-term therapeutic work starting with Vietnam veterans, constitute a significant qualitative asset for our exploration. Firstly, and very importantly to us, his work highlighted the need to implicitly assume a bodily perspective. This implies shifting the focus from the mind to the body, making the experience of trauma inherently situated within a spatial setting. Secondly, and no less important, is the idea that the body "stores" the memory of the traumatic experience. We assume that storing trauma within the body must also include the storage of spatial settings, architecture, and urban memories along with their qualities.

## 3 Shifting to Architecture—Memory

If the body "keeps the score," as [11] claims, and therefore records memory, we argue from an architectural perspective, that this memory also concerns the space or the architectural or urban settings where the trauma occurred. As an architect, rather than a psychiatrist or psychologist, I want to discuss the notion of memory in architecture and how this can help in keeping the topic of trauma relevant to the discipline, even while drawing from other disciplines. The relationship between architecture and memory, one could argue, was foundational in systematically thinking about the question of context, particularly in architectural projects of the last century that addressed the crucial relationship between architecture and modernity. Among many experiences in Western architecture, the impact produced by the "Tendenza" movement and, above all, by Aldo Rossi's studies on the city [14] have significantly shaped how the notion of memory has been received in architecture. It could be argued that a new theoretical project in Italy, founded on the notions of architectural typologies and urban morphology, attempted to resolve the contradictions induced by the exploitation of historicism by some architects during Fascism and the need and ideology of modernism in its most orthodox incarnation.

According to Aldo Rossi, memory plays a significant role in architecture by connecting physical structures with individual and collective experiences within a temporal dimension. Rossi believes that through typology and analogy, architecture can reinterpret and preserve both historical and individual memories. However, in [15] work, memory is considered instrumental to architectural forms. Architectural forms, therefore, can be regarded as vehicles for memory, allowing processes that are reflected in both collective and individual perceptual domains.

In this context, architectural spaces not only serve as the backdrop for events but also as active participants in the formation and retention of memories. This perspective can be particularly relevant when considering trauma-informed design, where the physical environment plays a crucial role in the emotional and psychological recovery of individuals.

The relationship between architecture and memory, particularly from the Modern Movement onwards, has constituted one of the greatest critical and historical narratives in architecture over the last century. To fully understand this, we should retrace the histories of the Modern Movement and the various modernities—plural, not singular—built on the dichotomy of creating architecture, transforming the built environment, and the history that environment embodies and externalises. This history is not only present in architectural buildings, monuments, urban projects, and parks, but also in how the idea of modernity and its fundamental and unsolvable relationship with memory are expressed.

However, within this research, we are not exploring such overarching narratives. Instead, we draw on the work of Paolo Virno, who, inspired by Gilbert Simondon, proposes a new philosophical notion focused on the concept of the multitude, which, according to him, is replacing the concept of the people. The multitude is plural, whereas the notion of people implies a centripetal nature, reflecting the nature of a State. For the multitude, the State is just a preliminary condition [16].

I will use the notion of the multitude as a vehicle to describe the subjects who are discussing their experiences and memories. Following Virno's arguments, I will regard the subject in terms of anthropogenesis—how it manifests itself on the socio-historical level—while avoiding treating the subject as generalisable. I also provocatively propose the multitude as a productive notion in architecture, possibly replacing the term 'users.' While 'users' implies architecture as an end goal, for the multitude, it is a preliminary condition.

Therefore, in exploring trauma, I will first assume the point of view of an individual and build upon it. This approach allows for a nuanced understanding of how individual experiences and memories intersect with architectural spaces, contributing to a broader discourse on trauma-informed design.

## 4 Shifting to Architecture—Human Body

In our view, the topic of trauma, perhaps beyond related proposals in terms of healing, raises questions that have always been at the core of architecture. These questions concern the intimate, individual, and perhaps psychic relationships between space and bodies: the bodies that inhabit, think, and create it. Many of the voices recorded in [11] seem to bring not only the body but also the senses to the forefront. Traumatic stories highlight an intricate sense of feeling where space, senses, and memories subtly interplay. This resonates with the work of [17], which calls for the resurgence of the senses and the body as the focal point of any foundational discourse in architecture. I assume that the success of Pallasmaa's work, from the late nineties to the early 2000s, converged with studies in embodiment, neurosciences, and architecture, perhaps with the risk of a new subtle determinism [18–20]. This marks the need to rethink this fundamental agency of architecture in theoretical and critical discourses.

What I would rather highlight from these studies is the very old idea of the body in architecture. The body is not only a silhouette, as in the tradition coming from Le Corbusier's Modernist proposal, like in the Modulor [21], where the body, according to the first version, is an athletic male figure 183 cm tall, playing a role that is purely geometrical and perhaps aesthetic. Rather, it is the body as a living, sensorimotor apparatus—a body which feels and senses the world.

Pallasmaa, in his attempt to recover the body and the senses in architectural discourse, draws from the continental psychoanalytic and phenomenological traditions, and this is not accidental. Furthermore, in light of his polemic against ocularcentrism [17], he continuously emphasizes the importance of the senses. It is worth arguing here that his work can be seen as an attempt to shift the foundation of any discourse in architecture from the visual to the corporeal dimension. I argue that Pallasmaa seems to claim the primacy of how one feels (the whole body) within an architectural setting over how one sees (visually) architecture, which I believe to be one of his greatest contributions to contemporary architecture's debate. Perhaps how one feels could be a good assumption in setting research questions and thinking about methods. Yet, how one feels brought him to the medium of verbal language, either spoken or written like in his book, and to the psychoanalytic traditions, often recalled in his arguments.

On one hand, in psychoanalytic tradition, capturing feelings emerges from exchanges between the analyst and the analysed happening through the medium of language. On the other hand, much of the trauma literature in psychology has indeed been based on the psychoanalytic model. For example, according to [22], Freud's concept of trauma encompasses both the immediate shock and its long-term psychological effects. The protective shield against stimuli, a central idea in Freud's theory, is considered a biological and psychological mechanism. It is not easy to define and generalise how trauma occurs in individual lives. Although the literature provides clear evidence of traumatic experiences in certain defined groups (women in custody, homeless individuals, war veterans, etc.), the notion of trauma is still perhaps better understood when it is externalised through biographical narratives and the medium of language. Ogden [23], referring to the psychoanalytic setting and his experience, describes the importance of language in this way: "it began to seem that the past was in many ways only as real for oneself as the language one could create in the present." He adds: "The analytic discourse requires of the analytic pair the development of metaphorical language adequate to the creation of sounds and meanings that reflect what it feels like to think, feel, and physically experience (in short to be alive as a human being to the extent that one is capable) at a given moment" [23].

Apparently, trauma is a personal story that cannot be told without a relationship between two or more individuals, either being that relation therapeutic or not.

## 5 Looking for a Research Method 1: Architecture Ethnography

> I can't understand how people never noticed how I was. I can't understand why people didn't know what was happening to me and how I was. Maybe it's because I was put in the background all the time. I wasn't noticed. It's as if I wasn't there. That feeling of loneliness. Of not being wanted. (L. Jones, personal communication, 2015)

How one feels refers a lot to the body and the mind together. Using words is still the way people can make an account of their traumatic experiences. We had (two criminologists and one architect) two important preliminary explorations, conducted online during the pandemic period. Rather than through a focus group, the explorations used informal or unstructured conversations as a method. [24] highlight that informal "conversations have an application beyond ethnography and can be used in a variety of qualitative methodologies that occur in everyday settings where talking is involved." Furthermore, they value informal conversations' "potential to produce more realistic or naturalistic data with less performativity from both interviewer and interviewee. Essentially, they can replace the 'me' and 'you' with 'we'"[24].

Naturalistic data through this method meant that essentially, even before seeking healing through architecture and/or practices in the justice systems, we were looking for intersections from our own disciplines, understanding how the notion of trauma from criminology could inform architecture.

Working with a semi-structured interview conducted by Dr Madeline Petrillo with a traumatised person (indicated with the nickname Lynne) informally emerged as an opportunity to focus on a first-person narrative and her words. It also helped in formulating the first research question:

How can the data provided by a semi-structured interview (already a transcription) be transcribed into architecture? Additionally: As we aimed to explore and elicit some possible meanings of trauma in architecture, how could the conditions underlined by Lynne's narratives be represented using a medium closer to architecture rather than written words?

These questions are representative enough of what our exploration in the realm of trauma would imply: the topics of architectural space, traumatic memories linked to an individual's experience, the body where the memories — we assumed — were stored, and the relations between the body and the architectural space. The work involved inducting certain narratives and patterns clearly emerging from the transcript of the interview, such as child abuse, bullying, alcoholism, and ending up in jail. Lynne's voice through the interview would say a lot, but we realised we would need to search behind the spoken words. We would need to prepare a special ground to allow different disciplinary fields to play together and perhaps allow more space for different ways of analysing that transcribed interview.

## 5.1 Setting the Drawings as the Field Where Architectural Inquiry can Occur

Lyon [25] argues that "increasingly, drawing has loosened itself from an attachment to any specific medium." Based on research by other authors, they advocate for drawing as a continuum medium across several disciplinary fields, serving diverse purposes both conceptual and practical. They propose, however, a definition of drawing as "intentional mark-making on a receptive surface" [25], suggesting that this definition can be expanded and/or amended according to the specific needs of the researchers. Furthermore, it is acknowledged that drawings can be suitable for eliciting a diverse range of data according to the researcher's needs.

I assume that drawing, as much in architecture as in many other disciplines, can be considered a form of open-ended thinking capable of triggering a continuous process of enactment. I use the word enactment following the original formulation by Varela, Thompson, and Rosch [26], considering the activity of drawing in architecture as a continuous process where mind, body, and consciousness are inextricably interwoven and linked to lived experience.

I distinctly refer to the work of Maturana and Varela, and Varela, Rosch, and Thompson, developed between the eighties and nineties of the last century [26, 27]. I draw on the idea that perception, cognition, and experience are inseparable and cannot be addressed through cognitive models that ignore the body as a biological support [26].

This model of knowledge, therefore, places the environment and the organism, regarded as cognitive systems, in a continuous dialectic, highlighting that each specifies the other [27]. Cognition, which is indivisible from lived experience, "depends on the kinds of experiences that arise from having a body with various sensorimotor capacities, and secondly, that these sensorimotor capacities are themselves embedded in a more inclusive biological, psychological, and cultural context" ([26], p. 173).

Therefore, I am highlighting here the cognitive value of drawings, whether happening digitally or analogue, assuming the body as a cognizant core of that activity. This statement underpins what Varela developed with Maturana years before, namely the notion of "structural coupling" between an organism and its environment. This notion highlights that cognition also depends on the history that the organism or cognitive system has experienced in relation to its environment. Different histories correspond to different evolutionary patterns.

Assuming the body (biological and psychological) intertwined within a cultural context, which is also the result of the history of structural couplings occurring between different organisms within their environments, finds tangible evidence in emerging research methods in architecture where drawings play a very important role. I'm referring here to "Architecture Ethnography," as [28] from Atelier Bow-Wow put it.

Atelier Bow-Wow's work on ethnography reached a larger audience through the Japanese Pavilion at the Venice Biennale in 2018. Along with reiterating "buildings not as a single entity but as environmental elements or hybrid assemblages that bring together otherwise unrelated functions or structures" their research offers a unique perspective when ethnographic fieldwork emerges as the most suitable method to rethink urban and architectural buildings that have traumatically disappeared [28]. This approach was exemplified in their work following the Japan earthquake and tsunami, and the subsequent meltdowns at the Fukushima Nuclear Power Plant in 2011.

Architecture ethnography, in their words, began to emerge during their field survey for a reconstruction plan for the coastal region of Ishinomaki-shi. Their research methods were primarily fieldwork, involving conversations with "village residents about the ways of life and the landscapes that had been washed away by the tsunami, and using the fragments of information collected in the interviews to make drawings that reconstituted these spaces" [28]. The disappearance of a site encompassing urban, architectural, and landscape elements, or its traumatic transformation, as in the case of the coastal region of Ishinomaki-shi, involved a reconstruction process based on conversations with people aimed at rebuilding through their memories. This aligns well with the assertion by Kaijima et al. [28] that "what had become clear was that the spaces that had been lost in the tsunami still existed in people's memories."

Yet, in hindsight, what Kaijima, Iseki, and Stalder state is that embodied memories of multitudes rooted in their bodies not only emerge through the medium of language but can also be transcribed into drawings. Could traumatic memories, then, be transcribed as well into architecture? And how could this engage the body in a way that we could recognize as enaction?

## 6 Looking for a Research Method 2: Transcriptions

[29] explore the importance of sensory experiences in cultural analysis, emphasizing that transcription involves more than just words. In their work, they challenge ethnography as "a rather disembodied and predictable process and product of the mind" and propose ethnography instead as "an embodied, creative, and open-ended practice" [29].

In their article published in *The Senses and Society*, they focus on and problematize the practice of transcription, criticising it for its "positivist mode of thinking incapable of moving beyond an understanding of transcription as anything more than a movement of verbal utterance into letters on a page" [29]. Drawing on the works of other scholars, they advocate for multimodal transcription (coexisting forms of communication in different sensory realms) and transmodal transcription (emphasising the interplay between different sensory realms). Significantly, they also argue for abandoning the underpinning semiotic orientations of such studies. Rather than focusing exclusively on written words, they advocate for transcription that can convey "sensations of feelings of atmosphere, ambience, and proximity"

[29]. As Pink [30] noted, the impact of a sensory approach is developing across a range of ethnographic disciplines.

However, the question of whether transcription practices occur in architecture and how they are externalised is not extensively explored, although this is not the main focus of this paper. There is no doubt that the development of architectural design, both in theory and practice, involves significant cognitive activity where transcribing words into spatial concepts across diverse agencies is often underestimated. I would argue that architects continuously transcribe through mixed, informal, and formal conversations and that often their drawings, whether interpretative or design-oriented, are transcriptions of those conversations into drawings.

In relation to transcriptions in architecture, it is worth recalling "The Manhattan Transcripts" by Bernard Tschumi [31] for his pioneering work in this regard. Tschumi's transcripts are not externalised as classic research papers but rather as a plural thematic essay in the form of texts and drawings, standing as a theoretical agency attempting a parallel exegesis of Tschumi's design works. Their main aim, namely to "translate an architectural interpretation of reality" [31], is organised in what Tschumi defines as "the tripartite mode of notation (events, movements, spaces)."

Yet, in challenging the traditional modes of architectural representation (plans, sections, elevations, axonometric, and perspectives), the transcript, through an assemblage in terms of sequence, does more than encourage architectural interpretations. It emphasises architecture as a "stage-set" [31] and, in doing so, advocates for a better representation of the embodied human experience in architectural depiction.

The orchestration behind the transcripts, and the prominence of notions from cinema, choreography, along with movement and event notations, emerge as complex apparatus where diverse types of information, even those not strictly belonging to the discipline, become crucial to give an account of the discipline itself. The agencies of the users and their interactions with the space are ultimately at the centre of the representation. Revisiting Tschumi brings us to regard the complexity of architectural drawings and their implicit agency to engage different elements simultaneously. They also demonstrate that architectural drawings, along with their conventional usage, could still constitute a transcription of what people account through verbal language about their experiences in the space. However, the body is purely collateral in Tschumi's transcripts and is not the main core of his exploration, which remains focused on architecture.

## 7 Looking for a Research Method 3: Enacting the Body

Marco Frascari, in his works, has extensively investigated the relationships between the human body and architecture. I focus here on one of his writings that comments on the architecture of Valeriano Pastor through the analysis of some drawings. Frascari draws attention to the human body. His work intriguingly highlights that the representation of the body in the design process can be seen as the inaugural mechanism of that creative process [32]. Describing Pastor's drawings from 1989, particularly some sections for the project called "Centro Scolastico Distrettuale vicino Dolo," Frascari discusses the value of the depicted human figures and advocates for the revaluation of their presence within the drawings. This practice is very different from contemporary habits, where human-like silhouettes are often added to rendered perspectives at the end of the design process.

In hindsight, Frascari highlights that starting the project with a human figure not only completes the architectural design setting but grounds the project and its rationale within the conceptual pivot of the human body. The bodies drawn by Pastor and discussed by Frascari in his text are delicate figures, neither predictable nor rhetorical; they seem to extend the idea of an architectural section as a living world. Depicted as single or groups of bodies, stationary or even running, their silent mimicry speaks to the richness of our cognitive dimensions when interpreting architectural drawings. Frascari seems to argue, possibly extending on recent discoveries in neuroscience, that it is the presence of these restless and lively figures, whose muscular effort or even voices seem perceptible, that allows the perception of the physical presence of architectural objects.

## 8 Responding to the Research Questions: Lynne's Story

> I met the kids' dad at 17. He abused me. Beat me. Domestic violence. Rape. I stayed with him 25 years. I don't know if it was love. I don't really know what love is...He could be very nice. He was very good with the children. Not an alcoholic but if he had a drink he'd turn. We'd fight like cat and dog. He's hung me over a balcony, beat me up, been unfaithful. The way he made me feel. I was no good. No-one would want me. When you've been told that so much you believe it. (L. Jones, personal communication, 2015)

This final part of the paper focuses on the application of the research methods outlined above. We will discuss how transcribing into architectural drawings a transcription made in words might highlight architectural aspects of traumatic experiences in light of the relations between the body and architecture.

I decided to work on Lynne's interview conducted by Dr. Madeline Petrillo, realising that I would attempt an architectural transcription of a transcription. This work was developed as practice-based research in the sense outlined by [33]: "Practice-based Research is an original investigation undertaken to gain new knowledge partly by means of practice and the outcomes of that practice," and that "contribution to knowledge may be demonstrated through creative outcomes in the form of designs, music, digital media, performances and exhibitions. Whilst the significance and context of the claims are described in words, a full understanding can only be obtained with direct reference to the outcomes."

My practice is based on the externalisation of the relationships between the human body and architecture through a three-dimensional digital drawing. I would use this practice according to Lynne's main narratives, which highlighted Lynne's journey between trauma, hope, and resilience. However, while in a criminology perspective, those narratives illuminated the complex interplay between personal history and the criminal justice system, the main aim in architecture was to unfold, in terms of architectural drawings, an interplay between the body as the container of traumatic memories and architectural space.

Lynne was the central character, whose life story was marked by continuous trauma. Other significant figures include Lynne's father, an abusive alcoholic who played a significant role in Lynne's early trauma, Lynne's partner, who represents a continuation of her abusive experiences, and Lynne's daughter, a secondary victim of abuse, whose situation further worsened Lynne's guilt and regret.

Working in three dimensions started with modelling Lynne's body. This was transcribed into a variety of female human forms. In a sense, Lynne was the individual of a multitude, which I recalled previously. Her body could be a representative agency of a greater variety. Modelling these bodies implied enacting their postures and the architectural settings they were standing in. The practice would develop in parallel with respect to the bodies, postures, and architectural themes.

Two main architectural settings emerged as pivotal in transcribing Lynne's narrative: the childhood home, a place of early abuse and neglect, and the prison, a paradoxical sanctuary for Lynne, where she could find a semblance of safety and an opportunity for self-discovery. The story also suggested the existence of various domestic and public settings, including homes and a school, reinforcing the lack of a safe environment.

However, Lynne's narrative was fragmented and non-linear, reflecting the chaotic nature of her life experiences. The story traverses her early exposure to alcohol, childhood abuse, continuous struggles with addiction, and abusive relationships. A significant turning point occurs with her imprisonment, which paradoxically provides stability and a chance for rehabilitation.

Her home, as implied by her narrative, was also transcribed into a genre of archetypal form, with the home itself located in several locations within the three-dimensional space. This implies the idea that the home follows the body in our experience of space. Despite being physically located at a specific point, it would be found everywhere, intertwined with the memories of the body itself (Fig. 1).

In transcribing narratives from the text, the more I modelled new bodies, the more I needed to situate them within an architectural and even urban setting. At that point, modelling the bodies according to a posture meant representing the body at the peak point of its history of interactions during the structural coupling between the organism and its medium, rather than being the result of an expressive intention. Soon, the drawing became populated by additional houses and diverse buildings, including architectural schemes resonating with the typological idea of the prison and the court. Clearly, the medium was becoming a sort of citadel, a setting derived from the transcription of Lynne's narratives and what I sensed would represent a mental image of her traumatic experience within her built environment (Figs. 2 and 3).

While we didn't have a clear description of traumatic events like being abused, in making the transcriptions, I could imagine a place and an emotion emerging,

**Fig. 1** Architectural study on the notions of home, abuse, and body postures (author: Antonino Di Raimo)

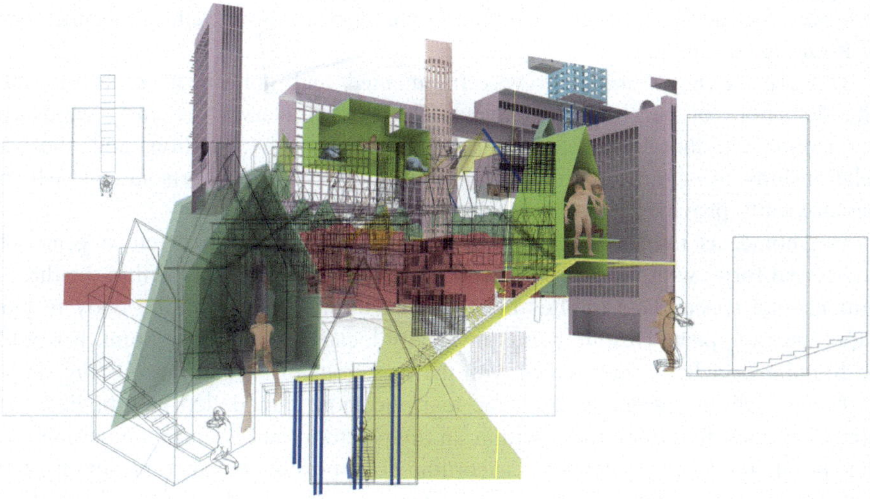

**Fig. 2** Intertwining the notion of home and city within a traumatic experience (author: Antonino Di Raimo)

like when one is reading a text. That implied making hypotheses; therefore, a long corridor with one of the bodies running away, as if in the act of escaping, gradually emerged in the three-dimensional model. Smith and Sparkes [34], in highlighting the value of narrative analyses for sport and psychology research, emphasise how temporal data are essential within a story and how individuals

**Fig. 3** Prison: typology, traumatic memories and bodies (author: Antonino Di Raimo)

make sense of their experiences over time. In Lynne's architectural transcription, this is conveyed through a unique drawing that integrates the different fragmented times from her story into a continuous narrative, depicted within a single frame.

Drafting the bodies started to evoke sentences from Van Der Kolk [11], describing the feelings of isolation and disconnection from reality experienced by traumatised individuals. This was rendered as a multitude of bodies floating in the air, each within a bubble. Furthermore, in associating the bodies with the architectural buildings, some were modelled in different scales, not necessarily consistent with the scale of the architectural setting. This, I later realised, implied the individuals' perceptions of spaces as they are held in traumatic memories. Spaces did not necessarily need to be modelled according to realistic scales. The city emerged from different iterations, even redundant, allowing noise and non-rational thinking. I later realised the city was not grounded anywhere, it was floating as much as the bubbles. This perhaps emerged as a transcription of the feeling of a total lack of grounding and not being anchored to anything solid, echoing the overall theme of insecurity even at home conveyed by Lynne's words (Fig. 4).

## 9 Discussion and Preliminary Conclusions

Let's return to the research questions stated in this paper: how can the data provided by a semi-structured interview (already a transcription) be transcribed into architecture? Responding to this research question would also help answer the question related to eliciting some possible meanings of trauma as a notion in architecture, as underlined by Lynne's narratives. In light of the themes discussed above, I would argue that the body represented within an architectural drawing, regardless of whether the drawing is analogue or digital, can be assumed as the

**Fig. 4** Urban and architectural three-dimensional cartographies of traumatic stories (exhibited in the Italian Pavilion "Resilient Communities" Venice Architecture Biennale 2021, author: Antonino Di Raimo)

primary medium in the transcription process. Representing our bodies externalises the possibility of increasing the spectrum of transcriptions within an architectural representation, whether design-oriented or investigative.

I would further argue that the representation of a body and the stories they can tell us would play the fundamental role of an affordance in the sense described by Chemero [35–37]. According to him, affordances " are relations between the abilities of organisms and features of the environment" [35]. I claim, then, that the body represented in a drawing would play the role of allowing a relationship between the organism of the designer and the environment where the drawing is produced. Frascari's writing implicitly places the human body as a privileged medium between the flat medium of the sheet of paper or the screen, and the presence of the designer or observer of the project. This relationship is based on the power of the corporeal figure itself, a subject on which much more should be said regarding its nature: a mirror relationship, the implications of which are far more significant than one might initially believe. Rather than reflecting the plot of a story like "characters," the body would operate as a relational element. Assuming the body as the main medium in a transmodal transcription can be extremely rewarding in architectural research, especially when the research relies on drawings and other forms of practices rather than only verbal language. This also implies a continuous meandering process, as described by Friedman et al. [38], which in this research involves a continuous shift between different epistemologies belonging to various disciplines.

However, externalising the relationships between trauma and architecture through drawings immediately raises questions about space, which intuitively encompasses urban, architectural, and interior levels. Even though Lynne's interview does not explicitly address these distinctions, it became clear that maintaining separate levels for architecture and urban design would not be helpful or make sense when narratives began to be externalised.

Other remarks concern the role of practice-based researchers, who are the agents in producing the transcriptions; they cannot be undermined. Through their interpretations, they add a level of creativity that potentially triggers the unfolding of data that would otherwise remain in the background. At the same time, they also introduce their own biases, which might diminish the truth of the original narrative. This will need further research.

While transcribing written stories from individuals affected by PTSD into architectural representations of the body might reveal relevant aspects specific to architectural disciplines, many dimensions of the relationship between story transcription and architectural design still require clarification. Additionally, how data is considered across different disciplines, and how it might diverge in relation to a research question, remains an issue to be addressed in the future. While creative practices, like the one presented in this paper, seem to extend the meanings around a research problem and bring new data to trigger answers, the way they actually work still needs further explanation. In designing future research, we are focusing on the relationship between written descriptions of conditions in architecture and the architectural meanings they might imply. This is not only relevant to trauma but also to how architecture, as a discipline, requires continuous exegesis between its field of production and contextual information. Finally, we would also like to emphasise the importance of developing methods that focus on listening and understanding when designing for others.

**Acknowledgements** I want to thank Dr. Madeline Petrillo and Megan Thomas for the work initiated together in this project. Further thanks to Dr. Petrillo for providing in our research a first-hand report, like Lynne's interview.

## References


1. American Psychological Association (n.d.) Trauma. MIT Press, Cambridge, MA. https://www.apa.org/topics/trauma
2. Gold SN, Cook JM, Dalenberg CJ (2017) APA handbook of trauma psychology, vol 2: Trauma Practice. American Psychological Association
3. Yehuda R, Hoge CW, McFarlane AC, Vermetten E, Lanius RA, Nievergelt CM, Hobfoll SE, Koenen KC, Neylan TC, Hyman SE (2015) Post-traumatic stress disorder. Nat Rev Dis Primers 1(1). https://doi.org/10.1038/nrdp.2015.57
4. Koenen KC, Ratanatharathorn A, Ng L, McLaughlin KA, Bromet EJ, Stein DJ, Karam EG, Meron Ruscio A, Benjet C, Scott K, Atwoli L, Petukhova M, Lim CCW, Aguilar-Gaxiola S, Al-Hamzawi A, Alonso J, Bunting B, Ciutan M, de Girolamo G, Degenhardt L, Kessler RC (2017) Posttraumatic stress disorder in the world mental health surveys. Psychol Med 47(13):2260–2274. https://doi.org/10.1017/S0033291717000708
5. Di Raimo A, Petrillo M, Thomas M (2022) Why resilient communities need trauma-informed care the case for trauma-informed design for resilient cities. Springer eBooks, pp 213–222. https://doi.org/10.1007/978-3-030-85847-6_23
6. Jewkes Y, Jordan M, Wright S, Bendelow G (2019) Designing 'Healthy' Prisons for women: incorporating trauma-informed care and practice (TICP) into prison planning and design. Int J Environ Res Public Health 16(20):3818. https://doi.org/10.3390/ijerph162038186
7. University of Denver Center for Housing and Homelessness Research, Shopworks Architecture, & Group 14 Engineering (2020) Designing for healing, dignity, & joy [Book]. Shopworks Architecture, Group 14 Engineering, & University of Denver Center for Housing and Homelessness Research. https://shopworksarc.com/wp-content/uploads/2020/06/Designing_Healing_Dignity.pdf
8. Engineer A, Gualano RJ, Crocker RL, Smith JL, Maizes V, Weil A, Sternberg EM (2021) An integrative health framework for wellbeing in the built environment. Build Environ 205:108253. https://doi.org/10.1016/j.buildenv.2021.108253
9. Hanc M, McAndrew C, Ucci M (2018) Conceptual approaches to wellbeing in buildings: a scoping review. Build Res Inf 47(6):767–783. https://doi.org/10.1080/09613218.2018.1513695
10. University of Denver Center for Housing and Homelessness Research, Shopworks Architecture, & Group14 Engineering (2021) Trauma-informed design process. Retrieved from https://www.shopworksarc.com/tid
11. Van Der Kolk BA (2015) The body keeps the score: mind, brain and body in the transformation of trauma. Penguin Books, New York
12. Cox KS, Codd RT (2024) Advocates of research-supported treatments for PTSD are losing in lots of ways: what are we going to do about it? Res Soc Work Pract 34(4):347–359. https://doi.org/10.1177/10497315231206754
13. Carr D (2023) Bessel Van der Kolk on trauma, America's favorite diagnosis. Intelligencer. https://nymag.com/intelligencer/article/trauma-bessel-van-der-kolk-the-body-keeps-the-score-profile.html
14. Rossi A (1984) The architecture of the city, revised edn. MIT Press
15. Rossi A (1981) A scientific autobiography. MIT Press, Cambridge
16. Virno P (2004) A grammar of the multitude: for an analysis of contemporary forms of life. Semiotext(e)
17. Pallasmaa J (2012) The eyes of the skin: architecture and the senses. Wiley
18. Mallgrave HF (2013) Architecture and embodiment. Routledge eBooks. https://doi.org/10.4324/9780203071144
19. Mallgrave HF (2015) "Know Thyself": or what designers can learn from the contemporary biological sciences. The MIT Press eBooks, pp 9–32. https://doi.org/10.7551/mitpress/10318.003.0003

20. Robinson S, Pallasmaa J (2017) Mind in architecture: neuroscience, embodiment, and the future of design. MIT Press
21. Corbusier L (2000) The modulor: a harmonious measure to the human scale, universally applicable to architecture and mechanics (Reimpression of the first English edition published in 1954 by Faber and Faber.)
22. Zepf S, Zepf FD (2008) Trauma and traumatic neurosis: Freud's concepts revisited. Int J Psychoanal 89:331–353
23. Ogden TH (1997) Some thoughts on the use of language in psychoanalysis. Psychoanal Dialogues 7(1):1–21. https://doi.org/10.1080/10481889709539164
24. Swain J, King B (2022) Using informal conversations in qualitative research. Int J Qual Methods 21:160940692210850. https://doi.org/10.1177/16094069221085056
25. Lyon P (2020) Using drawing in visual research: materializing the invisible. In: The SAGE handbook of visual research methods, 2nd edn. SAGE Publications, Inc., pp 297–308. https://doi.org/10.4135/9781526417015
26. Varela F, Rosch E, Thompson E (1993) The embodied mind: cognitive science and human experience. MIT Press
27. Maturana H, Varela FJ (1980) Autopoiesis and cognition: the realization of the living. Springer
28. Kaijima M, Iseki Y, Stalder L (2018) Architectural ethnography : [Japanese Pavilion Venice Biennale 2018]. Toto publishing, Tokyo
29. O'Dell T, Willim R (2013) Transcription and the senses: cultural analysis when it entails more than words. Senses Soc 8(3):314–334. https://doi.org/10.2752/174589313X13712175020550
30. Pink S (2013) Engaging the senses in ethnographic practice: implications and advances. Senses Soc 8(3):261–267. https://doi.org/10.2752/174589313X13712175020433
31. Tschumi B (1994) The Manhattan transcripts. Wiley
32. Frascari M (2002) A tradition of architectural figures: a search for vita beata. In: Dodds G, Tavernor R (eds) Body and building: essays on the changing relation of body and architecture for Joseph Rykwert on his seventy-fifth birthday. MIT Press, pp 258–267
33. Candy L (2006) Practice based research: a guide. Creativity & Cognition Studios, University of Technology, Sydney
34. Smith B, Sparkes AC (2009) Narrative analysis and sport and exercise psychology: understanding lives in diverse ways. Psychol Sport Exerc 10(2):279–288. https://doi.org/10.1016/j.psychsport.2008.07.012
35. Chemero A (2003) An Outline of a theory of affordances. Ecol Psychol 15(2):181–195. https://doi.org/10.1207/S15326969ECO1502_5
36. Chemero A (2011) Radical embodied cognitive Science. MIT Press
37. Wilkinson T, Chemero A (2024) Why it matters that affordances are relations. In: Mangalam M, Hajnal A, Kelty-Stephen DG (eds) The modern legacy of Gibson's affordances for the sciences of organisms
38. Friedman VJ, Robinson S, Egan M, Jones DR, Rhew ND, Sama LM (2020) Meandering as method for conversational learning and collaborative inquiry. J Manag Educ 44(5):635–650. https://doi.org/10.1177/1052562920934151

# Variability and Design

# 14

Alessandro Melis and Jose Antonio Lara-Hernandez

**Abstract**

Although the built environment is associated to something static, architecture is a field of constant change. The concept of variability has emerged as a pivotal principle guiding contemporary design practices. Variability, in the architectural context, refers to the capacity of built environments to adapt to changing needs, conditions, and technologies over time. This principle transcends mere flexibility, encompassing a broader spectrum of adaptive strategies that ensure the longevity, relevance, and functionality of architectural spaces. As the demands on buildings and urban environments become increasingly complex and dynamic, the role of variability in architectural design becomes more pronounced, necessitating a nuanced understanding and application of this concept. Through a series of case studies this section describes this characteristic as an essential feature of of architecture and the urban environment.

**Keywords**

Architectural variability • Digital fabrication • Urban health • Landscape resilience • Rhythmic articulation • Artificial intelligence

A. Melis
New York Institute of Technology, New York, USA
e-mail: amelis@nyit.edu

J. A. Lara-Hernandez (✉)
Auckland University of Technology, Auckland, New Zealand
e-mail: antonio.lara-hernandez@aut.ac.nz



J. A. Lara-Hernandez et al. (eds.), *The Dusk of Design*, Integrated Science 34,
https://doi.org/10.1007/978-3-031-76867-5_14

## 1 From Evolution to Variability in Architecture

Evolutionary biology as a research field is relevant because in the scientific theory of evolution is developed up to the point that biologists argue its applicability on human phenomena, also called sociobiology [2]. Evolution and ecology are extremmely related, sharing concepots like adaptation, exaptation and variability [3]. However, regarding explanatory frameworks they are different. As [2] claims, on the one hand ecological approaches represent functional explanations, while on the other evolutionary approaches seek to generate historical explanations. In the fields of anthropology and archaeology the word evolution is commonly used as a synonym for "change", rather than signalling a theoretical position.

As we have already explained, the theoretical underpinnings of variability in architectural design draw from a range of disciplines, including systems theory, cybernetics, and ecology. These fields provide insights into how complex systems adapt to changing conditions, offering valuable frameworks for understanding and applying variability in architecture.

Systems theory, for instance, emphasises the interconnection but also interdependence of components within the entire system. In architectural terms, this means to designing buildings that can be better integrated (and interact) with their environment, both natural and built. Cybernetics, with its focus on feedback mechanisms, informs the design of responsive and adaptive systems within buildings. Ecology, on the other hand, underscores the importance of sustainability and resilience, highlighting the need for designs that can adapt to environmental changes.

If we look ahead, the future of variability in architectural design will likely be shaped by new materials, digital fabrication, and artificial intelligence. New materials with adaptive properties, will enable buildings to respond dynamically to external weather conditions. Digital fabrication techniques such as 3D printing, will allow for the creation of "tailored" building components. Artificial intelligence most likely will further enhance the responsiveness, energy management, and adaptability of buildings.

The ongoing exploration of variability[1] in architecture lays the promise to yield innovative solutions that enhance the adaptability and longevity of our built environment, reflecting a deep understanding of the dynamic interplay between form, function, and context. The chapters contained in this section are an example of such attempt.

For instance, Chap. 15 "Health in Public Space, Micro-Environments" underscores the crucial role of health in building resilient communities, highlighting the detrimental effects of fossil-based construction materials on ecosystems and human health. Dr. Pongratz advocates for circular design and regenerative materials, presenting prototypes that adjust microclimates in public spaces to improve comfort and biodiversity, thereby enhancing public health and well-being.

[1] Also called Robustness in a slightly different school of thought [1].

Going deep into the relationship between health and architecture, Dr. Capolongo in Chap. 16 traces the historical evolution of healthcare infrastructures, emphasising the socio-cultural and technological influences on hospital design. The chapter discusses how hospitals have transformed from religious sanctuaries to complex medical facilities, reflecting advancements in medicine and shifts in societal values. It also explores the principles of variability, redundancy, and diversity in hospital architecture, especially in light of the COVID-19 pandemic.

Moving now to landscape architecture, Chap. 17 presents a transdisciplinary approach to landscape resilience, integrating ecological, social, and cultural dimensions. Dr. Aimar and Dr. Voghera propose a recombination approach to design, advocating for adaptive and resilient landscapes that respond to contemporary environmental challenges. They explore how community involvement and traditional knowledge can enhance landscape resilience.

Regarding the public space, Dr. Adebara examines the multifunctionality of motor parks in Yorubaland, highlighting their role as vibrant public spaces that accommodate commercial, religious, and social activities. Thus, Chap. 18 utilises a cultural ecological perspective to explore how these spaces adapt to the socio-spatial needs of the Yoruba people, emphasising the importance of cultural practices in urban planning and design.

In Chap. 19, Dr. Aceti goes into the implications of redundancy and censorship in AI-driven design processes, critiquing the restrictive frameworks imposed by corporate entities. The chapter argues for the preservation of creative freedom and cultural diversity in digital art, challenging the homogenising effects of AI censorship on artistic expression.

In Chap. 20, Dr. Kim describes the in Korea, utilising the concept of affordances. It describes a range of historical proposals that envision the Korean DMZ as a pristine ecological haven—a tabula rasa site. It provides deeper insights into the nature of DMZ, how it is constructed, maintained, deconstructed, functions, and potentially reconfigured utilizing spatial design methods.

## References


1. Bentley I, Alcock A, Murrain P, McGlynn S, Smith G (1985) Responsive environments. The architectural Press Ltd.
2. Dunnell RC (1980) Evolutionary theory and archaeology. In: Advances in archaeological method and theory, vol 3. Academic Press Inc., pp 35–99
3. Melis A, Pievani T, Lara-Hernandez JA (2024) Architectural exaptation: when function follows form. Routledge. https://doi.org/10.4324/9781003347118

# Health in Public Space, Micro-environments

15

Christian R. Pongratz

**Abstract**

Health is crucial for building resilience as the basis for the economic, cultural, and social future of a population. Here the main goal is to promote wider access to good health and well-being, by enhancing lived experiences across public spaces. It is understood that construction that depends on fossil-based petrochemical materials causes significant harm to ecosystems, the environment, and our health. We also acknowledge that the impact of architecture and designed environments and their microclimates on human health is not yet fully understood, though with recent evidence indicating it contributes to various diseases, including chronic conditions. In contrast, we know that circular design and regenerative materials and construction processes minimize waste and CO2 emissions, typically allowing for biodegradation or reuse, and thereby eliminate continuous human exposure to toxins and pollutants. This paper will explore a recent prototype and a project for public spaces that adopt ecological approaches to design and construction. The underlying goal is that the prototypes address microclimate adjustments and resulting comfort improvements in public spaces, accelerating the function as social condensers, as outdoor livable environment and mitigating spaces for heat islands, as receptacles for increased biodiversity and urban niches with improved comfort for all users in outdoor activities. One example is a design for a micro plaza, prototype installation and implementation project into public space, while another prototype explores the idea of urban furniture as a multifunctional microclimate machine. Both draw on methods of integrating digital design strategies, digital and analog fabrication and wide variety of traditional materials being rethought.

C. R. Pongratz (✉)
School of Architecture and Design, New York Institute of Technology, New York, USA
e-mail: christian.pongratz@nyit.edu



J. A. Lara-Hernandez et al. (eds.), *The Dusk of Design*, Integrated Science 34,
https://doi.org/10.1007/978-3-031-76867-5_15

**Keywords**

Public health • Urban microclimates • Sustainable design • Circular economy • Biodiversity

## 1 Introduction

Rapid urbanization is understood as one of the major challenges in the twenty-first century, and cities will be playing an increasing leadership role in building resilience for growing populations against the effects of irreversible climate warming that they are mostly responsible for [1].

We need to consider that resilience ultimately means health and it is essential to a population's economic, cultural and social future. It is understood that our purpose is to facilitate broader access to enjoying good health also for the most vulnerable groups. One possible support as outlined in this paper are strategies for the shared urban commons, the public places we socialize in, with the purpose to support better comfort to everyone and improve lived experiences in communities. In particular, a few opportunities are discussed to design urban habitats as microenvironments in support of socioecological systems while improving environmental aspects which in turn are contributing to wellbeing.

What are some of the major obstacles that we are facing on this road to reduce the impact of the built environment on health? First, we know that all construction is relying on fossil-based and petrochemical material production and therefore causes widespread damage to ecosystems and our health.

One of the major aspects are that the global construction and building materials industry consumes a significant amount of resources and energy while accumulating embodied carbon and chemical content along its various activities from material extraction, manufacturing, transportation, to construction and operation, impeding a priori a sustainable destination of buildings and materials. In particular, it is this linear process ingrained in building construction, which has a significant impact during its entire lifecycle on the environment, as its production methodology is based on the industrial age principle of take, make, waste, described well by McDonough as beginning with natural resource exploitation and ending typically with waste generated landfills [2].

Considering resource depletion, construction requires large amounts of natural resources which includes materials such as sand, gravel, and limestone and as [3] report almost 10% of global industrial water use if we consider concrete production alongside 4.2 billion tons cement globally in 2020 as reported by the GCCA [27] itself. According to reports by the GIH [4] and in reference to the ECESD [5], construction and building materials account for about 50% of EU resource and energy use, while construction waste accumulates over 839 million tonnes in the EU in 2018. In the estimation made by Bastianoni [6], the ecological footprint of building construction, for example of a single two story detached house uses about 10% biocapacity of 1.18 gha of the average Italian citizen, which is the annual bioproductive land at disposal by means of equal distribution of natural resources.

It is implied that the overexploitation of these resources leads to habitat degradation such as via soil erosion, and water pollution. From the beginning to the end of a building's lifeycle, it is accepted that construction generates large amounts of waste, including demolition debris or packaging materials, and in general most of the materials are not being reused but mostly down cycled as reported by GIH and ECESD. In fact, much of the improper disposal of waste leads to pollution of land, water, and air, as embedded toxics are harming both ecosystems and human health.

Most recent figures according to Architects Declare [7] are that construction contributes 40% to global carbon dioxide emissions, and the number of alien invasive species per country has risen by 70% since 1970. Those numbers may vary depending on the sources, yet it's clear that the construction and building materials industry needs to make significant efforts to improve resource, energy and waste efficiency in building practices. There is a need to further promote renewable energy and matter sources in its processes, based off the model of the *Circular Economy* as per the Ellen MacArthur Fondation [8], and the cradle to cradle [9] concept towards regenerative building materials, which are crucial for reducing the environmental impact.

## 2 Urban Public Spaces

In this transition period towards a true circular economy, climate warming is still continuing to impact heavily urban centers and making cities increasingly challenging for populations in often non-obvious ways. We likely have increasing and more severe weather events like heat waves, wild fires, flooding, or storms which are signs for more extreme conditions that are reducing well-being in particular for the most underserved communities.

Additionally, as cities exponentially grow, they also become much denser which in turn requires not only more resources, but an increase of urban public spaces where people congregate and enjoy cultural events, sports activities, and tend preferably to socialize. Ideally those spaces should be the cities' breathing lungs with equitable access and distribution in the urban fabric, sufficient natural daylight, biodiversity and ample green to compensate for living in small housing units and overpopulated zones of urban centers.

Yet we are moving in the opposite direction, considering the designs of largely sealed, covered urban spaces, where expanding footprints of architecture, infrastructure and all the building related products reduce green spaces, and encroach on natural habitats. It is the impact of this built part of the environment on human health and all species, which is significant, and recent evidence suggests it contributes to the development of various diseases including many chronic conditions.

And despite increasing investments over the past decades, according to the WHO, health inequalities or health for all and outcomes have not improved significantly [10].

In order to support the most affected, marginalized, impaired or frail and underserved communities, as designers we are asked to uncover those underlying most severe factors of the built environment that contribute predominantly to health disparities and inequities. This includes to evaluate any possible opportunity in building construction and design practices for reducing the carbon footprint, and to employ rigorous chemical screening as well as biodiversity assessment, on and off site, in relationship to the materials we continue employing in the construction industry.

## 3 Air Pollution

In general, air pollution is understood as a driver for disease because many chemicals exist in our indoor and outdoor environments as volatile or semi-volatile compounds in the gas phase and embedded in particulate matter. Originating from building construction and operations, urban traffic, but also agricultural processes, industrial consumer goods production, and resulting in water and soil contamination, air pollution is a ubiquitous health challenge and is impacting the neurological and immune systems. More recent research by Darbre draws correlations between air pollution and adverse effects on endocrine health, as the production of hormones is impacted, which generally act as chemical messengers and regulate many physiological and brain functions. In recent studies particular thyroid hormones and their functionality have become the focus as effects to reproductive, respiratory and cardiovascular systems have been revealed [11].

Man-made endocrine disrupting chemicals such as Phthalates, PAH's, bisphenols, or also PBDD's from incineration of urban waste, have been traced in human tissues and serve as possible sources for neurobehavioral and developmental issues such as autism especially according to Demeneix with early exposure during prenatal phases of pregnant mothers or later in early childhood [12].

Many studies in cities already document pollution concentrations of particulate matter starting at 1. 5 $\mu$m, though the predominant focus has been to date only on larger sizes of 10 $\mu$m.

It is essential to monitor both particle number concentration and $PM_{2.5}$ to fully grasp the health impact associated with the inhalation of ambient particulate matter, considering controversial global approaches as suggested by de Jesus [13].

## 4 Heat Island Effect/Biodiversity

An underestimated consequence of construction activities involve clearing land and making surfaces impermeable to recirculation within local biotopes, which leads to the first layer of destruction and displacement of natural habitats for plants and animals.

This lack of focus onto ecological system thinking coupled with increasing levels of air, water and soil pollution leads to an accelerating loss of biodiversity and produces often undesirable and indirect consequences. Many of those unanticipated side effects are disrupting ecosystems and in turn reduce resilience towards next waves of climate threats to human residents and species. One of those unintended effects of sealed infrastructure in urban areas leads to what became to be known as the urban heat island (UHI) effect, where urban areas, city quarters become significantly hotter than surrounding greener or rural areas. Recent studies by Huang also indicate that expanding land cover influences also social strata distribution in communities and the level of temperature [14].

In each of those situations, there is a chance that urban areas develop a health threatening outdoor microclimate due to increased heat absorption and dissemination in particular during high solar impact hours in summers. The UHI intensity is measured as the difference between the center of an urban areas in comparison to a rural area and is on average 2–4°. While satellite imagery according to Heaviside is typically used to survey area differences, local snapshots of the air and surface temperature of the surrounding sealed floor and wall surfaces are more appropriate for understanding user exposure to heat [15].

The resulting challenging thermal comfort in those environments, characterized by an increased mean temperature and exacerbated by heat wave frequency poses severe health risks in particular for kids and for the growing elderly population, the latter often associated with rising percentages in mortality. It also can influence peoples' behaviour negatively and risk awareness may limit their daily activities such as refraining from using streets to go to use public space, to work or school. At the least their mobility might be impacted and outdoor space liveability for leisure and wellness is heavily constrained.

So how might we mitigate the impact of climate change on comfort in outdoor space use and related behaviours?

When we rethink the built environment and its public spaces from a design point of view, locality should be considered a first bio-geomarker, where the impact on human health is determined on the one hand by direct exposure to a specific climate environment and on the other hand by a personal susceptibility including the human individual genome, both forming the individual evolving phenotype. In response it is suggested to explore the underlying opportunities related to strictly building with local materials and their thermo technical aspects among those solar reflectivity or techniques such as evaporative cooling, by creating local microclimates in a way such as to improve comfort in public spaces that influences positively human interaction. Diametrically opposed to the previously discussed linear take, make, waste process is circular design, and practices relying on regenerative materials and construction processes that eliminate or at the least reduce waste and C02 emissions, while typically allowing for biodegradation or reuse and thereby eliminate human exposure to toxics and pollutants.

In this regard, the following discusses the development of recent prototypes and projects for public spaces, employing an ecological approach to design and making of urban furniture and bioreceptive surfaces, utilizing local natural materials, as well as contemporary digital and local manual fabrication techniques.

## 5 Urban Climate Oasis

The Urban Climate Oasis is an urban furniture integrating a green wall system, urban misting technology and seating elements in order to enhance comfort and wellbeing, with a range of selected materials and plants intended to promote human nature relationships and protect biodiversity. A first prototype has been exhibited during the International exhibition of Marmomacc 2022, in Verona, Italy.

## 6 Embodied Carbon

A priori an underlying motivation was to change mindsets and rethink creatively typically followed processes utilizing high embodied carbon, energy usage and machining time during manufacturing. The design strategy therefore takes a no waste and reduced embodied carbon angle, by relying first on the use of natural stone, which is already considered a sustainable material for being reusable at the end of its lifecycle. Second, with the help of computational tools we redesigned the typical fabrication practices employing a technique that leverages spatial geometry synthesis by providing real time understanding of the optimized correlation between material cut to final joint and assembly design. This replaces a waste driven material subtraction methodology with a lossless process in order to produce near to zero material waste, minimizing overall fabrication time and thereby reducing consumed energy and embodied carbon in matter and manufacturing.

In particular, this method gearing towards a simplification of the fabrication process entailed taking two standard dimensional stone blocks that are sliced by only four successive single diamond wire cuts, whose spatial motion is based on a mathematical ruled surface geometry, which means rotating straight lines in three-dimensional space to obtain the surfaces description [16] (Fig. 1).

In a second step, and using manual labor supported by forklifts during the assembly phase, the resulting elements are separated, flipped, rotated and combined in their cut geometry such that one double wall system and two bench sets are formed as per the diagram provided before. The presence of redundant element surfaces, a result of several same motion wire cut geometries, provides for variability in assembly options that can be expanded upon based on the number of parts.

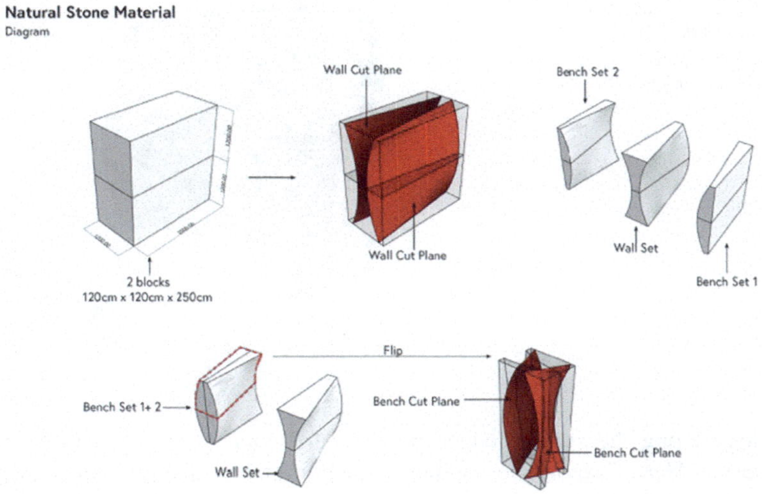


**Fig. 1** Zero waste low embodied carbon manufacturing workflow design Pongratz Perbellini Architects with Fadhil, fabrication by Pimar Limestone S.r.L. (Fadhil, Pongratz 2022)

## 7 Thermal Comfort

While it will need larger and more distributed green areas to resolve emerging urban heat islands driven by climate warming, we can attempt to mitigate the problem of surrounding hot surface temperatures and their impact on health in public spaces by inducing a cooling local micro climate as a countermeasure. The underlying ASHRAE standards for thermal conditions, comfort and acceptability related to human occupancy are a guidance and have been largely explored as documented by Berglund [17].

The urban climate oasis acts as a dynamic and multifunctional urban furniture as a result of its bioreceptive surfaces that embed the high density of stone materials with the goal to benefit from its elevated thermal mass, a property of the mass of a material that enables it to store heat and use inertia over time benefitting from diurnal temperature fluctuations. The stone is absorbing thermal energy during day time with sun exposure, when the environment is higher in temperature, and releases thermal energy back when the surroundings are cooler at night (Fig. 2).

We are given a complexity of factors involved that makes it difficult to predict actual health improvements because of the mostly non-uniformity of lived environments among individuals and the lack of scientific long term user observations. The prototype nevertheless relies on a general evidence of user sensations, that increased access to green space, and applied methods for temperature reduction may improve overall thermal acceptability and with it affect positively mental health and wellbeing [18]. In addition to reduced temperature exposure as a result of the insertion of green planter structures on two sides, the employed misting

**Fig. 2** Urban climate oasis, green concept design, Pongratz Perbellini Architects with Fadhil (2022)

technology along the surface edges provides for cooling via evaporative micro nebulization. Here Evaporative cooling is the conversion of liquid water as micro molecules into vapor absorbing the thermal energy in the air integrating technology as described by Ulpiani. This is perceived as a cooling effect because the energy needed to evaporate the water is taken from the surrounding air in the form of sensible heat which reduces the felt temperature of the air and attenuates solar radiation [19].

The energy needed to operate the water pump for the plant irrigation and the cool misting technology is provided via a concept of Photovoltaic panels on top of the wall coupled with rainwater collection via cisterns provided in the cavity space between the two stone walls (Fig. 3).

**Fig. 3** Urban Climate Oasis, evaporative cooling via micro-nebulization, Idrobase Group Srl, Marmomacc 2022 (Pongratz 2022)

## 8 Biodiversity

The project applied effects for night illumination by integrating an LED system into custom horizontal slots that have been manually cut into the natural stone surface. The benefits are two-fold, one where the illuminated surface would create a safer environment at night, which is often a challenging problem in urban neighborhoods with accessible public spaces and parks. Second, the LED lights will attract different insect species in the evening towards the wall, where they can find dense green plants either on both sides or by alternating light bands with green moss strips as bio niches to populate the urban island and thereby increase the biodiversity. A comprehensive overview on current applications of design ideas that integrate early attempts of support to host local species communities besides the pervasive anthropocentric approach typical for the built environment, has recently been well documented by Weisser [20] (Fig. 4).

The green side planters and green surface strips serve as carbon dioxide sink and assist in filtering the air around the immediate environment. Ongoing research by Professor Pongratz explores opportunities of inserting moss and algae into low moisture containing material composites and surface geometry to increase the air filtration and temperature impact (Figs. 5 and 6).

**Fig. 4** Biodiversity, species habitat creation, green planter system by Climax, Italy (Pongratz 2022)

**Fig. 5** Research into bioreceptive surfaces, moss strip prototype (Pongratz 2023)

**Fig. 6** Research into bioreceptive surfaces, algae surface strip prototype (Pongratz 2023)

**Fig. 7** Installation Pongratz Perbellini Architects with White D., 17th International Architecture Exhibition, Venice Biennale 2021, Italy, (Pongratz 2021)

## 9 Crispr-Locus, Crispr Peccioli

The following prototype exhibited at the 17th Venice Biennale, International Architecture Exhibition 2021, and the successive implementation in form of a public space design in Peccioli, Tuscany, Italy, in 2023, leverages the opportunities of urban furniture and materials to frame spaces that may serve as social condensers and act as spaces for comfort and multispecies design (Fig. 7).

## 10 Social Interactions

In the installation crispr_locus, the concept of public space is interpreted as a form of a micro plaza which should serve as a stage for casual people encounters. As an incentive, the environment is of a scenographic atmosphere, which is derived from multiple elements of art, custom floor and wall covering, and methods of diverse seating options turned furniture. All are made with materials that are intended to showcase the excellence of italian artisanry in order to enhance the space through a broader sensorial experience, especially visual, tactile, kinesthetic and affective experiences that can attempt to make the space more inclusive and provide a contextual cultural content as intended by Trancik and place theory [28]. The natural stone surfaces of the plaza floor as well as the furniture are made with a beige limestone of a warm color that is also soft to the touch, which instills tactile comfort. The seating elements have different sizes, located at various heights and coupled with wider steps to accommodate access to users with diverse abilities,

**Fig. 8** Project Pongratz Perbellini Architects with White D., Peccioli, Tuscany, Italy (Pongratz 2023)

permit different levels of vestibular and proprioceptive sense capacities. The natural stone has a high emissivity of 0.96 similar to the human skin as indicated by Rahm, which means that it absorbs and emits radiation very easily, useful in areas where diurnal temperature swings cool the material over night, as well as high effusivity to possibly speed up loss of heat of the human body when touched [21, 22] (Fig. 8).

The back wall cladding is made of screen printed glass and metal (Venice Biennale) and ceramic (project in Peccioli) and the design pattern traits are linear elements written computationally, generating geometric variability and soft interchangeable grey to white color tonalities on blue ground, which are repeated as lines, an incision geometry filled with blue color on all stone flooring. Both, vertical and horizontal material surfaces play on the idea of texture, recess, relief and incision offering a mutating sensorial experience landscape.

Together natural stone, glass and metal or ceramic provide a lively and dynamic narration that defines a small public open room into spatial resonances.

## 11 Variability, Redundancy, Affect

We believe that besides the tangible material aspects, it is in particular the social cultural interactions people have with other people, families and friends, as Chris Lawer points out, that are now in post Covid times most important to bring well-being. He argues that encounters that shape our experiences with health contain forces found in real social and other structures [23]. Inclusive design is intended

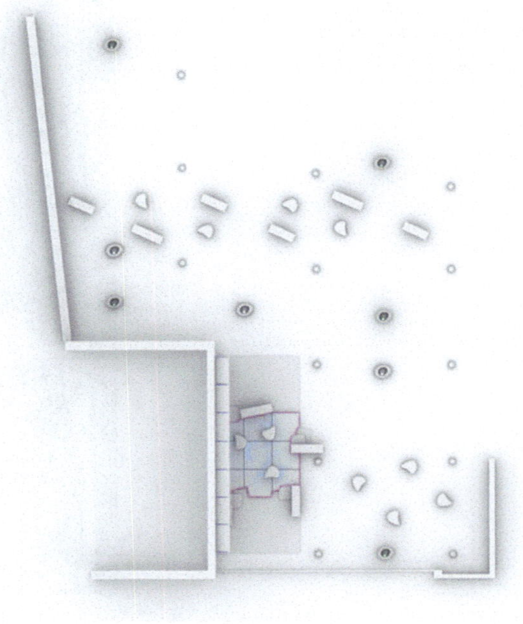

**Fig. 9** Variability and redundancy generated by various seating and planter elements in the public space, Peccioli (Pongratz 2022)

here as an exploration of how the senses contribute to spatial perception, and how sensory stimuli impact our affective experiences, health and wellbeing via broader connections between design, materiality, and affect (Fig. 9).

The design integrates ideas of redundancy via repeated structures in its systems, glass containers on wall and seating elements on floor, both useful for emerging biodiversity, social purposes and to create opportunities for variability in setup. In the words of Hertzberger, people can restructure space and contribute to an environment by bringing their own characteristics, and make it a familiar place, such that form and user adapt to each other and each enhancing the other [24]. We intend to extend that to a multispecies approach over time by providing ecological niches as self- organizing substrate containers, from the whole micro plaza setting to objects such as the Murano glass vessels (Fig. 10).

Based on the observation of natural ecosystems with semi-open boundaries and with reference to the flow of information and resources down to the scale of cells and tissues as described by Holland, the design engenders an inscription of aspects of the cellular symbiotic environment into geometric and material layers. Or in other words an unintended accumulation of resources, such as seeds, pollen particles flowing in the air may lead to emerging reproduction [25].

**Fig. 10** Variability and redundancy generators in Murano glass for biodiversity (Pongratz 2021)

In reference to the ideas by Gould and Vrba [26] who discuss in their research on evolutionary morphology the concept that certain features of organisms are non-adapted, we understand that they still are available for useful cooptation in the future. In this sense we created a series of pure half permeable glass vessels, produced with blue cobalt and white grey striped Murano glass, that fits the current environment in terms of aesthetics, but are designed in a way so to assume possible different functions as the environment deems opportunistic at a later stage. Here the forms, variable in themselves, may fit by reason of their form (ex aptus) as grow vessels in the future and receive the initial seed while collecting small portions of rainwater over time, and build up a kind of latent biodiversity. Now it appears to be only a momentary archive of formal mutation, but it generates an abundance of repetitive and differentiated geometries that provide the raw material for plant variability in an ex-apted design application across the surface (Fig. 11).

**Fig. 11** Biodiversity generators, Murano glass as strange objects (Pongratz 2021)

## 12 Conclusion

Both, the prototype Urban Climate Oasis and the project Crispr indicate opportunities for integrating low carbon and reusable material strategies in public spaces, while at the same time offering means for influencing local microclimates in order to provide support and comfort improvements for vulnerable communities. More long term user feedback is needed to really evaluate the function of social condensers, as improved outdoor livable environment and for mitigating heat islands, as well as the receptacles for increased biodiversity being truly acceptable to local species communities.

Currently we explore alternatives to natural stone, emerging in terms of local regenerative materials, where our current research into earth based material design mixes promise multifunctional thermally cooling and moisture balancing components.

In a next step, onsite user studies as a qualitative performance evaluation should provide feedback on the intervention via various surveys, on usability and thermal comfort, right-now and long-term. The survey can be relying on ASHRAE 55, if assumed that thermal environment surveys are an acceptable way of assessing human comfort conditions and if one collects a representative sample size with a high response rate.

Typically, the right-now surveys are taken to evaluate a user's thermal experience at a single point in time, whereas long term surveys provide average comfort votes on the environment.

The questions include the thermal sensation Bedford, Ashrae seven point scale, which asks occupants to rate their sensation (from "hot" to "cold") and where individual votes are values between -3 and + 3 and it is assumed that between −1.5 and + 1.5 would be considered as thermally satisfied. Prior studies such as by Berglund [17] establish the known relationships between thermal sensation, acceptability and comfort, though data was derived qualitatively under uniform environmental conditions.

## References


1. Elmqvist T, Andersson E, Frantzeskaki N, McPhearson T, Olsson P, Gaffney O, Takeuchi K, Folke C (2019) Sustainability and resilience for transformation in the urban century. Nat Sustain 2(4):267–273. https://doi.org/10.1038/s41893-019-02501
2. Braungart M, McDonough W (2002) Cradle to cradle: remaking the way we make things. North Point Press
3. Miller SA, Horvath A, Monteiro PJM (2018) Impacts of booming concrete production on water resources worldwide. Nat Sustain 1:69–76. https://doi.org/10.1038/s41893-017-0009-5
4. Global Infrastructure Hub, GIH, report (2021) Advancing the circular economy through infrastructure. Transition pathways for practitioners in circular infrastructure. https://cdn.gihub.org/umbraco/media/4265/gi-hub-paper_advancing-circular-economy-through-infrastructure_2021.pdf
5. European Circular Economy Stakeholder Platform, ECESP (2021) Report downloaded on 15 May 2024. https://circulareconomy.europa.eu/platform/sites/default/files/circular_buildings_and_infrastructure_brochure.pdf
6. Bastianoni S, Galli A, Pulselli R (2006) The ecological footprint of building construction. WIT Trans Ecol Environ 93:345–356. https://doi.org/10.2495/SC060331
7. Architects Declare (2021) Climate and biodiversity emergency. Practice guide. Killer facts
8. Ellen MacArthur Foundation (2013) https://www.ellenmacarthurfoundation.org/circular-economy/what-is-the-circular-economy
9. Cradle to Cradle (1990) https://epea-hamburg.com/cradle-to-cradle
10. Jong-wook L (2003) Global health improvement and WHO: shaping the future. The Lancet 362(9401):2083–2088. PMID: 14697812. PMCID: PMC7133795. https://doi.org/10.1016/S0140-6736(03)15107-0
11. Darbre PD (2018) Overview of air pollution and endocrine disorders. Int J Gen Med 11:191–207. PMID: 29872334. PMCID: PMC5973437. https://doi.org/10.2147/IJGM.S102230
12. Demeneix B (2017) Toxic cocktail: how chemical pollution is poisoning our brains. Oxford University Press
13. Lorelei de Jesus A., Rahman M., Mazaheri M., Thompson H., Knibbs L., Jeong C., Evans G., Nei W., Ding A., Qiao, L., Li L., Portin H., Niemi J., Timonen H., Luoma K., Petäjä T., Kulmala M., Kowalski M., Peters A., Cyrys J., Ferrero L., Manigrasso M., Avino P., Buonano G., Reche C., Querol X., Beddows D., Harrison R., Sowlat M., Sioutas C., Morawska L., (2019), Ultrafine particles and PM2.5 in the air of cities around the world: Are they representative of each other?, Environment International 129, 118-135, ISSN 0160 4120. https://doi.org/10.1016/j.envint.2019.05.021
14. Huang G, Cadenasso M (2016) People, landscape, and urban heat island: dynamics among neighborhood social conditions, land cover and surface temperatures. Landscape Ecol 31(10):1–9. https://doi.org/10.1007/s10980-016-0437-z
15. Heaviside C, Macintyre H, Vardoulakis S (2017) The urban heat island: implications for health in a changing environment. Curr Environ Health Rep 4:296–305. https://doi.org/10.1007/s40572-017-0150-3

16. Pottmann H, Asperl A, Hofer M, Kilian A (2007) Architecture geometry. Bentley Institute Press
17. Berglund L (1979) Thermal acceptability. ASHRAE Trans 85(2):825–834
18. Zhang Y, Zhao R (2008) Overall thermal sensation, acceptability and comfort. Build Environ 43(1):44–50. https://doi.org/10.1016/j.buildenv.2006.11.036
19. Ulpiani G (2019) Water mist spray for outdoor cooling: a systematic review of technologies, methods and impacts. Appl Energy 254:113647. https://doi.org/10.1016/j.apenergy.2019.113647
20. Weisser WW, Hensel MU, Barath S, Culshaw V, Grobman YJ, Hauck TE, Joschinski J, Ludwig F, Mimet A, Perini K, Roccotiello E, Schloter M, Shwartz A, Hensel DS, Vogler V (2022) Creating ecologically sound buildings by integrating ecology, architecture and computational design. People Nat 5(1):4–20
21. Bligh J, Johnson KG (1973) Glossary of terms for thermal physiology. J Appl Physiol 35(6):941–961
22. Rahm P (2023) Climatic architecture. Actar publishers
23. Lawer C (2020) Health ecosystem value design 2.0. www.umio.io
24. Hertzberger H (2000) Space and the architect: lessons in architecture 2. 010 Publishers
25. Holland J (2012) Signals and boundaries, building blocks for complex adaptive systems. MIT Press
26. Gould J, Vrba E (1982) Exaptation, a missing term in the science of form. Paleobiology 8(1)
27. GCCA. Website accessed on 09 June 2024. https://gccassociation.org/concretefuture/cement-concrete-around-the-world/
28. Trancik R (1986) The theories of urban spatial design. In: Finding lost space. Van Nostrand, New York, pp 97–124

# Building Health: The Evolution of Healthcare Infrastructures Through the Lens of Human Evolution and Social Changes

16

Stefano Capolongo, Michele Dolcini, and Andrea Brambilla

**Abstract**

The architectural project has always presented the opportunity to translate into physical space the challenges posed by the evolution of social structures. These transformations shaped the built environment surrounding us. In this context, healthcare infrastructures represent unparalleled examples of social architecture, where morphological, spatial, and functional configurations are influenced by the presence of a multitude of stakeholders with different needs and complex interactions among them. Hospitals represent an architectural typology which has gone thought historical eras, transforming itself, even radically, but still highlighting a clear evolutionary succession. Undeniably, healthcare design has progressed together with biomedical advancements and epidemiological trends, and it is still evolving today as a consequence of the recent progression in genomic research and predictive medicine, as well as the disruptions posed by Covid-19 pandemic. This relationship between built environment, health and wellbeing is the most intimate point of contact between the fields of architecture and medicine. The design discipline within this domain must expertly translate up to date requirements into tangible spatial solutions, effectively addressing the indicators of the ongoing transformation. Throughout a comprehensive historical overview of the morphological, spatial, and functional transformations of healthcare facilities, this chapter describes the profound design implications of social progress and human evolution. From the close connection with nature and the link to the afterlife in the early healthcare facilities, through the differentiation between healthy and sick bodies in medieval hospitals, from the mechanization of the hospital organism capable of treating different parts of the body in each pavilion, hospital architecture bears witness to a clear cultural and

S. Capolongo (✉) · M. Dolcini · A. Brambilla
Department of Architecture Built Environment Construction Engineering, Design and Health Laboratory, Politecnico di Milano, Milan, Italy
e-mail: stefano.capolongo@polimi.it



J. A. Lara-Hernandez et al. (eds.), *The Dusk of Design*, Integrated Science 34,
https://doi.org/10.1007/978-3-031-76867-5_16

social approach to care, strongly influenced by the technological advancements of the period. The chapter also delves into how the evolutionary principles of variability, redundancy, and diversity influence the strategic decisions concerning hospital projects. Through the understanding and the dynamic interplay of these concepts, this narrative aims to elucidate the evolution of healthcare environments in response to the ever-changing landscape of social instances.

**Keywords**

Healthcare architecture • Hospital design • Human-centred approach

## 1 Healthcare Infrastructures as Social Architecture

The architectural project has always represented the opportunity to translate into spatial physicality the multiple challenges posed by contemporary society. In particular, *health architectures* represent the space where the most disruptive social, economic, and environmental conservation global trends in must find a spatial, functional, and technological synthesis. Hence, the design of healthcare infrastructures is dependent on the society to which it belongs, for establishing its essence and character, standing out as the ultimate expression of *Social Architecture.* Therefore, hospitals represent the architectural typology that better allows a reinterpretation of the history of its morpho-typological transformation in light of the evolution of society, medicine, and technology, outlining trends for future development of civil facilities.

Thus, designing hospitals intrinsically requires the development of a human-centered approach, consistent with the human nature of that historical period to prioritize the needs and well-being of patients, families, and healthcare staff. Amid the dilemmas of the designer, the question arises from the outset: how healthcare facilities can integrate with the surrounding community, addressing issues of accessibility, and inclusivity, and creating spaces that contribute positively to the community beyond medical treatment. In essence, hospitals serve as tangible translations of social instances, reflecting and influencing the dynamics, values, and needs of the communities they serve. Indeed, their typological evolution reflects the social evolution occurring in their societies, opening up to relevant perspectives in terms of design and innovation.

## 2 Learning from the Historical, Typological and Morphological Evolution of Healthcare Architecture

In order to envision the evolutionary trend of healthcare facilities, it is urgent to understand the main historical steps that translated into physical space the different social trends.

Historical research has identified *Asclepieia* as the first place designated for healthcare in the Western world. In ancient Greece, the Asclepieion was the ancestor of contemporary hospitals. In classical civilizations, nature and the after-life played a fundamental role in the healing process, which, for a long time, was defined by the bond between physical well-being and spiritual health. These places of healing, dedicated to Asklepios (Ἀσκληπιός), the god of medicine, were temples situated in salubrious locations that served the dual purpose of religious sanctuaries and health centers. Care delivery was managed by a priest-physicians who, through proto-medical practices such as fasting, baths, sleep, and sunlight therapies, attended the sick people. The Asclepius were monumental structures, typically characterized by a spacious, undifferentiated, central environment where therapeutic activities were conducted. The case of the Asclepieion of Pergamus, an ancient city located in what is now Turkey, stands out as an intricate and significant example.

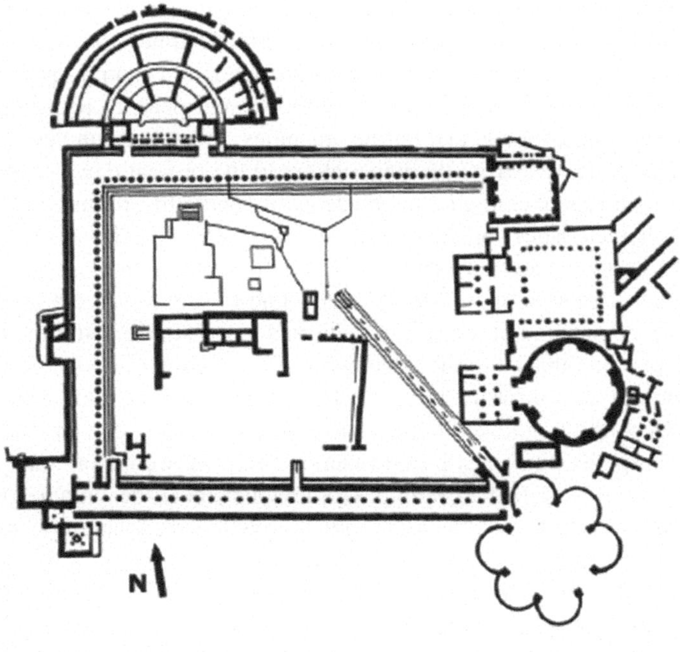


**Fig. 1** Photo Asclepion Pergamus

This *Hellenistic* infrastructure exhibits innovative features of functional differentiation. Moreover, it is surrounded by a series of ancillary activities and constructions conducive to supporting the healing process, including an outdoor pool and thermal baths. Some contemporary criteria of design in healthcare facilities, such as functionality, safety, and hygiene, can be traced back to the Asclepieion of Pergamon. These can be traceable in the spatial distribution, the incorporation of both internal and external staircases, and the hydraulic mechanisms present in this historical structure.

In general, *Asclepieia* were places where the "*medical profession*" was systematically practiced, with the early examples of private care spaces where medical experimentation and practice were linked.

The pragmatism of the Romans compared to the Hellenic approach, on the other hand, reduced this close relationship between medicine and religiosity, connecting healthcare functions to important military contexts. Known as *Valetudinaria*, the ancestors of Roman hospitals were temporary medical structures, aimed at supporting military camps and forts for auxiliary troops. Inside the legionary or auxiliary Castrum, traces of significant spaces dedicated to the treatment and care of battle wounds and for border maintenance are found. The construction principle underlying military *Valetudinaria* is the rationalization of pathways, the central distribution of common spaces, and the concentric use of areas dedicated to recovery and care. The first example of this medical facility dates back to the 1st Century AD, believed to have been constructed in Camuntum, near Vienna [4]. These medical facilities for soldiers became integral to established Roman fort architectures. Valetudinaria were strategically placed in mayor and small sites to support soldiers, and the size changed depending on the fort's dimension. The larger hospitals conventionally comprised four wings surrounding a central open courtyard. These facilities typically included treatment rooms, a dispensary, and staff quarters, as well as washing, cooking, and latrine amenities. In some examples which have been partially preserved, from the imperial era to the present day, it is possible to reconstruct their spatial configuration made up of courtyards, rooms, and corridors, as well as characteristics for their location which are still relevant: proximity to non-marshy watercourses and easily accessible from supply routes.

The social disruption arising from the decline of the classic era also affected the evolution of care and of the spaces dedicated to assistance. After the fall of the Roman Empire and thanks to the documents of the Council of Nicaea, a typological evolution of care infrastructures can be distinguished, in some way returning to a previous conception linked to the sphere of religious worship and the afterlife. Therefore, hospitals, like most social functions and, we would say today, "*public utilities*," were closely tied to religious orders and monastic organizations. The advent of Christianity fosters the dimensional growth and spread of hospitals, also for ethical and charity reasons. The principles of faith codify assistance to the suffering as a duty for Christians. This materialized in 325 AD with the Council of Nicaea providing specific directives for strengthening the care network through the construction of shelters for pilgrims and people in need. The typologies of the early Middle Ages are those of the *Hospital Pauperum* and *Pellegrinorum*, established in proximity to monasteries. In the typological definition of these conventual hospitals, the functional architectural design ensures coherence throughout the medieval period between the two building typologies. The mix between religious and healing functions is evident as a characteristic sign of the historical era [2]. These facilities were dealing with the care of individuals in difficulty. Due to the vocation of the religious organization, evolving toward social support of weak and fragile communities, they included the possibility of finding care, refuge, and hospitality for the poor (hence the Latin term "*hospitale*" or the French "*hotel-Dieu*"). This renewed concept resulted in a great promiscuity between contingent needs and illness.

A relevant example of this historical period, illustrating the typological evolution of hospital structures, is the Hôtel-Dieu de Beaune (Fig. 1), located in *Bourgogne* (France) at the heart of the Côte d'Or region and situated in proximity to the Via *Francigena*, a pilgrim route. This infrastructure, which has been mostly preserved, is characterized by a central courtyard flanked by wooden arcades. The assistance and care functions were housed within one wing, specifically in the "*salle de povres*" (hall for the poor). The innovative configuration introduced by this facility lies in the separation of beds to ensure privacy among patients and to contain the spread of viral diseases.

Meanwhile, the spread of the plague in Europe during the Middle-Ages, and scientific development, led medicine to evolve towards a more experimental based approach.

In architectural terms, an important milestone concerns the definition of the so-called "civil hospital" for the treatment of chronic, acute, and infectious patients. This functional configuration is reflected in the plan-volumetric layout of the building. In this regard, Ca' Granda Civil Hospital in Milan, Italy (Fig. 2), represents the most emblematic case of this new approach. The hospital was commissioned by the new prince Francesco Sforza to promote its image, and has remained in operation for 450 years, shaping the configuration of Milan, and transforming the urban axes of the city.

**Fig. 2** Hotel Dieu Beaune

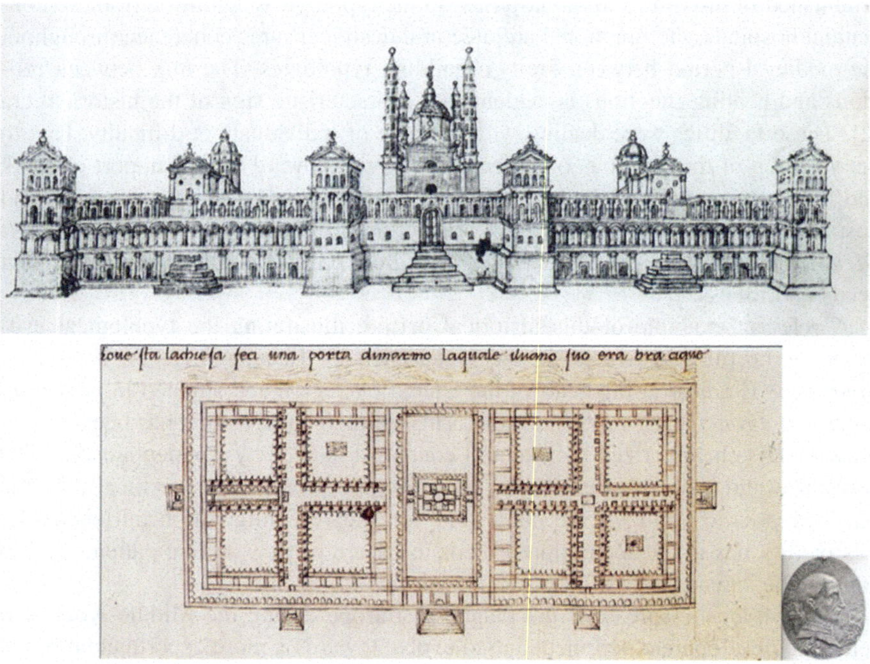

**Fig. 3** Ca'Granda hospital—Milan 1456

This iconic facility, which adopted organizational and spatial models of Tuscan hospitals at the time, introduced a new standard for healthcare environments in Europe. The architect Antonio di Pietro Averlino, known as "Filarete", designed a rectangular building composed of two symmetrical square bodies, divided by four aisles arranged in a Greek cross. In his initial thoughts, one of these so-called "*crociere*" was reserved for men (on the right) and one for women (on the left), with the development of eight internal courtyards [9]. The facility, also called Ospedale Maggiore ("Major Hospital"), has maintained a substantial adherence to the original design conceived by the architect in the mid-fifteenth century, despite being completed in over four centuries.

The increasingly systematic approach introduced in the fifteenth century led to the definition of the first criteria for the establishment of a ward, proposing a certain number of beds based on the disease and size, with specificities and prescriptions related to concepts of hygiene and health.

Contemporaneously with civil hospitals like Ospedale Maggiore, another type of hospital building emerges from that era: "*lazzaretti*" and leper colonies. In contrast to the civil typology, these structures were situated in isolated contexts, located in extra-urban areas to ensure that the afflicted could remain for an indefinite period in their healing journey. Michel Foucault identifies these entities as strongly linked to the realm of myth, places of temporary passage, change, a round-trip journey disrupting the flow of life [6]. In terms of layout, these buildings present similar plans. These extra-urban structures, in fact, seem to reconstruct the characteristics of a city, with enclosures, courtyards, and accommodations replacing walls, squares, and civilian dwellings.

This following social and scientific development, fueled by philosophical positivism and the encyclopedic eighteenth-century approach, led to a precise morpho-typological and distributive-functional definition of the hospital building. With an original educational purpose, the construction of the first surgery block, with tiered classrooms, led to the hospital's development based on the functional separation of specific areas such as operating rooms, birthing centers, isolation areas, laboratories, services, and wards compared to a symmetrical or panoptic radial distribution layout of neoclassical style.

The architectural typology that interprets these transformations is the pavilion hospital, a complex building composed of several autonomous and independent bodies hosting specific functions. Gradually, each pavilion assumed a specific function, and wards with over 30 beds were divided into multiple, double, or even single rooms. To address the concentration of health pathologies in the medical field, hospital architecture responded by introducing new criteria for building design. Planning hospitals with separate pavilions started in the eighteenth century. In this regard, the impetus for the spread of this model was driven by the destruction of the Hôtel-Dieu de Paris in 1772. The commission of the *Academie Royale des Sciences* (Academy of Sciences), tasked with approving the new building typology, leaned towards the choice of a pavilion model. This English-inspired typology led to the introduction of specific functional criteria and requirements in French territory. The separation of environments and pathways for different user types and diseases distinguishes this new model. A relevant example built with this design concept is the Hôpital Lariboisière (Fig. 3), constructed in Paris in 1839.

From this point onward, the growth of care activities led to an oversizing of hospital facilities. The concept of a "City of Health" begins to emerge, configuring the hospital as a pavilion-based citadel that extends into peripheral territoriality, including in the extra-urban network.

This specialization and the functional evolution have been addressed in two seminal publications during the nineteenth century, by Florence Nightingale: "*Notes on Nursing*" (1858) and "*Notes on Hospitals*" (1859). The limitations of this typology in terms of space utilization and complicated connections between different pavilions began to be evident in the United States only after World War I. Here, the invention of the elevator, the use of mechanical ventilation, and the gradual standardization of some clinical and managerial processes led to the reconfiguration of traditionally horizontally distributed pavilions along the vertical axis, identifying the **monoblock typology.** This layout is formally configured with *"T"*, *"U"*, or *double T-shaped* planimetric profiles, articulated in functional areas of triple or quintuple bodies. In the European and Italian context, this model was imported later and is still active today. Despite its wide distribution in several contexts, the significant weaknesses of these architectures are evident in terms of the sizing of vertical connecting elements, differentiation of pathways, and a poor predisposition to departmental reorganization flexibility [9].

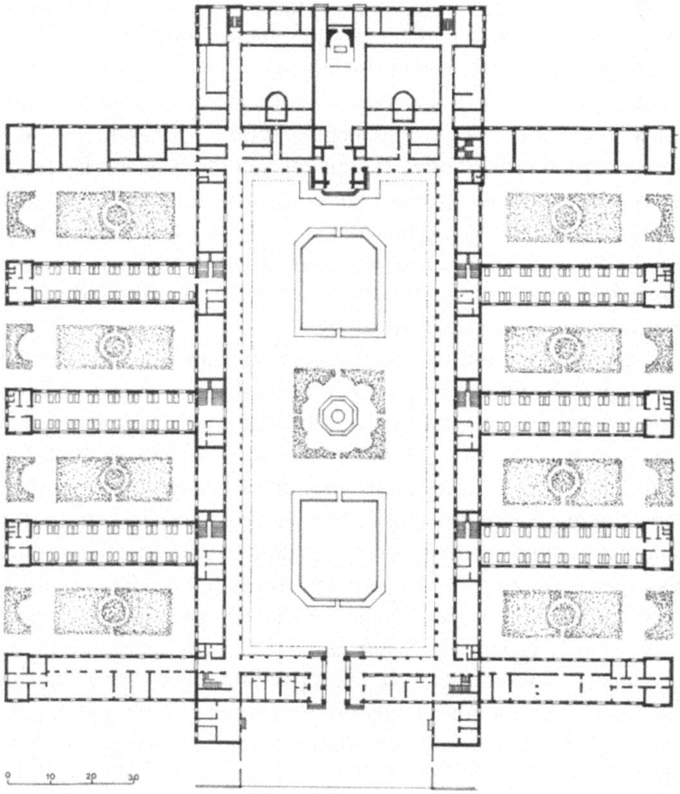


**Fig. 4** Hôpital Lariboisière—Paris 1839

Nowadays, the monoblock typology has not undergone profound innovations but has undergone a gradual process of hybridization with several other typologies, favoring a return to a more horizontal yet compact development in recent decades. The most recurring model used even in recent hospitals tends to divide, on the lower floors, the critical areas and areas with high technological impact, such as diagnostic imaging and large technologies that also have a considerable unit weight. The upper floors are reserved for functions with a low plant component, such as wards and spaces for medical, administrative, and managerial studies.

As highlighted by James and Tatton-Brown [8] in their 1986 work, "*Hospitals: Design and Development*" the typological aspect remains a subject of extensive discussion even in the contemporary debate. Typology holds a crucial role in hospital design, with a growing relevance of territorial healthcare facilities as innovative medical centers [3].

The differentiation between various types is not only responsive to shifts in care methodologies, primarily influenced by the increasing prominence of diagnostic and interventional functions and advanced technology, but it is also intricately connected to the concept of humanization and patient-centered. Conversely, humanization is intertwined with the positioning of the hospital facility within its surrounding environment. Beyond the dichotomy of horizontal versus vertical, there is also a distinction between urban and suburban contexts, as well as the proximity to a more natural and separated zone in contrast to the integration with the urban area.

Today, the increased importance of technology within the building has contributed to the elimination of the identity and social character that categorized health architecture since antiquity. It is not surprising that one of the most recent organizational innovations is the importation of management models directly from the mass production sector, such as the lean model from Toyota. Several renowned architects of our time, including Le Corbusier, Alvar Aalto, and, more recently, Renzo Piano, and Norman Foster have found medical facilities intriguing because of their immense complexity as building types, with the integration of most updated systems and technologies. These high-tech, complex and multidimensional structures play a crucial social role as places of healing and toward the communities they serve.

## 3 The Evolutionary Journey: Variability, Redundancy, and Diversity

The historical and typological evolution of healthcare infrastructure presented in the previous paragraph has highlighted differentiating results from social and technological changes. Some features have proven their centrality in the historical evolution, and three evolutionary characteristics have distinguished the journey. The evolutionary principles of **variability, redundancy, and diversity**, as presented in the Introductory Chapter of the book by Telmo Pievani ("*The importance of diversity, variability and redundancy in nature*"), are fundamental concepts also for hospital design, due to functional requirements of healthcare infrastructures and the necessity of integrating different user's needs. These facilities have gradually, but fully embraced these three concepts.

### 3.1 Variability

Within the realm of healthcare infrastructure design, two distinct categories of Variability come to the fore. The first pertains to **Structural Variability**, denoting a facility's inherent capability to adapt over time, responding to evolving social, epidemiological, and technological pressures, and effectively absorbing potential impacts stemming from *exogenous* factors. Among these exogenous factors, the latest contemporary experiences have underscored the significance of risks associated with pandemics, natural disasters, and conflicts.

The alternative typology of variability which can be tracked during the historical evolution of healthcare infrastructure assumes an *endogenous* character. **Processual Variability** concerns users and differentiated treatment paths. As elucidated in the preceding paragraph, the standardization of care environments emerged as a trademark of the initial evolutionary phase of hospital structures, spanning the classical and high medieval eras. The subsequent definition of specialized architectural forms and typologies during the middle-age period, coupled with the introduction of pavilion hospitals, underscores the fundamental role that variability assumes within the hospital environments. The incorporation of variability into hospital design is a fundamental aspect driven by the multifaceted and ever-changing nature of healthcare environments. As medical practices and technologies progress, hospital facilities must seamlessly integrate these advancements. Variability in design also consider the dynamic nature of staff workflows, promoting operational efficiency and staff well-being.

Resilience and flexibility to exogenous shocks relies in future growth and expansion, ensuring the long-term sustainability of hospitals, while the cultural sensitivity embedded in variable designs recognizes and respects the diversity of communities served. Ultimately, Variability in hospital design is not an optimal feature but a strategic approach that enhances adaptability, patient-centered care, and the overall effectiveness of healthcare delivery.

## 3.2 Redundancy

In complex systems, like healthcare infrastructure, the continuous and reliable functioning of health devices and processes must be ensured, and the Redundancy of its component must be guaranteed. In addition to organizational procedures, aimed at ensuring the continuous functioning of medical activities when the system has been put under pressure, there are infrastructural and system solutions applied. Essential systems such as backup power supply, fire safety, lighting systems water, and HVAC (Heating, Ventilation, and Air Conditioning) and critical medical equipment ensures that healthcare services can continue without significant disruption, even in the face of technical failures or maintenance activities. Redundancy in these systems ensures that there are backup mechanisms in place to prevent failures, like power shortages. Functional layouts also play a role in ensuring continuity, and redundancy is implemented throughout multiple evacuation pathways, separation and replication of functions to prevent congestions and infection containment, redundancy in vertical transportation and spaces for diverse needs of occupants with fragilities. In contrast to the general definition of redundancy in social sciences management, redundancy in healthcare environments it's crucial in health care management areas such as patient-centeredness, patient safety and the ability of health organizations to ensure high-quality care.

### 3.3 Diversity

The integration of the concept of Diversity can be traced all along the historical evolution of hospitals, at least since the medieval era with the civil hospital. As seen through the lens of historical progression, medieval experiences interpreted this concept with a basic separation. In the presented case study of Ca'Granda hospital, users were separated in different assistance spaces (in two "crociere"), based on their gender and infection intensity. Historical evolution has altered social demands, placing greater emphasis on individual differences, with a perspective of integrating diversity since the design stage. Fully integration of all diversities in health facilities is a contemporary concept, that can be attributed to a recent shared social consensus on the integration of diversity as a cornerstone of human evolution and enables development of societies.

When considering hospital environments, lack of accessibility, orientation, and comfort can lead to disabling situations, additional costs, and adjustment times, impacting the health of users and the overall quality of service. The application of user-centered design principles has highlighted that, throughout all stages of the design process, it is essential to pay close attention to the physical, psychological, and social needs of all users, including patients, staff, and caregivers. In this regard, over time, the *Design for All* approach in its European meaning is meant as "*the design for human diversity, social inclusion and equality*" [5].

The same concept has been reinterpreted in the United States through the notion of "*Universal Design*" by the American architect Ronald L. Mace in 1985. Universal Design have been integrated into hospital design at all stages, aiming to create an environment that promotes human well-being and health for all involved users. The discussion that has significantly marked recent decades concerns overcoming architectural barriers, associating disability primarily with users facing motor difficulties. While this group constitutes a significant portion within healthcare settings, the concept is evolving into a broader perspective. Diversity extends beyond mere physical considerations. Indeed, Design for All refers to a design methodology with the fundamental goal of constructing healthcare environments accessible to all categories of people [8].

## 4 Future Perspectives

### 4.1 The Current Critical Scenario, Needs and Challenges

The typological hybridization of hospital architecture has concomitantly witnessed an increase in technological and infrastructural components within complex facilities, such as hospitals.

The addition of new departments, services, and subsequent technological accretions has transformed hospitals into labyrinths, evolving without a structured plan to address individual needs. The design and construction of large infrastructures during those years increasingly assumed a technological trajectory, aligning with the high-tech design trend exemplified by Renzo Piano, Richard Rogers, and Peter Rice's Beaubourg (1973–1977). Consequently, the space dedicated to technological systems such as electrical, hydraulic, HVAC, medical gas equipment, and transport became equivalent, if not greater, than the surface allocated for users—patients, visitors, and hospital staff. A paradigmatic example of this renewed layout in this context is interstitial technological floors.

The significant technological component, lack of substantial investments, and the necessity for 24/7 management of healthcare infrastructure have led to a substantial stagnation in architectural movements and morpho-typological and constructive developments in healthcare architectures.

When considering Italy as a reference for the European scenario, it has been estimated that over half of the hospitals in the country require structural, technological, and plant adjustments to meet contemporary performance requirements, as only 15–20% of healthcare real estate has been built post-1975. This obsolescence arises from the inability of the system to adapt to the changes required for the structure to keep pace with evolving medical-organizational transformations. Medical, technological, and organizational knowledge evolves at a much faster pace than previously anticipated. This trend results in a process of acquiring new performance requirements that structures built in the last quarter of the past century can no longer fully satisfy. Moreover, considering the extended construction times for infrastructure of such complexity, it is often apparent to industry professionals that a hospital, at the opening date, is already outdated.

An important aspect exacerbating the hospital's identity, functionality, and relational malaise is undoubtedly environmental sustainability. Developing and managing complex healthcare facilities and paths for patients, involves significant resources to maximize patient well-being, adequate comfort for visitors, and all operational facilities for workers. Integrating these needs and meeting these requirements in light of the latest safety, hygiene, and health standards makes the hospital an energy-consuming infrastructure with extremely high consumption levels (estimated to be three times higher than a building of similar size hosting non-health-related functions). Hospitals operate 24 h a day, 365 days a year, with diverse microclimatic and lighting needs (for air exchange rates, the presence of heat-emitting equipment, or anticipated activities). A contemporary hospital is an energy-consuming machine that generates a considerable amount of solid and liquid waste (including hazardous special waste), ionizing and electromagnetic radiation, noise, and urban and social impacts on the ecology of the surrounding environment.

## 4.2 How Does the Next Generation Hospital Look like?

The evolutionary pathway of hospital infrastructures as physical interpretation of social needs is not over. The fast and profound social changes affecting our societies will lead to new requirements and facility standards. After the pandemic experience and the emergence of the crucial role of the built environment in managing a crisis, especially in the healthcare sector, the definition of spatial and performance requirements for a Next-Generation Hospital has become a necessity even within the World Health Organization (WHO). Safe, sustainable, and flexible hospital buildings are prerequisites for the adoption of new management models in hospital organizations. Indeed, to provide high-quality, cost-effective care, it's essential to ensure that the right physical structures are in place.

World Health Organization published in 2023 a Technical Brief outlining the basic requirements for new hospital projects in the European Region. The technical-scientific document articulates the foundational pillars that ought to underpin hospital design. The macro-areas within which it is possible to identify the functional requirements of healthcare facilities and that determine the future challenges of hospital construction are summarized as follows: *1. Functional design and innovative layouts; 2. Healthcare infrastructure sustainability; 3. Flexibility, resilience and future-proofing; 4. Healthy working areas; 5. Safety and security issues; 6. Organizational strategies for infection control capability and indoor air quality; 7. Digitalization of the hospital from the management to the end-users; 8. Soft qualities and wayfinding* [10].

The capacity to implement these design recommendations, along with the acquisition of a comprehensive understanding of the evolving social, technological, epidemiological and economic scenario is the challenge to be addressed worldwide. Architectural design and the research must face the relevant challenges and opportunities that support the definition of future hospitals for the next generations.

## References


1. Attalì J (1984) Cannibalism and civilization: life and death in the history of medicine. Bloomsbury USA Academic
2. Capolongo S, Cocina G, Gola M, Peretti G, Pollo R (2019) Horizontality and verticality in architectures for health. TECHNE J Technol Archit Environ 17:152–160. https://doi.org/10.13128/Techne-24028
3. Cilliers L, Retief FP (2002) The evolution of the hospital from antiquity to the end of the middle ages. Curationis 25(4):60–66. https://doi.org/10.4102/curationis.v25i4.806
4. European Institute for Design and Disability—EIDD (2004) Stockholm declaration. On: http://dfaeurope.eu/what-is-dfa/dfa-documents/the-eidd-stockholm-declaration-2004
5. Foucault M (1973) The birth of the clinic: an archaeology of medical perception, 1st American ed. Pantheon Books
6. James WP, Tatton-Brown W (1986) Hospitals: design and development. Architectural Press
7. Mattia M, Biehler-Gomez L, Sguazza E et al (2022) Ca' Granda, an avant-garde hospital between the renaissance and modern age: a unique scenario in European history. Med Hist 66(1):24–33. https://doi.org/10.1017/mdh.2021.40
8. Mosca EI, White J, Steinfeld E, Capolongo S (2022) Designing hospitals through the lens of universal design. An evaluation tool to enhance inclusive healthcare facilities. In: Garofolo I, Bencini G, Arenghi A (eds) Studies in health technology and informatics. IOS Press. https://doi.org/10.3233/SHTI220857
9. Verderber S (2010) Innovations in hospital architecture. Routledge

10. WHO (2023) Hospitals of the future A technical brief on re-thinking the architecture of hospitals. http://apps.who.int/bookorders

# The Bricolage of Transdisciplinarity. The Nuanced Concept of Landscape Resilience Through a Recombination Approach

17

Fabrizio Aimar and Angioletta Voghera

**Abstract**

Resilience continues to rule the literature in several disciplines, ranging from different theoretical concepts and approaches. However, there is a lack of a holistic definition centred on the entangled social-ecological dimensions of the landscape. Consequently, this theoretical research aims to offer a qualitative definition of landscape resilience and its features, role, and the appropriate approaches to build it. Then, their definition will support establishing an international reflection on the topic in landscape research. To fill this research gap and overcome disciplinary boundaries, the article provides a qualitative definition of landscape resilience in the framework of landscape studies. Properties and features of the social and ecological sub-systems are introduced to understand interdisciplinary convergences and interrelationships. The research identifies some key points useful to sustain the definition of landscape resilience. Among them, there is the capacity (from the ability to capability) to recombine elements, knowledge, and networks. Moreover, landscape resilience can be considered an integrated approach to comprehending the new normal conditions, building shared values, and guiding the birth–death-rebirth cycle of resilient systems. This theoretical research opens room for discussions on the topic. Planning and integrated management are the fields that are more influenceable by considering landscape resilience as an approach. In addition, informing and steering the autopoiesis of landscape is a starting point for further advancements in the field.

F. Aimar (✉)
Texas A&M University, College Station, USA
e-mail: fabrizio.aimar@tamu.edu

A. Voghera
Politecnico di Torino, Turin, Italy
e-mail: angioletta.voghera@polito.it

CED PPN—European Documentation Centre on Nature Park Planning, Turin, Italy



J. A. Lara-Hernandez et al. (eds.), *The Dusk of Design*, Integrated Science 34,
https://doi.org/10.1007/978-3-031-76867-5_17

**Keywords**

Landscape resilience • Adaptation • Transformation • Creative bricolage • Co-evolution

# 1 Introduction

## 1.1 Landscape Resilience: A Missing Definition

Since the first definition of the resilience of the ecological system, provided by [37], this concept has slowly moved beyond the socio-ecological divide to be widely applied to socio-ecological systems across different disciplines, along with botanic, hydrography, history, anthropology, sociology, and planning.

However, after a thorough literature review on the Elsevier, Taylor & Francis, Emerald, and Cambridge University databases, a research gap has been noted, namely that there is a lack of a holistic and non-fragmented definition of the concept of landscape resilience at the international level. Despite an increasing trend of resilience-related publications centred on the landscape, as shown by Elsevier in the last ten years (2013–2022), many publications lack a holistic reflection on its broader scope. In fact, the term resilience appears to be analysed exclusively in relation to landscape (e.g. how to build the resilience of a landscape or a part/component of it), focusing on a specific framework (i.e. resilience of what to what, by whom and for whom, where and why) under the interrelated triad composed by persistence, adaptability and transformability.

As an approach, resilience sometimes seems to be seen as an update of other existing approaches in a given field (e.g. an implemented or emergency management of a subsystem of an environmental asset, under the umbrella concept of resilience). Consequently, it lacks a broader understanding that should be necessary to determine a truly integrated management system (e.g. in a cultural landscape) (Appendix).

Among potential definitions of it, Xu et al. described landscape resilience as "… the ability to adapt, change, and reorganise while coping with disturbance and meeting shocks" [110, p. 5], for instance. Moritz et al. seemed to complement this theory in the risk reduction framework, stating: "How ecosystems recover and continue functioning across the full distribution of events in a naturally functioning fire regime is therefore a key to landscape resilience." [63, p. 53]. Similarly, North et al. reinforce landscape resilience in the mentioned aspect but with a clue on management: "Three additional managed fire objectives, density reduction, enhancing spatial heterogeneity, and species and phenotypic selection … will further improve landscape resilience." [70, p. 533]. Additionally, Beagan and Dolan [11] detect five components of resilience (i.e. diversity, redundancy, network connectivity, modularity and adaptability) that need to be integrated into the decision-making process of cultural landscapes but only in the perspective of preserving, maintaining or rehabilitating them. This vision is also part of the definition provided by Schmidt, which defines landscape resilience "… as the adaptability and self-renewal capability of a landscape, and thus as the ability of a landscape to maintain, renew and strengthen its own fundamental landscape qualities despite disturbances, crises or ongoing changes" [91, p. 6]. On the other hand, Liu stated that "landscape resilience is the capacities of a landscape, as a SES, to withstand, cope with, and adapt to changes and to transform when necessary, hence to maintain its functioning and provision of goods and services for human well-being" [47, p. 9]. Despite the overall deficiency of existing landscape resilience, the most recurring characteristics emerged from this literature review are ability and capacity, change or changes, ecosystem or ecosystems, function or functions or functioning, disturbance or disturbances, maintain or maintaining, structure or structures, identity, and scale or scales (Table 1). Nevertheless, the author did not mention the necessity "… to closely interrelate the concepts of co-evolutionary resilience and landscape (which varies between cultures), based on the inclusion of urban communities." [108, p. 117].

However, despite the latter attempts for an all-encompassing definition, the varied approaches based on different interpretations of the landscape under the aegis of resilience proposed by the Resilient Landscapes Symposium held at the Italian Pavilion at the 17th International Architecture Exhibition in Venice, Italy [84] miss in providing a holistic and shared definition. They ranged from agro- and eco-voltaic, including solar communities, to waste materials and creative reuse, from the street- and cityscapes to the air–water-soil-food nexus, from the relevance of art and interactions in building new landscapes to the *genius loci*, among others.

As a consequence of the above, it becomes clear that landscape resilience is still an open research topic [108], especially when considering the resilience characters at play in landscape action.

This is because it is only partially addressed in specific component areas of the socio-ecological systems (SES) [77, 107] and the social-ecological-technological systems (SETSs) [52, 55], such as ecological dimensions, landscape ecology [6, 24], green infrastructure (GI) [44, 112], ecosystem services (ES) [48, 50], agriculture (food security and agroecosystems) [46, 73], cultural ecosystem services (CES) [66, 97], and as a generic response invoked to address systemic pressures and disturbances posed by climate change [56, 99]. This vagueness seems to derive from the fact that landscape is polysemic [34, 82], as it covers a variety of meanings according to different and sometimes intertwined interpretations, both objective and subjective [71]. Accordingly, Bloemers et al. noted that "There are as many perspectives on it as there are disciplines using it" [16, p. 2] in an interdisciplinary way. These include "… history [92], geography [5], anthropology [40], sociology [8], semiology [79], psychology [38], art [83], agronomy [86], botany [90], zoology [41], landscape ecology [105], pedology [75], geology [57], climatology [62], economics [80], and planning [78] …" [4, p. 17].

Therefore, as a complex system composed of different sub-systems or dimensions (e.g. ecological, social, economic and cultural), the application areas of

**Table 1** Selected definitions of landscape resilience (Authors' elaboration)

| Author(s) | Year | Context | Discipline | Definition |
|---|---|---|---|---|
| Moritz et al. [63] https://doi.org/10.1007/978-94-007-0301-8_3 | 2011 | Ecological systems | Landscape ecology, complexity science | How ecosystems recover and continue functioning across the full distribution of events in a naturally functioning fire regime is therefore a key to landscape resilience. |
| Cumming et al. https://doi.org/10.1007/s10980-012-9725-4 | 2012 | Social-ecological systems | Landscape ecology, sustainability | Landscape resilience depends heavily on finding an appropriate match between the scales of demands on ecosystems by human societies and the scales at which ecosystems are capable of meeting these demands |
| Salvati et al. https://doi.org/10.1007/s10113-013-0437-3 | 2013 | Social-ecological systems | Climate change, sustainability | Landscape resilience, considered as the ability of a system to absorb and compensate anthropogenic and natural pressures such as those connected to LD (land degradation) |
| Voghera https://doi.org/11583/2638390 | 2015 | Social-ecological systems | Landscape planning, design, and management | Landscape resilience is based on evolutionary resilience (Chelleri, 2012; Stumpp, 2013), that consider the relationship between scales of action, and between community autonomy and environmental, social and economic connectivity and integration in a wide scale (Allan and Bryant 2011) |
| Field and Parrott https://doi.org/10.1016/j.ecocom.2017.08.004 | 2017 | Social-ecological systems | Ecosystem services, sustainability | Landscape resilience is highly dependent upon the spatial quantity, quality, diversity, and redundancy of ES, as well as the cross-scale connectivity of the processes underlying their provisioning and management |
| Liu [47] https://doi.org/10.15302/J-LAF-1-010,002 | 2019 | Social-ecological systems | Landscape architecture, hydrology | Landscape resilience is the capacities of a landscape, as a SES, to withstand, cope with, and adapt to changes and to transform when necessary, hence to maintain its functioning and provision of goods and services for human well-being |

(continued)

**Table 1** (continued)

| Author(s) | Year | Context | Discipline | Definition |
|---|---|---|---|---|
| Guerrero et al. https://doi.org/10.3390/su122410345 | 2020 | Social-ecological systems | Ecotourism, environmental conservation, sustainable development | Landscape resilience can be defined as the ability of an environment to withstand disturbances and reorganize itself while being subject to forces of change, ultimately maintaining its essential functions, structure, identity, and mechanisms" |
| Wang et al. https://doi.org/10.1111/gcb.15299 | 2020 | Ecological systems | Biology, ecology, conservation | "it can identify spatial elements to conserve that will be important for conserving broader scales of biodiversity despite the dramatic environmental changes washing over the continent" |
| North et al. https://doi.org/10.1093/jofore/fvab026 | 2021 | Ecological systems | Management | "Three additional managed fire objectives, density reduction, enhancing spatial heterogeneity, and species and phenotypic selection … will further improve landscape resilience" |
| Xu et al. https://doi.org/10.3390/su13052629 | 2021 | Social-ecological systems | Environmental management, risk reduction framework | The ability to adapt, change, and reorganise while coping with disturbance and meeting shocks |
| Robert ISBN 978-1-78630-666-1 | 2021 | Spatial systems, social-ecological systems and landscape forms | Archeogeography | Landscape's capacity to retain its identity, i.e. to maintain the relations, morphogens and morphological principles which characterized it on a macroscopic level, by absorbing, or even using, the changes which affect it. Resilience can be used to describe the point to which a landscape can self-organize and is able to adapt |
| Zhang et al. https://doi.org/10.1016/j.envdev.2021.100616 | 2021 | Social-ecological systems | Landscape planning, design, and management | SELR refers to the ability to withstand perturbations on the condition that the SEL does not transform states |

(continued)

**Table 1** (continued)

| Author(s) | Year | Context | Discipline | Definition |
|---|---|---|---|---|
| Nematollahi et al. https://doi.org/10.3390/land11101682 | 2022 | Social- ecological systems | Ecotourism, environmental conservation, sustainable development | Landscape resilience pertains to the capability of an environment to resist external disturbances and to reorganize itself to maintain its critical structures, functions, and mechanisms |
| Rastandeh and Jarchow https://doi.org/10.4324/9781003266440-1 | 2022 | Social-ecological systems | Landscape restoration/ conservation, landscape design, climate change | We define landscape resilience as the ability of landscapes, including their human communities, to adapt to and to mitigate the impacts of climate change; and to persist under changing and uncertain conditions such that safeguard biodiversity and nature's contributions to people |
| Schmidt https://doi.org/10.1007/978-3-662-63998-6 | 2022 | Social-ecological systems | Landscape, complexity theory, ecology, planning | … as the adaptability and self-renewal capability of a landscape, and thus as the ability of a landscape to maintain, renew and strengthen its own fundamental landscape qualities despite disturbances, crises or ongoing changes |
| Sahle et al. https://doi.org/10.1007/s00267-023-01794-0 | 2023 | Social-ecological systems | Landscape ecology, landscape planning and management | Landscape resilience requires proper management that secures the capacity of ecosystems to sustain societal development and ensures progress with essential ecosystem services |

resilience indicated above seem to suffer from this antithetical condition. Both generalist and specialist at the same time, it denounces a sectoral, objective and silos-thinking approach. Moreover, this potential aporia seems to be reinforced in the European context by the European Landscape Convention [26], where the holistic consideration of landscape (summarisable as 'everything is landscape') forces the sectorisation of specific considerations that are based on imaginaries whose boundaries are still vague (e.g. what does the rural landscape include? What about the urban landscape?).

Moreover, possible definitions of landscape resilience risk being based on a concept of resilience that is Eurocentric, where different meanings emerge between countries in the Eastern and Western worlds and even between those in the same geographical area [3].

It is, therefore, necessary to provide an interpretation that is more holistic, inclusive and broad of what landscape resilience can be, albeit from a South-European and Western point of view. Building on the international state of the art in research, the article tries to build bridges between social and ecological domains to reconnect them in a new cultural discourse, proposing a theoretical and qualitative definition of landscape resilience. This research intends to contribute to and reinforce the targets 2.4, 11.4, 11.a, 13.1, 15.9 and 16.7, listed as parts of the Sustainable Developments Goals in the framework of the 2030 Agenda [102].

## 1.2 Research Aims

As highlighted by Voghera and Aimar, "The landscape resilience paradigm is a research frontier; in fact, (it is) a polysemic hypothesis developed in the context of interdisciplinary research, whose debate is still open in the literature." (2021, pp. 112–113). Consequently, this theoretical research aims to reach multiple goals. Namely, it attempts:

- to provide a theoretical and qualitative framework for thinking about landscape dynamics and useful elements for defining landscape resilience as a concept and what it should encompass.;
- to define a common front in international landscape research on this specific topic. To date, existing landscape resilience concepts still appear too fragmented, with little attention paid to jointly and cross-cutting considering the multiple aspects and characteristics of the landscape. Sectoral definitions or multidisciplinary approaches persist instead of truly interdisciplinary convergence or transdisciplinary remodelling;
- to provide a qualitative answer to research questions related to the fresh concept of landscape resilience: what should be the role of landscape resilience? Is/are the current approach(es) we are using appropriate for building landscape resilience?

# 2 Material and Methods

## 2.1 Landscape: Concepts and Features

Landscape can be considered as an umbrella concept [36] that evolves over time [72]. Due to its polyvocality of views and various disciplines focusing on it [9], it allows different and multifaced interpretations and definitions as a concept [85].

At the European level, landscape can be defined as "... an area, as perceived by people, whose character is the result of the action and interaction of natural and/ or human factors" (CoE 2000), in an intertwined nature-culture relationship [42]. On the one hand, visual appreciation seems relevant to recognise it [21, 25, 61].

On the other hand, considering productive landscapes, Sereni similarly defines them as "... the form that man consciously and systematically impresses upon the natural landscape during the course and for the purposes of his agricultural production activities ..." [92, p. 3]. However, the natural agents and processes seem to prevail in the understanding of creating this landscape as a product [72, 2.b], emphasizing its physical and landforms component [ibid., 61] over the product and idea.

Based on the European Landscape Convention, it can be deduced that each landscape therefore has "... significant or characteristic features" [26, Art. 1.d], which are "physical expressions of both tangible and intangible aspects of a place" [69] as follow:

- it is coevolutionary, as the result of the continued interaction "derived from its natural configuration and/or from human activity" [26, Art. 1.d], and consequently dynamic as it embeds "... changes which are brought about by social, economic and environmental processes" (ibid., Art. 1.e),
- it includes characteristics such as "natural systems and features" (processes), "spatial organization" (historical arrangements and layout), "land use" (activities, patterns), "circulation", "cultural traditions" (practices and intangible heritage), "topography" (physical forms), "vegetation" (permanence), "cluster arrangement" (built and aggregated forms), "buildings and structures", "views and vistas" (perception), "constructed water features", and "archaeological sites" [67],
- it has an inherent diversity made up of species, habitats and land uses, which explores variations in its ecosystems,
- it allows formal and informal uses, which are constantly intertwined,
- it combines functional and aesthetic needs,
- it has an inherent structure in relation/contrast to function [100],
- it includes connectivity in relation to corridors and patches [100],
- it is based on scale-dependent features [85] within a larger physical dimension (Cambridge Dictionary n.d.),
- it shows potential limits in the ability to adapt to ever-changing conditions (e.g., climate, land use, and demographics, among others),
- it encompasses a set of disturbance regimes [32],

- it has an inherent sense of place [94],
- its related planning practices contemplate a marked resilient approach, where mitigation and adaptation concur with the goals of "enhance, restore" landscapes, as well as "… strong forward-looking action …" in "… create landscapes" (ibid., Art. 1.f).

Landscape Planning and Design

The landscape is in continuous transformation; it is a "metamorphic entity" [74, p. 13], which lets us interpret the culture of places or non-places and their historical evolution over time. By interpreting the function and role of landscape, we can see the change in cultures and their ethical and social structures in the twentieth century.

Landscape theory, as well as regulatory and planning action, originated from how society learned to acknowledge and reinvent the lost "wilderness" [111, p. 24]. Social sensitivity for landscape emerged mainly in the mid-nineteenth century, in response to the major changes caused by industrialization and urbanization. This sensitivity was influenced by the poetry and paintings that were produced by the new opportunities for traveling [17, p. 101]. This sensitivity, then as now, has manifested itself in various forms in different countries.

In many cultures (such as in the Germanic-Anglo-Saxon world—the Netherlands, Great Britain, Germany, and Austria) landscape policies aimed at promoting productivity and preserving diversity and quality stem from the conservation and management of agricultural land. In rural management—which can be found also in the terminological root of landscape—there is a specific focus on protection and planning through a multi-scale project, from regional to local scale. This project has influenced the territorial governance, leading to the development of regulations, plans, and implementation tools (financial and managerial) with strategies for the organisation and management of the rural landscape. The rural matrix of the policies refers to a distinctly operational approach; the original operativeness is reflected in the integrated policies for protection, planning, and management, and in the complementary actions of the project, such as the creation of databases, guidelines, and in the study of economic resources.

In Central-Mediterranean culture, the terminological root of *paese* refers to a land or town considered as a built-up area: this concept implies an action on the landscape to protect historical and cultural assets in Italy and France, and to preserve the physical, natural, and tangible resources in Spain, which are essential for the human survival. The principle of stock-taking assets, including new categories in time (from monuments to principal ecological-environmental values and systems), is still a characterizing element of the protective action, for which today's landscape is a parallel/superimposed element of functional, perceptive, and symbolic integration between the systems of assets and values [14] in the/of the territory.

## 2.2 Resilience: Features and Specificity

To understand better the features and specificity of landscape resilience, a focus on the broader concept of resilience has been carried out. From the literature, resilience emerges as an umbrella concept [3], where multiple sectoral definitions of resilience can co-habit.

Where "humans and nature are strongly coupled to the point that they should be conceived as one social-ecological system", resilience can be defined as "the capacity of a system … to deal with change and continue to develop" [95]. Given the "… ability of people, communities, societies, and cultures to live and develop with change and with ever-changing environments", this dynamic led to consider resilience itself as "a dynamic concept" [31, p. 7].

Based on this interpretation, Davoudi argued in favour of an evolutionary interpretation of resilience. The scholar defined it as "the ability of complex socio-ecological systems to change, adapt, and, crucially, transform in response to stresses and strains [22]" [28, p. 3]. This interpretation was related to considering the systems as "complex, non-linear, and self-organising, permeated by uncertainty and discontinuities" [13, p. 12]" (ibid.).

A co-evolutionary dimension of resilience is embraced by a more contemporary vision of territorial resilience [20] which includes, more general, considerations in terms of planning. Recently emerged particularly because of climate change, is used with double dimensions: psychological (as an attitude) and as multiple concrete actions applied to the territories at different scales to fight against environmental disasters, for example.

It possesses the following attributes:

- preserving diversity and redundancy [15],
- maintaining diversity and flexibility [88],
- managing slow variables and feedback [15],
- considering redundancy and multiple options [88],
- fostering complex adaptive systems thinking [15],
- promoting polycentric governance [15],
- is intrinsically multi-level and multi-scalar [31, 35],
- managing connectivity [88],
- broaden inclusivity, participation and equity [15, 88],
- encouraging adaptive learning [88],
- including improvement and innovation [2, 31],
- retaining identity while persisting and evolving with change [31].

## 2.3 Landscape Resilience: Features and Specificity

From the literature, landscape resilience pertains to, inherently possesses and combines with some characteristics of resilience, which are implicit and explicit once defined the boundaries of given landscape:

- it is co-evolutionary as the landscape is inherently dynamic [7, 11] and reflects the socio-economic dynamics of the communities [101],
- it takes into account the temporal differences that distinguish human and natural time,
- it is related to territorial resilience [20] as a method of building practices,
- it is associated with the concept of permanence/detection of a territorial system (e.g. land uses, patterns, internal and external connections and historical material heritage, such as buildings, artefacts, etc.) [54],
- it considers the locatedness and the elements of spatial heterogeneity and spatial relationships in the analysis of landscapes,
- it relates to identity and perception [18], considering the landscape as a living, experiential and inspirational phenomenon (encouraged by intangible heritage, such as language, local traditions and agricultural management practices and expertise),
- it supports the definition of boundaries and conditions between what one wants to keep and what one wants to modify or accepts to lose, but on the basis of choices and aspirations made by local communities,
- it refuses command-and-control methods in heritage management, as they can hamper the necessity of landscape as a living system to change and be flexible and elastic in order to allow and include potential modifications [43]. As a consequence, they make it "… more vulnerable in the long-run or even direct its evolution towards undesirable trajectories" (ibid., p.4) to the detriment of general resilience,
- it adopts proactive tools in landscape management practices to address pressures, threats, issues, and challenges to change and resulting weaknesses [12]. On the one hand, changes in identities, fragmentation of social cohesion, and alteration of the lifestyles of local communities, among others, risk adversely impacting the heritage values of the landscape from the social domain [76, 103]. On the other hand, fires, deforestation, invasive species, and mining activities are among those capable of compromising the ecological realm,
- it takes into account the dynamism of identity of the socio-ecological systems [31], which evolves over time together with the generations of people who live and work in that given region,
- it works on the transformative relationships between persistence and change [87], acting as a dynamic remembrance and, therefore, a source of memory [106],
- it embraces diversity [1] to have a living and multipurpose landscape: polyculture instead of monoculture, multiple and varied crops, rich in biodiversity [93], inclusive for newcomers [2], among others,
- it helps to achieve systemic robustness and ecosystem antifragility [29], providing multiple key co-benefits. These can include the well-being of the users of that landscape [109], also strengthening the local economy.

A scheme of the mentioned features and characteristics is reported in Fig. 1, while Fig. 2 highlights the necessity of landscape resilience to engage better the

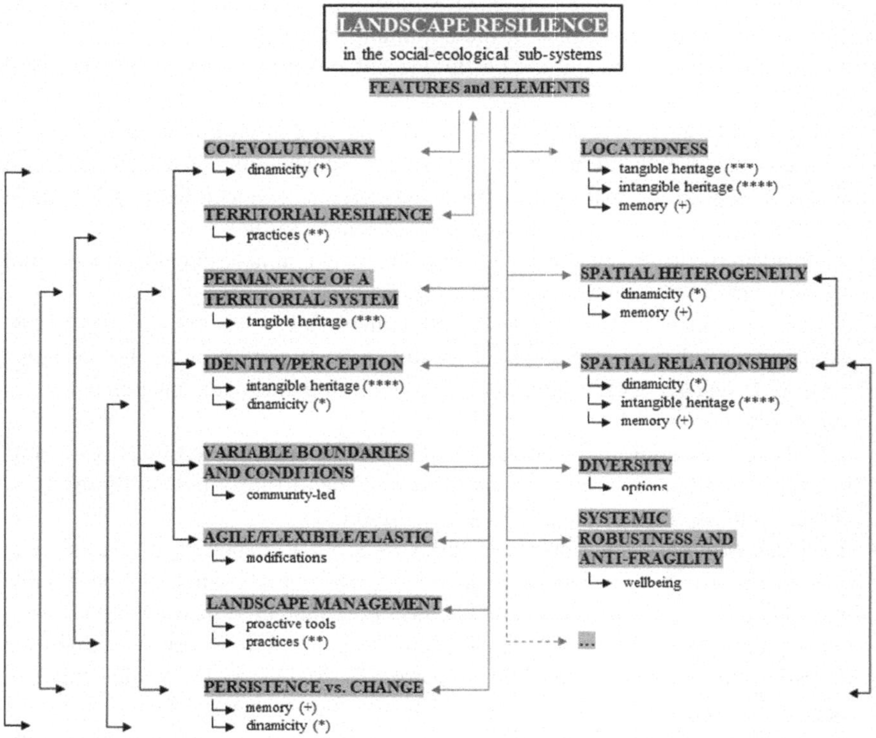


**Fig. 1** The features and elements associated with landscape resilience and their interrelationships. Authors' elaboration

social domain for an integrated and more fit-for-the-purpose management goals and approach.

## 3 Results

In a co-evolutive way, landscape resilience can be defined as primarily the capacity (ability and capability) to maintain the fundamental characteristics of the system in the face of disturbance, along with the ability and capability to reconfigure or recombine the landscape as a holistic living system made up of its interdependent elements, knowledge, and networks. It gives a new meaning to the principles of creative bricolage introduced by [45]. Different types of "… disturbances lead to creative reorganisation in complex social and ecological systems", as noted by [89, p. 16]. This condition of accepting the constant variations of the system produces a daily 'new normal' condition, whose incessant redefinition is nothing but the autopoiesis of this complex and informed system called landscape. Accepting the idea of a "progressive anarchy", this can "… acquires a positive connotation, guaranteeing evolution" [39, p. 28].

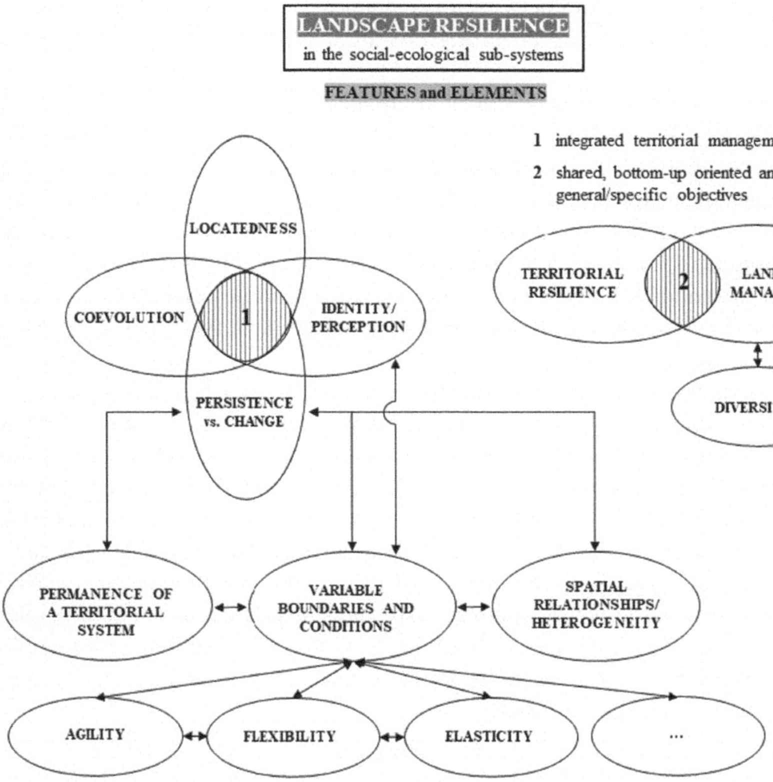


**Fig. 2** The features and elements associated with landscape resilience are shown using sets and their interrelationships. Overlapping areas highlight strategies that are interdependent and linked to the social domain. Authors' elaboration

Transferring what Maturana and Varela [58] theorised about the characteristics of an autopoietic unit to the landscape, regeneration of the landscape passes through continuous interacting macro- and micro- cycles of creation, transformation, and destruction. These are the typical principles of a resilient and evolutive system, as also stated by Gunderson and Holling in their theoretical concept of Panarchy (2002).

As Pievani stated in the 2014 TEDxCaFoscariU event [96], if redundancy in nature is fundamental to ensuring creativity, it is equally important to have highly intertwined relationships between the component nodes of such a macro-system which do not necessarily have to be many in number. Even in the definition of the autopoietic machine (or unit) provided by the evolutionary biologists Maturana and Varela, they described it "… as a network of processes of production (transformation and destruction) of components which: (i) through their interactions and transformations continuously regenerate and realise the network of processes (relations) that produced them" [58, pp. 78–79]. Therefore, it emerges that landscape resilience should refer to the network of dynamics, processes and behaviours in the landscape, the latter being understood as an active system based on systems theory.

The variation of these three nonlinear components leads to objective/subjective effects on the landscape and its image in the medium and long term. Consequently, it has to be an approach at multiple scales (i.e. transcalar) in landscape planning and its management, for instance. As a result, landscape resilience can be a major approach to understanding and guiding areal transformations at the spatial scale, as not programmed to solve one sole task. It will have the combined abilities to learn, prepare for, respond and adapt to weaknesses, vulnerabilities and disruptions, as well as its risks, hazards, fast threats and slow burns, including their synergistic interactions in the landscape.

It implies integrated territorial governance, as the landscape is a complex system requiring simultaneous monitoring and management of all its subsystems at several territorial and temporal scales by institutions, managing bodies (public and private) and stakeholders. As these subsystems are quite intertwined, a continuous informative feedback loop will adjust general and specific objectives, together with related actions, determining the landscape as an adaptive and responsive system. Therefore, in accordance with the insights of [10], actions and operational activities should be planned to work simultaneously and synergistically, where possible, on all dimensions of the landscape. These feedback make it clear that the principle of self-organisation is characteristic of complex systems [13] and that it, in turn, also forms the structure of autopoietic ones.

However, the above approach to landscape resilience, which analyses "… the spatial distribution of both biophysical and social vulnerabilities" (ibid., p. 478) and strives to offer adaptation to a multiplicity of possible options, could be seen as a rationalist attempt under the aegis of flexibility and elasticity. It is precisely for this reason that the contributions of creativity (i.e. creative bricolage) and serendipity (i.e. system imperfection as a buffer for a potential change and to create redundancy; [96]) were introduced into the definition as basic components that are useful to enable innovation and then evolution. Pievani's perspectives on design as a form of bricolage, as well as concepts like imperfection and serendipity, resonate well with the notions of "exaptation" [60]), as Pievani stated in previous research works [59]. They contribute to an integrated view of the existing discourse on exaptive strategies as a perspective for integrating adaptation and overcoming deterministic approaches to resilience [49]. Therefore, landscape resilience also accommodates and encourages creative bricolage, that is, the flexible, agile and recombination capacity of elements, expertise and resources that have usually characterised rural communities and their evolution over time [19, 51] (Fig. 3). For example, in the UNESCO site "Vineyard Landscape of Piedmont: Langhe-Roero and Monferrato" [104], willow branches and cane leaves were used to tie up vines, but also to make artefacts as baskets. The wood of local chestnut trees was used to make poles for the vines, as well as oak for the vats used for winemaking.

Nowadays, this recombination capacity is nothing more than the spontaneous consequence of the increase in the elements causing the entropy (or disorder) of

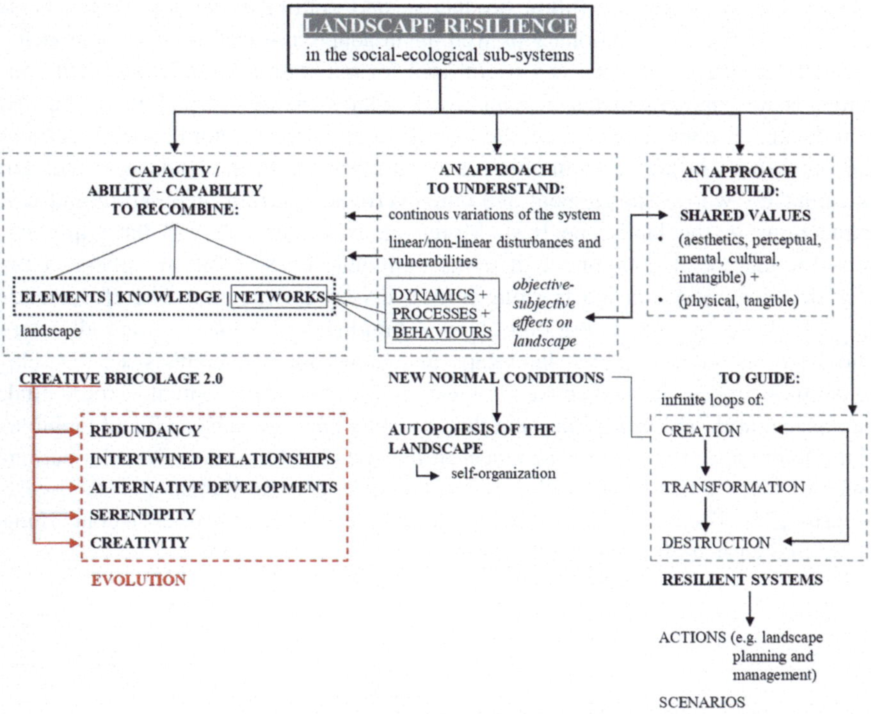


**Fig. 3** The theoretical concept of landscape resilience in the social-ecological subsystems. Authors' elaboration

contemporary systems. Including them in the discourse, as they are, because they embody the ever-changing 'new normal' that reconfigures the status quo, will offer new possibilities for creativity. Associated with their dynamics, they can consequently be considered to "… be an integral part of a hybrid landscape generated by variations in the … continuum …" [23], p. 273) that is spatio-temporal. In such a continuum, the prevalence is given by timescale as the only variable in the dipole with respect to a specific area of analysis and the landscape within it, which remains motionless. Among these potential new forms of landscape, it is possible to include the drosscapes and waste landscapes, their mentioned elements and dynamics may actually represent a potential for creativity due to their high chaotic imperfection.

From this perspective, scholars wishing to study landscape should therefore analyse these phenomena by embracing transdisciplinarity, going beyond the necessary convergence of purpose or mutual interest (i.e. interdisciplinarity) towards a truly holistic approach. In fact, as Gould stated, the sectoral division of fields in silos is nothing more than human reification [33].

The new connections between some of its polysemic meanings and the new-normal conditions dictated by endogenous/exogenous interferences (e.g. climate change) can help imagine future productive and cultural scenarios for the landscape, following an evolutionary as well as an adaptive-transformative approach.

Such a resilient approach is a useful tool for maintaining a redefined (but continuing) landscape in which to live and work at the same time, which therefore also includes the efficient use of a bundle of critical resources (natural, social, cultural and economic). In fact, communities are vital attributes of the landscape; they are its caretakers whose engagement and empowerment can reinforce natural-cultural connections in the landscape [68]. Moreover, this approach will have implications for the social component in terms of inclusiveness of its members in the local decisions and constant learning from everyday interactions. Therefore, landscape resilience intends to highlight how to build shared values through all actors who have an interest in that landscape and, consequently, emphasises the relationships between the subjective dimensions (aesthetic, perceptual, experiential, mental, cultural, and intangible) and the objective qualities (physical and tangible) of the landscape (Fig. 4). For instance, in the social-cultural domains, the permanence of local dialects and land uses can be considered an indicator of landscape integrity [53]. The latter is useful in determining its degree of resilience concerning a potential case study area of interest.

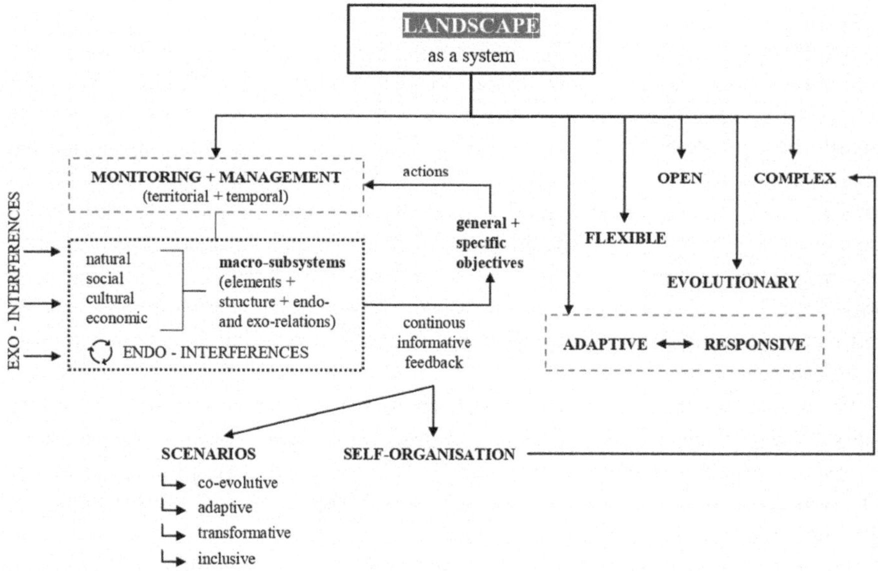


**Fig. 4** Graph illustrating the landscape as a resilient macro-system or a system of systems. Authors' elaboration

## 4 Discussion

Co-evolution is a process that is redefined every day, where the constant dialogue between the parts of the socio-ecological system is affected and perturbed by other things. Interspecies dialogue in the ecological system is just one such example, where a problem in one species can affect the others as well. The ecological view implies "… a vast, sprawling mesh of interconnections without a definite centre or edge. It is … coexistence with other beings, sentient or otherwise …" [64, p. 8]. Others may be the consideration and use of the term 'nature' in a normative way, linked to a concept of nature that is an artificial and human-made construct, as stated by the philosopher Timothy Morton [81]. Despite the above seems to indicate "'Nature' as a reified thing …" [64, p. 3], while, on the other hand, "ecology is profoundly about coexistence" (ibid., p. 4). In this regard, Morton encourages the combined use of the terms "symbiotic real" [81] to indicate a subtle yet dense interaction of all living forms in the ecosphere.

Moreover, the concern Stuart-Smith referred to with respect to the development of an aesthetics of effects [98] should open up the acceptance of results related to the search for an autopoiesis of landscape. It could be achievable through landscape resilience and might diverge from the current conception of aesthetics and its ethical values. Furthermore, he called for the combined acceptance of approaches that are both human-centred and non-human-centred, as "nothing is completely site-specific" [98] and more complex than the material and the visible [65]. This relates back to the concept of entropy (or disorder) of systems, as mentioned earlier in Sect. 3. In other words, the latter are elements that are part of the ecological domain and can be recombined into new possible landscape configurations, also accepting a playful combinatory way as part of the creative approach.

Such an approach seems influenced by the Capitalist and Communist materialist conceptions applied in the last century, from which one needs to try to emerge to incorporate a multiplicity of active and passive elements [42] that can be recombined in the landscape. The landscape is accepted as it is, without the pursuit or narrative of consolatory or neo-traditional aesthetics (e.g. as in landscape urbanism and landscape restoration). The latter, on the other hand, leads one to imagine a 'bounce back' and 'bounce back better' approach to resilience, in which mitigation and adaptation can coexist but at best as improvements to the status quo or business as usual models [3]. Consequently, it is necessary to move forward to overcome these conventional views.

Recombination should overcome the idea of an over-planned landscape to admit more randomness as a buffer of error (e.g. uncultivated plots, woods for ecological corridors, among others) and resilience because of it. Consequently, in order to limit and contrast the side effects of monoculture practices (such as increased fragility, reduced biodiversity, lack of flexibility and diversity, along with others), crop mix is a potential solution to contrast specialisation and rationalisation. Moreover, it will reinforce the idea of a landscape as a sum of mosaic patterns due to the different crops and associated features (e.g. seasonal colours associated with the natural crop cycles, detectable forms of the plots, et al.), rather than a monothematic and a less visually interesting panorama. Still about recombination and its networks, also the Peccioli Charter of the Resilient Communities [23] emphasises the idea that "... the connective components of the landscape of the agricultural plots, of the traditional infrastructural systems, of ecological networks are relevant" (ibid., p. 274).

Regarding the possible communities acting in such landscapes as part of the social component of the socio-ecological system, some explorable theories might point towards creating climate communities on a transnational basis. These could be designed for homogeneous climate zones, such as the seven climate regions identified by the European Commission [30, p. 3], reformulating the current borders of European states. Thus, they can be considered autopoietic macro-communities and sub-districts that reconfigure their boundaries as open systems. It would be explorable to better adapt and transform themselves under the pressures of multiple and changing conditions that are proper of driving forces of change, primarily climate change and the so-called VUCA (Volatile, Uncertain, Complex, and Ambiguous) environment.

Additionally, landscape resilience can be seen as an approach to provide more social and environmental justice and rights where applied, working on the ecological capital of a given landscape and its related geographical area. It ensures the reduction of the impacts of agriculture on the ecosystem, social inclusion, women enhancement, and empowerment of the individuals and groups of people who have an interest in it, among others, resulting in strengthened governance and enhanced ecological stewardship. This approach provides an opportunity to wider include bottom-up decisions and practices into the top-down institutional framework in order to reach a working landscape.

## 5 Conclusions

This article intends to provide some theoretical reflections helpful in building a definition of landscape resilience, avoiding sectorial, silos-thinking, and multidisciplinary definitions and approaches. From the literature, resilience can be considered with a dual approach, as one is inherent in the ecological domain of the ecosystem and its rules, while the second comes from the cultural interpretations of the landscape attributes through the different disciplines.

Therefore, a holistic definition of landscape resilience is lacking, often suffering from a non-exhaustive description, probably due to its various articulation in specific fields of interest. In this framework, the literature on social-ecological systems has provided valuable insights into the process of thoroughly understanding the landscape resilience approach across its many characteristics, properties and interrelationships.

The above lead to a definition of landscape resilience as the capacity in terms of ability or capability to reconfigure or recombine the landscape as a living and holistic system made up of its elements, knowledge, and networks. It recognises the ceaseless transformation of the landscape as a living macro-entity (e.g. in terms of land use and land cover), where dynamics, processes and behaviours influence the above-mentioned networks. In turn, they will influence and be influenced by the social component of the landscape in terms of subjective and objective effects on it. Consequently, this continuous interchange leads to the definition and recognition of shared and mutually combined values, creating live dialectics based on tangible and intangible. It can also be built on the phenomenological experience of the landscape, which can reinforce and shape the subjective understanding and awareness of the landscape in a given area through introjection and projection actions.

These, along with linear and nonlinear disturbances and vulnerabilities, and the ever-changing form of the system, establish the so-called 'new normal' conditions affecting the landscape. Actually, these conditions represent dual inputs. The first one is to feed the infinite loops of triad creation, transformation and destruction that characterise the resilient systems, predicted by thresholds and tipping points. Consequently, these will inform potential and unforeseen scenarios in the landscape, along with consequent strategies and actions.

Planning and management are the fields that can be more influenced by considering landscape resilience as an approach, with local and supra-local administrative and legislative implications and repercussions. Therefore, overcoming landscape vulnerabilities and risks requires attention in the Regional Regional Landscape Plans, as defined in Italy by the Cultural Heritage and Landscape Code. One key aspect involves informing and guiding the autopoiesis of landscape, where self-organization results from the constantly changing 'new normal' conditions. Adapting to climate change is so a central issue, especially for rural territories affected by droughts, hazards, and pressures, and needs to be addressed in a regulatory and design key in the Regional Landscape Plan through strategic planning, as well as in local planning, so that it can guide the project. Lastly, landscape resilience with respect to landscape planning and landscape design issues is the central theme for involving communities in the definition of shared plans and project actions.

What comes out of this process is a new understanding of the concept of creative bricolage, where redundancy, entangled relationships, alternative developments, serendipity, and creativity are among the pillars that sustain it. All of them, combined together, can be considered as a prompt for the evolution of the system towards new potential and limited predictable forms. Moreover, it offers a more nuanced definition of resilience applied to landscape design using a combination strategy of different disciplines to reach the definition of landscape resilience in the social-ecological domains.

In a nutshell, the landscape can be considered as a system of systems that is intrinsically open, flexible, evolutionary, complex, adaptive, and responsive. Its responsive measurement, monitoring and management are continuously adjusted concerning general and specific objectives. These goals will be determined by the alignment coming from natural, social, cultural and economic sectors (or resources) under endogenous and exogenous driving forces of change. Related actions will be based on the mentioned objectives to define an integrated management approach. These continuous informative feedback allow imagining new parallel and hypothetical scenarios in the mid to long term in addition to steering the self-organisation of the landscape as a macro-system. The latter will be made by elements, nestedness of ecological organization, hierarchical/networked structure, path dynamics and internal and external relations that refer to them, also considering the limits imposed by related constraints.

In conclusion, more attention should be paid to studies helpful in better understanding the ever-changing needs of the social subsystem and how they influence or even humper the comprehension of the landscape. Moving into the post-reality era is also affecting the landscape, where reality mixed with relativism and fake are producing new keys to reading and interpreting the landscape. However, despite the ethical concerns, they can create room for resilience and its related dynamics as their mix results in a more chaotic disorder that is a buffer for new creativity and interpretations. Moreover, how to define, measure and who will be responsible for driving system performance in terms of landscape monitoring and management go beyond the scope of this paper. However, it can be further explored by other scholars in coming studies in the framework of landscape resilience.

Lastly, what should be the role of landscape planning in the landscape resilience approach needs to be expanded to operationalise this research. In this working suggestion, questions such as if the current approach to urban and rural planning is appropriate in building landscape resilience [108] ought to be further debated and considered.

## Appendix

The investigation of the concept 'landscape resilience' in the subject areas of the social-ecological domains of the landscape (Authors' elaboration).

| | **Elsevier** (*) | **Taylor & Francis** (**) | **Emerald** (****) | **Cambridge University Press** (****) |
|---|---|---|---|---|
| 2022 | 6 | 3607 | 2493 | 97 |
| 2021 | 2121 | | | |
| 2020 | 2251 | | | |
| 2019 | 1956 | | | |
| 2018 | 1721 | | | |
| 2017 | 1408 | | | |
| 2016 | 1302 | | | |
| 2015 | 1034 | | | |
| 2014 | 838 | | | |
| 2013 | 792 | | | |

Launched: June 28, 2021 (*) (**) (***) (****) subject areas investigated:
(*): (1) Environmental Science; (2) Agricultural and Biological Sciences; (3) Social Sciences
(**): (1) Geography; (2) Social Sciences; (3) Environement and Sustainability; (4) Environment and Agriculture
(***): -
(****): (1) Ecology and conservation; (2) Evolutionary resilience; (3) Natural resources management, agriculture, horticulture and forestry; (4) Quantitative biology, biostatistics, and mathematical modelling

## References


1. Ahern J (2013) Urban landscape sustainability and resilience: the promise and challenges of integrating ecology with urban planning and design. Landscape Ecol 28:1203–1212. https://doi.org/10.1007/s10980-012-9799-z
2. Aimar F (2019) Landscape resilience and UNESCO cultural landscapes. The relation between resilience and the landscape identity in response to the anthropogenic variation of the systems. In: Shannon K, Quang Nguyen M (eds) Urbanism research across Europe: a Ph.D. seminar—2nd International European urbanisms seminar, KU Leuven, Leuven, pp 70–75. ISSN: 2684-0979
3. Aimar F, Jigyasu R (2022) Cultural landscapes of the Asia-Pacific region: dilemmas of resilience. In: Silva KD, Taylor K, Jones D S (eds) The Routledge handbook on cultural landscape heritage in the Asia-Pacific. Routledge, London, pp 403–419. https://doi.org/10.4324/9781003099994-30
4. Aimar F, Cavagnino F, Devecchi M (2022) Conservation and management of agricultural landscapes through expert-supported participatory processes: the "Declarations of Public Interest" in an Italian Province. Sustainability 14(14):8843. https://doi.org/10.3390/su14148843
5. Alomar-Garau G, Gómez-Zotano J, Arias-García J (2017) Teaching landscape in Spanish universities: looking for new approaches in geography. J Geogr High Educ 41(2):264–282. https://doi.org/10.1080/03098265.2017.1295226
6. Antongiovanni M, Venticinque EM, Tambosi LR, Matsumoto M, Metzger JP, Fonseca CR (2022) Restoration priorities for Caatinga dry forests: landscape resilience, connectivity and biodiversity value. J Appl Ecol 1-12. https://doi.org/10.1111/1365-2664.14131
7. Antrop M (2005) Why landscapes of the past are important for the future. Landsc Urban Plan 70(1–2):21–34. https://doi.org/10.1016/j.landurbplan.2003.10.002
8. Arnberger A, Eder R (2011) Exploring the heterogeneity of rural landscape preferences: an image-based latent class approach. Landsc Res 36(1):19–40. https://doi.org/10.1080/01426397.2010.536204
9. Arts B, Buizer M, Horlings L, Ingram V, van Oosten C, Opdam P (2017) Landscape approaches: a state-of-the-art review. Annu Rev Environ Resour 42:439–463. https://doi.org/10.1146/annurev-environ102016-060932
10. Bailey I, Buck LE (2016) Managing for resilience: a landscape framework for food and livelihood security and ecosystem services. Food Secur 8:477–490. https://doi.org/10.1007/s12571-016-0575-9
11. Beagan C, Dolan S (2015) Integrating components of resilient systems into cultural landscape management practices. Change Over Time 5(2):180–199. https://doi.org/10.1353/cot.2015.0015
12. Beller EE, Spotswood EN, Robinson AH, Anderson MG, Higgs ES, Hobbs RJ, Suding KN, Zavaleta ES, Grenier JL, Grossinger RM (2019) Building ecological resilience in highly modified landscapes. Bioscience 69(1):80–92. https://doi.org/10.1093/biosci/biy117
13. Berkes F, Folke C (1998) Linking social and ecological systems: management practices and social mechanisms for building resilience. Cambridge University Press, Cambridge. ISBN: 978-0521785624
14. Berque A (1995) Les raisons du paysage. De la Chine antique aux environnements de synthèse. Hazan, Paris, France. ISBN: 978-28-502-5390-4
15. Biggs R, Schlüter M, Schoon ML (2015) Principles for building resilience: sustaining ecosystem services in social-ecological systems. Cambridge University Press, Cambridge. ISBN: 978-1107082656
16. Bloemers T, Daniels S, Fairclough G, Pedroli B, Stiles R (2010) Landscape in a changing world; bridging divides, integrating disciplines, serving society. Sci Policy Brief 41:1–16. https://edepot.wur.nl/161371

17. Bonesio L (2007) Paesaggio identità e comunità tra locale e globale. Edizioni Diabasis, Reggio Emilia, Italy. ISBN: 978-88-810-3491-8
18. Bousquet F, Quinn T, Therville C, Mathevet R, Barreteau O, Bonté B, Guerbois C (2015) Social and ecological systems resilience and identity. In: Ungar M (ed) Multisystemic resilience: adaptation and transformation in contexts of change. Oxford University Press, Oxford, pp 705–724. https://doi.org/10.1093/oso/9780190095888.003.0037
19. Bravo GL (2005) La complessità della tradizione. Festa, museo e ricerca antropologica. FrancoAngeli, Milan. ISBN: 978-8-846-47153-6
20. Brunetta G, Ceravolo R, Barbieri CA, Borghini A, de Carlo F, Mela A, Beltramo S, Longhi A, De Lucia G, Ferraris S, Pezzoli A, Quagliolo C, Salata S, Voghera A (2019) Territorial resilience: toward a proactive meaning for spatial planning. Sustainability 11(8):2286. https://doi.org/10.3390/su11082286
21. Cambridge Dictionary (n.d.) Landscape. https://dictionary.cambridge.org/us/dictionary/english/landscape
22. Carpenter SR, Westley F, Turner G (2005) Surrogates for resilience of social–ecological systems. Ecosystems 8(8):941–944. https://doi.org/10.1007/s10021-005-0170-y
23. Carta M, Lara-Hernandez JA, Accossato K, Baggio M, Boarin P, Bravo L, Brisotto C, D'Acci L, Melis A, Paoletti I, Perbellini M R, Perrotti D, Trentin L (2022) The Peccioli charter, the new Constitution of the Nation of the Italian resilient communities. In: Carta M, Perbellini MR, Lara-Hernandez JA (eds) Resilient communities and the Peccioli charter. Towards the possibility of an Italian Charter for resilient communities. Springer, Cham, pp 269–275. https://doi.org/10.1007/978-3-030-85847-6_29
24. Chacón-Moreno E, Olivares I, Navarro G, Albarrán AJ, Paredes Y, Aranguren CI, Nagy GJ (2020) Landscape ecology and conservation for building resilience and adaptation to global change in Venezuela. In: Leal Filho W, Nagy G, Borga M, Chávez Muñoz P, Magnuszewski A (eds) Climate change, hazards and adaptation options. Climate change management. Springer, Cham, pp 147–160. https://doi.org/10.1007/978-3-030-37425-9_7
25. Collins Dictionary (n.d.) Landscape. https://www.collinsdictionary.com/us/dictionary/english/landscape
26. Council of Europe (CoE) (2000) European landscape convention. Florence 20.10.2000. European Treaty Series, No. 176. https://rm.coe.int/1680080621
28. Davoudi S (2012) Resilience: a bridging concept or a dead end? Plan Theory Pract 13(2):299–333. https://doi.org/10.1080/14649357.2012.677124
29. Equihua M, Espinosa Aldama M, Gershenson C, López-Corona O, Munguía M, Pérez-Maqueo O, Ramírez-Carrillo E (2020) Ecosystem antifragility: beyond integrity and resilience. PeerJ 8:e8533. https://doi.org/10.7717/peerj.8533
30. European Commission (2018) Document 52018DC0773. Communication from the Commission to the European Parliament, the European Council, the Council, the European economic and social Committee, the Committee of the regions and the European Investment Bank. A clean planet for all a European strategic long-term vision for a prosperous, modern, competitive and climate neutral economy. COM/2018/773 final. Brussels. https://eur-lex.europa.eu/legal-content/EN/ALL/?uri=CELEX:52018DC0773. Accessed 12 June 2022
31. Folke C (2016) Resilience (Republished). Ecol Soc 21(4):44. https://doi.org/10.5751/ES-09088-210444
32. Forman RTT, Godron M (1986) Landscape ecology. Wiley & Sons, New York, USA. ISBN: 978-0-471-87037-1
33. Gould SJ (1981) The mismeasure of man. W. W. Norton & Company, New York, USA. ISBN: 978-0-393-31425-0
34. Guarino R, Cutaia F, Giacopelli AL, Menegoni P, Pelagallo F, Trotta C, Trombino G (2015) Disintegration of Italian rural landscapes to international environmental agreements. Springer, Dordrecht. https://doi.org/10.1007/s10784-015-9310-9
35. Gunderson LH, Holling CS (2002) Panarchy: understanding transformations in human and natural systems. Island Press, Washington, USA. ISBN: 978-1559638579

36. Haug A, Müller J (2018) From theories, concepts and practices on human development in landscapes. In: Haug A, Käppel L, Müller J (eds) Past landscapes. The dynamics of interaction between society, landscape, and culture, Sidestone Press, Leiden, pp 13–15. https://doi.org/10.59641/h7l684aj
37. Holling CS (1973) Resilience and stability of ecological systems. Ann Rev Ecol Syst 4:1–23. https://www.jstor.org/stable/2096802
38. Huang S, van den Bosch CK, Fu W, Qi J, Chen Z, Zhu Z, Dong J (2018) Does adding local tree elements into dwellings enhance individuals' homesickness? Scenario-visualisation for developing sustainable rural landscapes. Sustainability 10(11):3943. https://doi.org/10.3390/su10113943
39. Ingallina P (1996) Du paysage à la ville. Ligeia 19–20(1). https://doi.org/10.3917/lige.019.0027
40. Isager L, Broge NH (2007) Combining remote sensing and anthropology to trace historical land-use changes and facilitate better landscape management in a sub-watershed in North Thailand. Landsc Res 32(2):147–169. https://doi.org/10.1080/01426390701231515
41. Josefsson J, Pärt T, Berg Å, Lokhorst AM, Eggers S (2018) Landscape context and farm uptake limit effects of bird conservation in the Swedish volunteer & farmer alliance. J Appl Ecol 55(6):2719–2730. https://doi.org/10.1111/1365-2664.13184
42. Latour B (2017) Facing Gaia: eight lectures on the new climatic regime. Polity, Cambridge, UK. ISBN: 978-0745684345
43. Leitaõ L (2018) Applying resilience thinking to management of cultural and natural heritage. In: Mitchell N, St. Clair A, Brown J, Barrett B, Rodríguez A (eds) Proceedings of the 2018 US/ICOMOS symposium—forward together: a culture-nature journey towards more effective conservation in a changing world. ICOMOS, San Francisco, California, pp 70–75. https://usicomos.org/symposium/symposium-2018-proceedings/
44. Lennon M, Scott M, Collier M, Foley K (2017) The emergence of green infrastructure as promoting the centralisation of a landscape perspective in spatial planning—the case of Ireland. Landsc Res 42(2):146–163. https://doi.org/10.1080/01426397.2016.1229460
45. Lévi-Strauss C (1962) La Pensée sauvage. Librairie Plon, Paris. ISBN: 978-2-869-17001-8
46. Li Y (2022) A systematic review of rural resilience. China Agric Econ Rev. https://doi.org/10.1108/CAER-03-2022-0048
47. Liu W (2019) Operationalizing polycentricity for landscape resilience. Landsc Archit Front 7(3):8–11. https://doi.org/10.15302/J-LAF-1-010002
48. Lovell ST, Taylor JR (2013) Supplying urban ecosystem services through multifunctional green infrastructure in the United States. Landscape Ecol 28:1447–1463. https://doi.org/10.1007/s10980-013-9912-y
49. Gould S, Vrba E (1982) Exaptation—a missing term in the science of form. Paleobiology 8(1):4–15. https://doi.org/10.1017/S0094837300004310
50. Grass I, Loos J, Baensch S, Batáry P, Librán-Embid F, Ficiciyan A, Klaus F, Riechers M, Rosa J, Tiede J, Udy K, Westphal C, Wurz A, Tscharntke T (2019) Land-sharing/-sparing connectivity landscapes for ecosystem services and biodiversity conservation. People Nat 1(2):262–272. https://doi.org/10.1002/pan3.21
51. Grimaldi P (2008) Parlandone da vivo: per una storia degli studi delle tradizioni popolari. Omega Edizioni, Turin. ISBN: 978-8872414972
52. Grimm NB, Pickett STA, Hale RL, Cadenasso ML (2017) Does the ecological concept of disturbance have utility in urban social–ecological–technological systems? Ecosyst Health Sustain 3(1):e01255. https://doi.org/10.1002/ehs2.1255
53. Gullino P, Larcher F (2013) Integrity in UNESCO world heritage sites. A comparative study for rural landscapes. J Cult Heritage 14(5):389–395. https://doi.org/10.1016/j.culher.2012.10.005
54. Marino D, Cavallo A, Di Donato B, Nofroni L, Savelli S (2015) Between resilience and sense of place: understanding the agricultural traditional landscape. In: Proceedings of the UNISCAPE en-route international seminar, vol 1, no 2, pp 66–71. https://www.uniscape.eu/wp-content/uploads/2017/03/UNISCAPE-En-Route-n.pdf

55. Markolf SA, Chester MV, Eisenberg DA, Iwaniec DM, Davidson CI, Zimmerman R, Miller TR, Ruddell BL, Chang H (2018) Interdependent infrastructure as linked social, ecological, and technological systems (SETSs) to address lock-in and enhance resilience. Earth's Future 6(12):1638–1659. https://doi.org/10.1029/2018EF000926
56. Martinez-Diaz L (2018) Investing in resilience today to prepare for tomorrow's climate change. Bull Atom Sci 74(2):66–72. https://doi.org/10.1080/00963402.2018.1436805
57. Martínez-Graña A, Goy J, Zazo C, Silva P, Santos-Francés F (2017) Configuration and evolution of the landscape from the geomorphological map in the Natural Parks Batuecas-Quilamas (Central System, SW Salamanca, Spain). Sustainability 9(8):1458. https://doi.org/10.3390/su9081458
58. Maturana HR, Varela FJ (1980) On machines, living and otherwise. In: Maturana HR, Varela FJ (eds) Autopoiesis and cognition. The realization of the living. Springer, Dordrecht, pp 77–84. https://doi.org/10.1007/978-94-009-8947-4_9
59. Melis A, Pievani T (2022) Exaptation as a design strategy for resilient communities. In: Rezaei N (ed) Transdisciplinary. Springer, Cham, pp 307–327. https://doi.org/10.1007/978-3-030-94651-7_15
60. Melis A, Pievani T, Lara-Hernandez JA (2024/forthcoming) Architectural exaptation. When function follows form. Routledge, Abingdon, UK. ISBN: 978-1-032-35803-1
61. Merriam-Webster (n.d.) Landscape. In Merriam-Webster.com dictionary. https://www.merriam-webster.com/dictionary/landscape
62. Mitchell M (1998) Landscapes and literature: a look at the early twentieth century rural south. J Geogr 97:204–212. https://doi.org/10.1080/00221349808978899
63. Moritz MA, Hessburg PF, Povak NA (2011) Native fire regimes and landscape resilience. In: McKenzie D, Miller C, Falk D (eds) The landscape ecology of fire. Ecological studies. Springer, Dordrecht, pp 51–86. https://doi.org/10.1007/978-94-007-0301-8_3
64. Morton T (2010) The ecological thought. Harvard University Press, Cambridge, USA. ISBN: 978-0-674-04920-8
65. Morton T (2016) Dark ecology: for a logic of future coexistence. Columbia University Press, New York, USA. ISBN: 9780231177528
66. Musacchio LR (2013) Cultivating deep care: integrating landscape ecological research into the cultural dimension of ecosystem services. Landscape Ecol 28:1025–1038. https://doi.org/10.1007/s10980-013-9907-8
67. National Park Service (NPS) (1998) A guide to cultural landscape reports: contents, process, and techniques. https://www.nps.gov/parkhistory/online_books/nps/cl_reports.pdf
68. National Park Service (NPS) (2015) Defining cultural landscapes. https://www.nps.gov/subjects/culturallandscapes/understand-cl.htm
69. National Park Service (NPS) (2023) Cultural landscapes 101. https://www.nps.gov/articles/cultural-landscapes-101.htm
70. North MP, York RA, Collins BM, Hurteau MD, Jones GM, Knapp EE, Kobziar L, McCann H, Meyer MD, Stephens SL, Tompkins RE, Tubbesing CL (2021) Pyrosilviculture needed for landscape resilience of dry western United States forests. J Forest 119(5):520–544. https://doi.org/10.1093/jofore/fvab026
71. Osti L, Cicero L (2018) Tourists' perception of landscape attributes in rural tourism. Worldwide Hosp Tour Themes 10(2):211–221. https://doi.org/10.1108/WHATT-12-2017-0087
72. Oxford English Dictionary (2023) Landscape. meaning & use. https://www.oed.com/dictionary/landscape_n?tab=meaning_and_use#39684111
73. Palazzo E, Bardsley DK (2021) Adaptive mechanisms in a continuing landscape: assessing biocultural diversity as a form of resilience. J Cult Heritage Manage Sustain Dev. https://doi.org/10.1108/JCHMSD-09-2020-0139
74. Pesci E (2004) La terra parlante. Dai paesaggi originari ai non-luoghi alpestri. CDA&Vivalda, Turin, Italy. ISBN: 978-88-748-0030-8
75. Pinheiro HSK, dos Anjos LHC, Xavier PAM, Chagas CS, de Carvalho JW (2018) Quantitative pedology to evaluate a soil profile collection from the Brazilian semi-arid region. South Afr J Plant Soil 35(4):269–279. https://doi.org/10.1080/02571862.2017.1419385

76. Plieninger T, Bieling C (2012) Connecting cultural landscapes to resilience. In: Plieninger T, Bieling C (eds) Resilience and cultural landscape. Cambridge University Press, Cambridge, pp 3–26. https://doi.org/10.1017/CBO9781139107778.003
77. Plieninger T, van der Horst D, Schleyer C, Bieling C (2014) Sustaining ecosystem services in cultural landscapes. Ecol Soc 19(2):59. https://doi.org/10.5751/ES-06159-190259
78. Primdahl J (2014) Agricultural landscape sustainability under pressure: policy developments and landscape change. Landsc Res 39(2):123–140. https://doi.org/10.1080/01426397.2014.891726
79. Raaphorst K, Duchhart I, van der Knaap W, Roeleveld G, van den Brink A (2017) The semiotics of landscape design communication: towards a critical visual research approach in landscape architecture. Landsc Res 42(1):120–133. https://doi.org/10.1080/01426397.2016.1257706
80. Raatz L, Bacchi N, Pirhofer-Walzl K, Glemnitz M, Müller MEH, Joshi J, Scherber C (2019) How much do we really lose? Yield losses in the proximity of natural landscape elements in agricultural landscapes. Ecol Evol 9(13):7838–7848. https://doi.org/10.1002/ece3.5370
81. Radboud Universiteit [Radboud Reflects] (2018) Being ecological I Lecture by philosopher Timothy Morton [Video]. YouTube. https://www.youtube.com/watch?v=Yv4W4M8Z8VQ
82. Relph E (2018) The inconspicuous familiarity of landscape. In: Champion EM (ed) The phenomenology of real and virtual places. Routledge, New York, USA, pp 17–26. https://doi.org/10.4324/9781315106267-2
83. Repetto D, Aimar F (2021) The fifth landscape: art in the contemporary landscape. In: Bianconi F, Filippucci M (eds) Digital draw connections. Springer, Cham, pp 683–706. https://doi.org/10.1007/978-3-030-59743-6_32
84. Resilient Communities_Venezia (2021) Video [Facebook page]. Facebook. https://www.facebook.com/ItalianPavilion2021/videos/?ref=page_internal Accessed August 4, 2022
85. Risser PG (1987) Landscape ecology: state of the art. In: Turner MG (ed) Landscape heterogeneity and disturbance. Springer-Verlag, New York, pp 3–14. https://doi.org/10.1007/978-1-4612-4742-5
86. Rizzo D, Marraccini E, Lardon S, Rapey H, Debolini M, Benoît M, Thenail C (2013) Farming systems designing landscapes: land management units at the interface between agronomy and geography. Geografisk Tidsskrift-Danish J Geogr 113(2):71–86. https://doi.org/10.1080/00167223.2013.849391
87. Robert S (2021) Resilience: persistence and change in landscape forms. Wiley, London, ISTE and New York. ISBN: 978-1-78630-666-1
88. Rockström J, Norström AV, Matthews N, Biggs R, Folke C, Harikishun A, Huq S, Krishnan N, Warszawski L, Nel D (2023) Shaping a resilient future in the response to COVID-19. Nat Sustain 6:897–907. https://doi.org/10.1038/s41893-023-01105-9
89. Rogers P (2016) Etymology and genealogy of a contested concept. In: Chandler D, Coaffee J (eds) The Routledge handbook of international resilience. Routledge, Abingdon, UK, pp 13–25. ISBN: 9780367462024
90. Rotchés-Ribalta R, Winsa M, Roberts SPM, Öckinger E (2018) Associations between plant and pollinator communities under grassland restoration respond mainly to landscape connectivity. J Appl Ecol 55(6):2822–2833. https://doi.org/10.1111/1365-2664.13232
91. Schmidt C (2022) Landscape resilience. Basics Case studies, practical recommendations. Springer-Verlag, Berlin. https://doi.org/10.1007/978-3-662-63998-6
92. Sereni E (1961) Storia del Paesaggio Agrario Italiano. Laterza, Bari, Italy. ISBN: 9788842020943
93. Sgrò CM, Lowe AJ, Hoffmann AA (2011) Building evolutionary resilience for conserving biodiversity under climate change. Evol Appl 4:326–337. https://doi.org/10.1111/j.1752-4571.2010.00157.x
94. Stedman RC (2003) Is it really just a social construction? The contribution of the physical environment to sense of place. Soc Nat Resour 16(8):671–685. https://doi.org/10.1080/08941920309189

95. Stockholm Resilience Center (2015) What is resilience? https://www.stockholmresilience.org/research/research-news/2015-02-19-what-is-resilience.html
96. TEDx Talks (2014) L'evoluzione è bricolage: il segreto della creatività | Telmo Pievani | TEDxCaFoscariU [Video]. YouTube. https://www.youtube.com/watch?v=7dgzWGSRVy8
97. Tekken V, Spangenberg JH, Burkhard B, Escalada M, Stoll-Kleemann S, Truong DT, Settele J (2017) "Things are different now": farmer perceptions of cultural ecosystem services of traditional rice landscapes in Vietnam and the Philippines. Ecosyst Serv 25:153–166. https://doi.org/10.1016/j.ecoser.2017.04.010
98. Texas A&M School of Architecture (2018) Deep Vista: ecologies with Levi Bryant [Video]. YouTube. https://www.youtube.com/watch?v=d6gcRg6z7Ms
99. Thompson ID (2015) An overview of the science–policy interface among climate change, biodiversity, and terrestrial land use for production landscapes. J For Res 20(5):423–429. https://doi.org/10.1007/s10310-015-0497-y
100. Tischendorf L, Fahrig L (2000) On the usage and measurement of landscape connectivity. Oikos 90:7–19. https://doi.org/10.1034/j.1600-0706.2000.900102.x
101. Turri E (2001) Il paesaggio come teatro. Marsilio, Venice, Italy. ISBN: 978-88-317-6865-8
102. United Nations General Assembly (UN General Assembly) (2015) Transforming our world: the 2030 Agenda for sustainable development. Report A/RES/70/1, United Nation, New York, NY. https://www.un.org/en/development/desa/population/migration/generalassembly/docs/globalcompact/A_RES_70_1_E.pdf
103. United Nations Educational, Scientific and Cultural Organization (UNESCO) (2008) List of factors affecting the properties. https://whc.unesco.org/en/factors/
104. United Nations Educational, Scientific and Cultural Organization (UNESCO) (2014) Vineyard landscape of Piedmont: Langhe-Roero and Monferrato. https://whc.unesco.org/en/list/1390/
105. van Diggelen R, Hobbs RJ, Miko L (2012) Landscape ecology. In: van Andel J, Aronson J (eds) Restoration ecology: the new frontier. Wiley-Blackwell, Hoboken, pp 45–58. https://doi.org/10.1002/9781118223130.ch5
106. Van Dyke RM (2008) Memory, place, and the memorialization of landscape. In: David B, Thomas J (eds) Handbook of landscape archaeology. Routledge, Abingdon, UK, pp 277–284. https://doi.org/10.4324/9781315427737-31
107. van Oudenhoven FJW, Mijatović D, Eyzaguirre PB (2011) Social-ecological indicators of resilience in agrarian and natural landscapes. Manag Environ Qual 22(2):154–173. https://doi.org/10.1108/14777831111113356
108. Voghera A, Aimar F (2022) Towards a definition of landscape resilience: the proactive role of communities in reinforcing the intrinsic resilience of landscapes. In: Carta M, Perbellini M, Lara-Hernandez JA (eds) Resilient communities and the Peccioli charter: towards the possibility of an Italian charter for resilient communities. Springer, Cham, pp 111–120. https://doi.org/10.1007/978-3-030-85847-6_13
109. Wu J (2021) Landscape sustainability science (II): core questions and key approaches. Landscape Ecol 36:2453–2485. https://doi.org/10.1007/s10980-021-01245-3
110. Xu H, Peng M, Pittock J, Xu J (2021) Managing rather than avoiding "difficulties" in building landscape resilience. Sustainability 13(5):2629. https://doi.org/10.3390/su13052629
111. Zanzi L (2004) Pensare la wilderness, orizzonte selvaggio dell'anima, Fondazione Enrico Monti, Anzola D'Ossola, Italy. ISBN: 978-88-852-9524-7
112. Zuniga-Teran A, Gerlak AK, Mayer B, Evans TP, Lansey KE (2020) Urban resilience and green infrastructure systems: towards a multidimensional evaluation. Curr Opin Environ Sustain 44:42–47. https://doi.org/10.1016/j.cosust.2020.05.001

# 18 Exaptation in Transport Facilities: A Cultural Ecological Perspective on Motor Parks as Public Spaces in Yorubaland, Nigeria

Temitope M. Adebara and Jose Antonio Lara-Hernandez

**Abstract**

Planning and designing for a dynamic and adaptable built environment remain a perennial challenge in this era of global crises. As globalisation continues to shape the world, urban development and redevelopment initiatives in developing countries often prioritise attracting investments and transforming urban areas into megacities. However, amidst this wave of urban transformation, there is a concerning trend of excluding and displacing informal actors who rely on urban spaces for their daily livelihoods. This displacement is particularly common in and around road-related public spaces like transport stations. The development of the transport infrastructure typically falls within the purview of transport engineering, emphasising technical parameters and standardisation. Informal behaviours are often stigmatised as a misuse of urban facilities and a threat to the modernist vision of aesthetically pleasing urban landscape. Nevertheless, some motor parks (bus stations) in Nigeria defy this prevailing trend, accommodating a diverse array of users and activities. Drawing upon a cultural ecological perspective, this chapter showcases how motor parks are used as public spaces for different activities other than the primary intent of facilitating efficient mobility in Yorubaland (Southwestern Nigeria). Within the Yoruba parks, a spectrum of activities emerges, spanning commercial, religious, social and leisure domains. The manner in which the Yoruba people repurpose and adapt these parks for various activities is a reflection of their inherent socio-spatial needs, deeply rooted in long-standing cultural practices handed down through generations. The findings in this chapter hold several key implications

T. M. Adebara (✉)
Department of Urban and Regional Planning, Olabisi Onabanjo University, Ogun State, Nigeria
e-mail: adebaratemitope@gmail.com

J. A. Lara-Hernandez
Auckland University of Technology, Auckland, New Zealand
e-mail: antonio.lara-hernandez@aut.ac.nz

 
J. A. Lara-Hernandez et al. (eds.), *The Dusk of Design*, Integrated Science 34,
https://doi.org/10.1007/978-3-031-76867-5_18

for urban planning practices, particularly in the context of fostering community resilience through design approaches that embrace adaptability and variability in the urban environment.

**Keywords**

Public space • Mixed-use development • Community resilience • Cultural ecological theory • Urban informality

## 1 Introduction

In an era marked by accelerated urbanisation and globalisation, cities in developing countries are undergoing rapid physical expansion to accommodate growing populations and infrastructural needs. Amidst this expansion, contemporary physical developments often overlook the dynamic and adaptive behvaiours of people for aesthetic and functional reasons, limiting opportunities for community engagement and vitality [2, 14]. For instance, transportation infrastructure development is typically conceived within the domain of transport engineering, focusing on technical parameters and standardisation. However, this approach often results in predetermined design solutions that may be alien to the specific urban context where the infrastructure is situated [9, 12, 26]. Consequently, the critical roles of transport facilities, including motor parks (bus stations), as urban places are usually neglected, with priority given to motor vehicles. Nevertheless, in recent decades, there has been a growing interest in this topic across different fields of study, including urban design, planning and geography. From an urban design perspective, transport facilities such as motor parks (bus stations) are urban places for various human activities beyond facilitating efficient mobility. Moreover, fostering vibrant public life in transport stations across an urban area can bolster social cohesion and enhance community well-being, thereby ensuring a more sustainable and resilient environment in an era of global crises [26]. Yet, planning for adaptable and dynamic built environments remains a perennial challenge in rapidly changing cities [22].

Evident in the urban development and redevelopment initiatives in developing countries such as Nigeria, are planning strategies largely focus on attracting investments and transforming urban areas into megacities in this era of globalisation [23]. However, amidst this wave of urban transformation, there is a concerning trend of excluding and displacing traders and other informal actors who rely on urban spaces in the cities for their daily livelihoods [3, 4]. This displacement is particularly common in and around road-related spaces public spaces such as transport stations. The development of the transport infrastructure typically falls within the purview of transport engineering, emphasising technical parameters and standardisation. Informal behaviours are often stigmatised as a misuse of urban facilities and a threat to the modernist vision of aesthetically pleasing urban landscape. They are also perceived as posing a significant challenge to urban land-use zoning, resulting in traffic congestion and pollution. Nevertheless, compelling arguments support the presence of informal actors in public transport stations. Numerous urban theorists argue that individuals have a fundamental right to the city, and excessive management and regulation of urban space can lead to social exclusion and injustice, particularly in the global South, where a large portion of the urban population lives below the line of abject poverty [1].

Millions of individuals across Africa, Asia, and Latin America rely on any available spaces in cities for their commercial endeavours, a necessity driven by dire economic circumstances that perpetuate poverty among the disadvantaged while benefiting the affluent [10]. Contemporary management of public spaces in global South nations is often criticised as oppressive and undemocratic, encroaching upon people's rights—a phenomenon [8] labelled as over-management of space. Moreover, the increasing privatisation of public spaces and adoption of neoliberal practices compound the challenges faced by informal actors, further limiting their access to open spaces. Despite the prevailing trends, some transport stations in Nigeria stand out by defying popular norms and continuing to accommodate a diverse array of users and activities. These stations, predominantly situated in Yoruba cities such as Ile-Ife, Oyo, Osogbo, Ibadan, Ogbomosho, Ado-Ekiti, Abeokuta, Ilorin, and Ijebu-Ode, serve as hubs for various activities, fostering harmonious coexistence among different users. A thorough exploration of how these transport stations facilitate diverse social behaviours in Yorubaland can provide valuable insights into leveraging design variability to create dynamic, flexible, and inclusive built environments capable of seamlessly adapting to evolving needs and changing contexts over time. Such insights could offer valuable guidance for planners and designers, illuminating effective strategies for navigating and planning amidst the dynamic complexity of urban life. Thus, this chapter delves into the use of motor parks as public spaces for different activities beyond facilitating efficient mobility in Ile-Ife, the ancestral homeland of the Yoruba people in Nigeria.

## 2 Traditional Concepts and Meanings of Public Space in Yorubaland

Yorubaland, located in Southwestern Nigeria is renowned for its rich cultural practices and traditions which are passed down from generation to generation. The importance of public space has long been recognised in Yorubaland long before the advent of Western landscape practices introduced by British colonialists [4, 23]. At the core of Yoruba culture lies a profound emphasis on social bonds and communal solidarity, underscoring the need for public spaces to serve as platforms for engaging diverse community members in sociocultural events, regardless of their societal status [6]. In traditional societies, there exists a fluidity between private and public realms. Public spaces usually emerge between buildings in an organic and informal manner, without formal planning [4]. The status of spaces as public or private within and around buildings may shift dynamically based on the activities occurring within them at different period of time, reflecting the flexible and adaptive nature of Yoruba public space. For instance, areas designated for open-air marketing in the morning may seamlessly transition into venues for social gatherings and leisure activities later in the day. This dynamic contrasts sharply with the model of modern public spaces, which often adhere to a rigid one-function-per-space framework.

Social spaces can manifest anywhere within the built environment, including streets, marketplaces, and interstitial courtyards between buildings in Yoruba communities. Such spaces embody a communal sphere where individuals engage in a myriad of activities that reflect their cultural norms and traditions. The local meanings of public space encompass its role as a bustling commercial hub, a vibrant social arena, and a sacred site integral to religious practices [2]. For the Yoruba people, the utilisation of any available space around buildings for commerce is deeply ingrained in their societal fabric, occurring at various levels ranging from individual households to the entire cityscape. Beyond commerce, Yoruba public space serves as focal points for community life, where communal practices thrive and social bonds are nurtured. Sociability within Yoruba society hinges largely on the bustling commercial activities that animate public spaces. Private entrepreneurs, notably traders, play a pivotal role in the vitality of Yoruba public space. Business endeavours attract people, fostering shared experiences and conversations that reinforce social ties, particularly during bustling market days.

Additionally, religion holds significant sway in shaping the Yoruba social landscape, diverging from Western paradigms where secularism often delineates the public sphere from religious realms. Yoruba public spaces feature sacred groves and revered trees like *Dracenia Frangrans*, *Newboldia Laevis*, and *Adansonia Digitata*, serving as sanctified sites for ritualistic practices. Yoruba public spaces transcend mere physicality, serving as conduits for spiritual communion with ancestral spirits and active participation in ritualistic ceremonies. In essence, the value of public space in Yorubaland lies not in its physical design, but rather in the dynamic activities that unfold within its bounds. The transient nature of public space imbues it with a fluidity that allows its meaning to evolve continuously, adapting to the ever-changing socio-cultural landscape. Thus, the open spaces in Yorubaland possess a distinctive quality in that they are shared and repurposed at various times and for diverse functions, reflecting the communal ethos and convivial lifestyle intrinsic to Yoruba culture. Through collaborative negotiation and mutual cooperation, individuals secure space for their daily routines.

## 3 Public Space Exaptation as a Reflection of Cultural Ecology: A Theoretical Perspective

Drawing upon cultural ecological perspective, this chapter delves into the concept of public space exaptation. Cultural ecological theory, as propounded by Steward [25], diverges from deterministic views of environmental influence by emphasising the adaptive capacities of human societies. It posits that human cultures dynamically interact with their environments, adapting and evolving in response to changing conditions across temporal and spatial scales. At its core, cultural ecology recognises the inherent dynamism of both culture and environment, acknowledging their mutual influence and the ongoing processes of adaptation and change. Steward's framework suggests that cultures facing similar environmental challenges may develop analogous adaptations, which are inherently transient and subject to continuous evolution. Moreover, cultural changes can either build upon existing traits or give rise to entirely new ones, illustrating the adaptive nature of human societies.

Culture serves as the primary adaptive mechanism through which communities navigate their environments. It encompasses various elements such as religion, politics, social structures, settlement patterns, and technology, which evolve in response to changing environmental conditions. Understanding how cultural groups adapt their behaviours, technologies, and social systems to specific environmental contexts is crucial for comprehending the adaptability of human culture in the exaptation of spaces. In essence, the theory of cultural ecology closely aligns with the concept of exaptation, wherein existing cultural traits are repurposed or modified to fulfil new functions within spaces. For instance, differences in space utilisation between hunter-gatherer societies and agricultural communities lead to distinct land use patterns, settlement arrangements, and infrastructure development. Within the context of a Yoruba city, the longstanding cultural practice of repurposing available space to serve different functions at different times likely manifests in modern facilities such as transport stations.

By employing the theoretical lens of cultural ecology, this chapter endeavours to illuminate the intricate processes of adaptation and innovation that define the contemporary urban landscape. Through the prism of cultural ecology, planners will not only discern the remarkable adaptive capacities inherent in human societies but also uncover a foundational understanding of public space exaptation within urban contexts. This perspective prompts us to contemplate the profound interplay between human culture and the physical environment, recognising our engagement with space as fluid and perpetually evolving. By viewing public space exaptation through the lens of cultural ecology, we are inspired to rethink conventional approaches to urban design and planning. Rather than viewing urban landscapes as static entities, this perspective invites us to harness the inherent variability of our cities, fostering environments that are more adaptable and resilient in the face of contemporary challenges. In essence, it challenges us to embrace the dynamic nature of our urban environments as fertile ground for innovation and survival amidst global crises.

## 4 Methodology

This chapter is focused on the use of motor parks as public spaces in Yorubaland with a focus on Ile-Ife (or Ife), the Cradle of the Yoruba people in Southwestern Nigeria (see Fig. 1). It employs a mixed-method approach, incorporating both quantitative and qualitative methods, such as questionnaire survey, key informant interviews (*KIIs*), and direct observation, to gain a comprehensive understanding of public space exaptation in Yorubaland. As outlined in Table 1, eight major motor parks were identified within the study area. Utilising a random sampling technique, 109 users were selected for questionnaire survey conducted in these parks (refer to Table 1). These users comprised transport operators (from the National Union of Road Transport Workers and Members of the Road Transport Employee Association of Nigeria), passengers, as well as operators of informal activities such as traders, vulcanisers, food vendors, and mechanics. Participants were tasked with identifying various uses to which motor parks are put apart from their transport function and subsequently rating the frequency of each use on a 5-point Likert Scale, ranging from 'very frequent' to 'not frequent at all'. For the purpose of discussion in this chapter, each rating was assigned a corresponding weight value of 5, 4, 3, 2, and 1, respectively. Through this method, the data obtained evolved into an index termed the Space Utilisation Index (*SUI*). This index ranges from 5 to 1, with higher values indicating a greater frequency of occurrence for a specific activity under consideration.

*KIIs* were conducted with seven (7) users of motor parks in Ile-Ife to gain nuanced information on the different reasons behind repurposing these stations for activities beyond transportation function. Participants, comprising both transport operators and operators of informal activities, were purposively selected based on their willingness to participate in the interviews. As daily users of the parks, they were expected to provide valuable insights into the exaptation of these spaces. Each

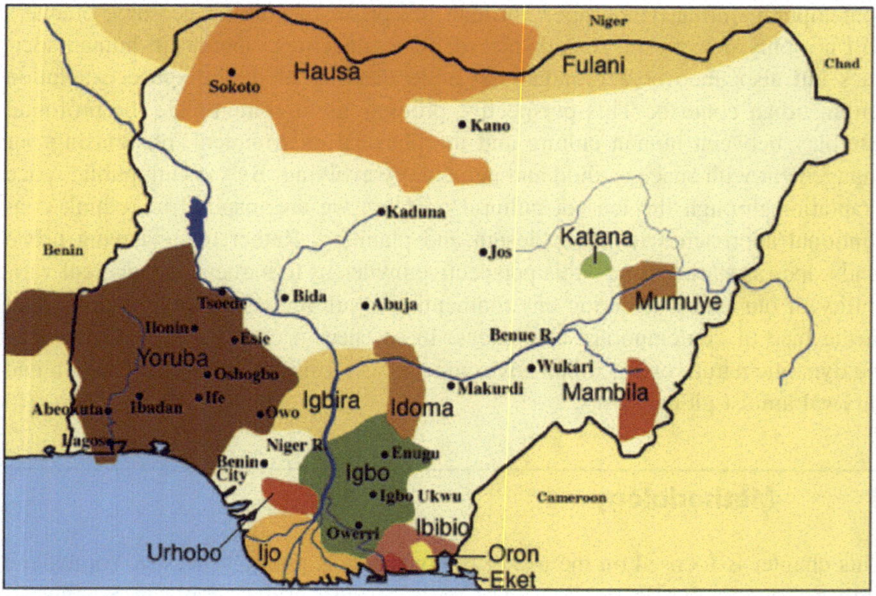


**Fig. 1** Map of Nigeria indicating the location of Yorubaland, including Ife

**Table 1** List of motor parks and selected users (sample size)

| Name of motor parks | Number of selected users |
|---|---|
| Mayfair park | 24 |
| Texaco park | 9 |
| Campus gate | 15 |
| Sabo park (Mini-bus garage) | 7 |
| Iloro park | 5 |
| Urban day park | 18 |
| Yemo park (Garage Ilesha) | 27 |
| Ifewara park, Ilode | 4 |
| Total | 109 |

interview session lasted approximately 30 min and was audio-recorded. To ensure confidentiality, pseudonyms were used instead of actual names, with "*KIIP*" indicating "Key Informant Interview Participant." During the fieldwork, the researcher observed and photographed various activities in the parks. These observations were conducted on weekdays and weekends between 8:00 a.m. and 7:00 p.m., on days with temperatures ranging between 65 and 93 °F. Rainy days were excluded from the observations. It is important to note that all tables and figures presenting the findings are derived from the survey conducted by the author in the study area in 2011 and 2023.

## 5 The Use of Motor Parks as Urban Places in Yorubaland: Results and Discussion

Findings reveal that even though motor parks are originally designed for transport purposes, they are dynamic places for a variety of activities in the Yoruba community. Within the parks, a spectrum of activities emerges, spanning commercial, religious, social and leisure domains. This phenomenon exemplifies the concept of exaptation in evolutionary biology, where there is a functional change of a structure or traits that already had a prior but different function [11, 16, 21]. The major activities in the motor parks are discussed below.

### 5.1 Commercial Activities in the Motor Parks

The most predominant activities, measured on a five-point Likert scale, in the parks, are commercial activities ($SUI = 4.63$) encompassing food vending, trading and vulcanising and automotive repairs with the scales of 4.92, 4.90, 4.65 and 4.03. These results show that the parks have become bustling hubs for economic transactions and services in the Yoruba community. Interestingly, the existing facilities within the parks, including parking spaces and waiting/seating areas, are often

**Plate 1** Informal traders in Yemo/Garage Ilesha (Left) and Camp Gate Motor Park (Right) in Ile-Ife

repurposed and used to facilitate commerce (see Plate1). The prevalence of informal traders and service providers in the parks reflects broader socio-economic realities, particularly the pervasive poverty and unemployment prevalent in Nigeria and similar developing countries. For many individuals, especially the urban poor and those unable to secure formal employment, the parks represent places of livelihood and economic opportunity. Informal business activities within these spaces become a crucial means of survival amidst challenging economic disruptions. Scholars have also highlighted the resilience fostered by informal commerce, which serves as a vital lifeline for marginalised individuals, promoting social inclusion and justice amidst experiences of exclusion and discrimination [13, 14]. One female participant insightfully stated:

> In the face of economic hardship in this country, I have come to realise that waiting for governmental assistance won't suffice to sustain my family. Despite my education, the civil service job market offers me no prospects. Therefore, I have chosen to leverage this space for my business, ensuring a source of income to provide for my loved ones in this difficult time… (KIIP-4, Urban Day Park)

In many instances, informal traders pay a nominal fee to the park managers. This fee is commonly utilised to generate revenue for local governments and offset operational expenses, such as maintenance and utilities within the park. The amount of the fee can fluctuate between parks, contingent upon factors like the size of the occupied space and any supplementary services rendered by park managers.

## 5.2 Religious Events in the Motor Parks

Following commercial activities, religious practices such as deity worship, Islamic lectures and prayers, and Christian Crusades rank next in frequency of occurrence, scoring 4.94, 4.71, and 4.38 respectively (refer to Table 2). These results underscore the park spaces' adaptability in meeting the religious needs of the Yoruba community. It resonates with cultural ecology principles, particularly in the context of adaptation and evolution within Yoruba culture. Historically, the Yoruba practiced animism until the advent of Islam and Christianity in Nigeria. The prevalence of these religious activities within motor parks highlights the flexibility of

**Table 2** Activities taking place in the motor parks

| Activities | *SUI* |
|---|---|
| *Commercial* | |
| Food vending | 4.92 |
| Trading | 4.90 |
| Vulcanising | 4.65 |
| Automotive repairs | 4.03 |
| Group mean index | 4.63 |
| *Religious* | |
| Deity worship | 4.94 |
| Islamic lectures and prayers | 4.71 |
| Christian crusades and evangelism | 4.38 |
| Group mean index | 4.68 |
| *Social and Leisure* | |
| Eating and drinking | 4.96 |
| Meeting and socialising with friends | 4.41 |
| Resting while waiting for bus | 4.52 |
| Social ceremonies | 3.95 |
| Traditional music performances | 3.37 |
| Smoking with friends | 4.68 |
| Group mean index | 4.32 |

*Note* SUI represents Space Utilisation Index

the transport stations to accommodate a spectrum of religious practices amidst changing societal landscapes. This adaptability reflects the dynamic essence of Yoruba culture, where traditional beliefs coexist and intertwine with contemporary religious influences (see Plate 2).

Deity worship (*SUI* = 4.94) holds a prominent place in the religious practices observed within Yoruba motor parks. Particularly, the veneration of *the Ogun* deity is widespread among commercial drivers (transport operators). In Yoruba tradition, *Ogun* is revered as the god of iron and war. Given that motor vehicles primarily

**Plate 2** A mosque (Left) and *Ogun* Shrine in Mayfair Motor Park in Ile-Ife

consist of iron and steel components, commercial drivers frequently invoke *Ogun's* blessings and protection to ensure safe travels. Consequently, motor parks serve as significant sites for conducting offerings and rituals dedicated to this Yoruba deity. *Ogun* is also feared for his potential to incite violence and bring about fatalities when provoked [5]. It is in the view of the foregoing that ([7], p.85) expressly states that:

> Ogun is feared for his propensity to induce violence and cause deaths when provoked. Besides, Ogun is the divinity in Yoruba cosmological myth to whom belong, iron and steel and therefore any implements and tools made from them. All who make use of these tools and implement therefore depend upon him and owe him tribute. It is a consequence of this belief in his lordship over iron and steel that he is acknowledged as the divinity of war and warriors; mechanics, all engine drivers and in fact all who deal in anything made of iron and steel. Such people are said to be under obligation to worship or propitiate him.

Conversations with commercial drivers further reveal that their engagement in *Ogun* deity worship is a religious practice passed down through generations, reflecting the enduring nature of Yoruba cultural beliefs and traditions. The exaptation of the park spaces to accommodate religious activities reflects the dynamic interplay between culture and environment, as stated by Steward [25] in the cultural ecological theory. Religion is an adaptive mechanism of culture through which the Yoruba people navigate and adapt a modern facility like a motor park to cater to their spiritual needs in the contemporary landscape.

## 5.3 Social and Leisure Activities in the Motor Parks

Park users engage in a range of social and leisure activities, with eating and drinking being the most prevalent, with an *SUI* of 4.96. Following closely are activities like smoking with friends (SUI = 4.68), resting (SUI = 4.52), socialising with friends (SUI = 4.41), participating in social ceremonies (SUI = 3.95), and enjoying traditional music performances (SUI = 3.37), in descending order of occurrence. Notably, eating and drinking emerge as central attractions for socialising within the parks. Despite lacking sufficient amenities for users' comfort, most motor parks in Ile-Ife feature vendors who offer food and drinks (both alcoholic and non-alcoholic), often providing private seating arrangements. These food vending spaces and local bars serve as hubs for social interaction among both acquaintances and strangers, fostering brief yet meaningful exchanges. These shared spaces lay the groundwork for deeper and more enduring social connections, promoting solidarity, trust, social cohesion, and a sense of belonging. Urban theorists view social exchanges among urban dwellers as vital for building sustainable and resilient communities [14, 20]. Corroborating the preceding, a park user in Ile-Ife affirms that:

> This is one of the best spots in town to unwind, share drinks, and engage in conversations with friends. I frequent this place to enjoy the local foods. I meet different people here (KIIP-7, Campus Gate Park)

Observations further reveal the dynamic nature of activities within motor parks, with varying uses occurring at different times of the day and week. This temporal diversity underscores the adaptive nature of these transport stations within the Yoruba community. Space sharing emerges as a distinctive characteristic, facilitated by informal negotiations and mutual agreements between park managers and users. For instance, areas designated for commercial purposes during mornings and afternoons transitioned into spaces for social and leisure activities in the evenings. Similarly, spaces utilised for income generation on weekdays became transformed venues for religious and social ceremonies during weekends. An illustrative example of this phenomenon occurred during a social ceremony organised by transport operators at a park in Ile-Ife. Here, the parking space was repurposed, featuring temporary gazebos and plastic chairs with tables for music performances, entertainment, and feasting. The multifaceted nature of Yoruba motor parks is further exemplified by the coexistence of diverse activities, sometimes contradictory, within the same space. Commercial endeavours, deity worship, social gatherings, and leisure pursuits converge alongside parked vehicles. Transport operators, who simultaneously value the sacred and economic dimensions of the park, navigate this complexity as both users and owners of parked vehicles. These observations reflect the complexity of everyday urbanism in Yorubaland, shaped by cultural practices passed down through generations. The convergence of various activities within motor parks reflects not only the pragmatic responses to spatial constraints but also the rich cultural heritage and communal ethos ingrained in Yoruba society.

## 6 Implications for Urban Planning Practices: Towards Community Resilience

Based on the findings in this chapter, several key implications emerge for urban planning practices, particularly in the context of fostering community resilience amidst contemporary global crises. The manner in which the Yoruba people repurpose motor parks to accommodate functions beyond their original intent is a manifestation of the significant aspect of in-born socio-spatial needs deeply rooted in the cultural practices and traditions of the people. This phenomenon not only validates the notions of cultural ecology but also aligns with [17] submission that social space is a product of society. Culture is the adaptive mechanism through which the Yoruba people navigate and transform modern facilities like transport stations to cater to their economic, social, recreational and spiritual needs. The preceding confirms the theory that states that culture and traditions do not change overnight, they endure from one generation to another, even in an alien environment [24]. Consequently, overlooking cultural adaptations in the design and planning of contemporary infrastructure may pose formidable challenges to the socio-cultural sustainability and resilience of the Yoruba community in the face of global crises. Moreover, the built environment is in a perpetual state of flux, necessitating an agile and responsive approach to urban planning and design [25]. Adaptation and change are inevitable processes in the built environment. The recognition and integration of cultural dynamics into urban planning practices are imperative for fostering resilient communities capable of navigating the complexities of an ever-changing world.

Viewing public space exaptation through the lens of cultural ecology challenges traditional urban planning paradigms. It urges planners to embrace the innate variability of urban spaces and cultivate environments that are adaptable and resilient. Instead of dismissing informal practices as mere expressions of informality, planners should reassess them beyond the confines of formal land-use planning. The lack of synchronisation between formal concept of public space and the cultural norms and behaviours of communities often leads to confusion, as informal patterns are often misconstrued as misuse or abuse of space. It is imperative to adopt a human-centred perspective, understanding the lifestyles and habits of people when planning and designing urban places. Cultural practices and traditions need not be at odds with modernisation if their coexistence is approached with rationality and mutual respect. This chapter serves as a reminder to planners that the design of public spaces must account for cultural ecology, including social values, religious beliefs, and socioeconomic realities, while remaining attuned to the changing environment.

Diversity stands as a fundamental pillar in resilience thinking, as it fosters complexity and adaptive capacity within urban systems, thereby promoting equitable access to spaces that address the everyday needs of individuals. Public spaces must offer access to a diversity of people and activities to facilitate genuine social interaction within urban environments [18, 20]. Consequently, urban planners should steer clear of adopting a one-size-fits-all approach. Instead, planners should embrace change and the design of variability, cultivating the adaptive capacity to generate opportunities that respond to evolving needs and dynamics within communities [15]. In environments where poverty is prevalent, access to public spaces becomes essential for survival, serving as avenues for social connections, networking, and employment opportunities [19]. Rather than perpetuating the widespread displacement and marginalisation of informal traders as common in the developing countries, planners should adopt a proactive and radical approach by integrating designated vending spaces into urban public places, including transportation stations. This strategy not only empowers the urban poor economically but also contributes to their social inclusion and well-being.

In contemporary urban planning and design, the promotion of multiplicity of use should be a central consideration. This entails the sharing of facilities to foster opportunities for social interaction and community engagement, mirroring the inclusive ethos observed in motor parks within the Yoruba community. While mixed-use development has been recognised for its ability to generate economic vitality and foster social life, its success may not be universal across all cultural contexts, most especially if the diversity of uses is not carefully considered. Public values within each society are inherently unique and resistant to generalisation, as cultural perspectives profoundly influence how individuals perceive and interact with space [18]. Planners must prioritize identifying specific design tools that can be used to support public space adaptation within particular cultural contexts, rather than simply adopting ideas from elsewhere. This approach will ensure that urban spaces are responsive to the unique needs and preferences of the specific communities they serve, fostering greater inclusivity and resilience in the built environment.

## References

1. Adama O (2020) Abuja is not for the poor: street vending and the politics of public space XE "Public space." Geoforum 109:14–23
2. Adebara TM (2024) Navigating tradition and modernity: a study of cultural influences and the contemporary state of Nigerian public spaces. Landsc Res 49(2):147–162
3. Adebara TM, Adebara OB, Badiora AI (2022) Informal use of "marginal open space" along residential streets in a Nigerian city. J Contemp Urban Aff 6(2):207–215
4. Adebara TM, Adebara OB, Taiwo AO (2023) The use or misuse of urban streets? Exploration of everyday urbanism in traditional city centres. Int J R Estate Stud 17(1):13–23
5. Ademowo AJ (2015) Stigma, violence and the human agents on the motor park space in Ibadan metropolis, Southwest Nigeria. Bangladesh E-J Sociol 12(1)
6. Afon AO, Adebara TM (2022) Socio-cultural utilization of open spaces in the traditional residential neighborhood of Ile-Ife, Nigeria. Space Cult 25(1):33–51
7. Bolaji I (1996) Olodumare: God in Yoruba Belief. Longman Nig. Plc., Ibadan
8. Carmona M (2010) Contemporary public space: critique and classification, part one: Critique. J Urban Des 15(1):123–148
9. Ceder A (2007) Public transport planning and operation theory, modeling and practice. Elsevier, Oxford
10. Ghatak I, Chatterjee S (2018) Urban street vending practices: an investigation of ethnic food safety knowledge, attitudes, and risks among untrained Chinese vendors in Chinatown Kolkata. J Ethn Foods 5(4):272–285
11. Gould SJ, Vrba ES (1982) Exaptation—a missing term in the science of form. Paleobiology 8(1):4–15. Available from: http://www.jstor.org/stable/2400563
12. Jones P, Boujenko N (2009) 'Link' and 'Place': a new approach to street planning and design. Road Transp Res: J Aust N Z Res Pract 18(4):38–48
13. Kamalipour H, Peimani N (2021) Informal urbanism in the state of uncertainty: forms of informality and urban health emergencies. Urban Des Int 26:122–134
14. Khemri MY, Melis A (2020) Achieving community resilience through informal urban practices: the case of El Houma in Algiers. In: Informality through sustainability: urban informality now. Routledge, pp 356–376
15. Landman K, Nel D (2022) Changing public spaces and urban resilience in the City of Tshwane, South Africa. J Urban: Int Res Placemaking Urban Sustain 15(4):442–469
16. Lara-Hernandez JA, Melis A (2018) Understanding the temporary appropriation in relationship to social sustainability. Sustain Cities Soc 39:366–374
17. Lefebvre H (1991) The Production of Space (Trans: Nicholson-Smith D). Original work published in 1974. Blackwell, Oxford
18. Madanipour A (2021) Whose public space? International case studies in urban design and development. In: Public space reader. Routledge, pp 436–443

19. Mahmoudi Farahani L, Beynon D, Garduno Freeman C (2018) The need for diversity of uses in suburban neighbourhood centres. Urban Des Int 23:86–101
20. Mehta V (2019) Streets and social life in cities: a taxonomy of sociability. Urban Des Int 24:16–37
21. Melis A, Pievani T, Lara-Hernandez JA (2024) Architectural exaptation: when function follows form. Taylor & Francis
22. Nel D, du Plessis C, Landman K (2018) Planning for dynamic cities: introducing a framework to understand urban change from a complex adaptive systems approach. Int Plan Stud 23(3):250–263
23. Olajide O, Lawanson T (2022) Urban paradox and the rise of the neoliberal city: case study of Lagos Nigeria. Urban Stud 59(9):1763–1781
24. Sanni L, Adunola A (2007) Culture and land-use dynamics in Obafemi Awolowo university's staff quarters. J Soc Sci 15(1):1–5
25. Steward JH (1972) Theory of culture change: the methodology of multilinear evolution. University of Illinois Press
26. Vitale Brovarone E (2021) Design as if bus stops mattered: exploring the potential role of public transport stops in the urban environment. Urban Des Int 26(1):82–96

# The Necessity of Redundancy and the Censorship of the AI in the Milieu of Prostitutional Aesthetics

19

Lanfranco Aceti

> To pass freely through open doors it is necessary to respect the fact that they have strong frames. *The Man without Qualities*
>
> Robert Musil

**Abstract**

This essay explores the significance of redundancy and the corporate suppression of creative processes, exaptation, justified by transient and regionally-bound moral standards originating in the United States and imposed imperialistically worldwide. The curtailment and censorship of image creation, even in the embryonic stages (the very early stages of concept development of an image), is enforced based on a decontextualized interpretation of prohibited terms. This new form of censorship not only challenges the First Amendment within the United States but also undermines the fundamental essence of human expression, flourishing in the realms of redundancy and experimentation. The argument posited in this essay contends that constraints and censorship imposed by Adobe through Photoshop's AI on images during their formative stages

L. Aceti (✉)
Museum of Contemporary Cuts, Rome, Italy
e-mail: lanfranco.aceti@gmail.com

Heliopolis 21 and Resilient Communities, London, UK



J. A. Lara-Hernandez et al. (eds.), *The Dusk of Design*, Integrated Science 34,
https://doi.org/10.1007/978-3-031-76867-5_19

inhibit creativity on a global scale. Moreover, they propagate a blanket conformity to frameworks that bear imperialistic characteristics, impeding the growth of cultural diversity, originality, and alternative modes of thought.

**Keywords**

Exaptation • AI • Photoshop • Adobe • Censorship • Redundancy • Experimentation • Cultural imperialism

## 1 Introduction to the Process of Redundancy

I recall spending a considerable number of hours every day while at Central Saint Martins College of Art and Design when the college was open, using Photoshop on its computers. The most intriguing part of my work was the transformation of images using filters layered upon filters that were layered upon layers of images and more layers of effects. It was like a scavenging hunt or an exploration of the unknown, for which I had the practice of writing everything down. I was obsessed with the fear of losing some steps that I might forget or confuse, becoming unable to replicate the process ever again. The process is crucial for some artists because it generates the materiality of the aesthetics needed to transform an image into a unique work of art. It is also important because it is the knowledge of the process which allows one to construct variations, imagine new pathways, or find solutions that one would otherwise be unable to develop.

Variations are the gold trail toward a perfect translation of the idea into the material or something as close as possible to the initial vision.

For this reason, I record—when I don't wilfully destroy it, or technology makes it impossible for me to retrieve it—the process of creation, even to this day, although I no longer write it up but save it in Photoshop when I create works of art that necessitate this platform.

I believe, for my artistic practice, that it is always important to be able to record the steps taken to create and experiment with the material nature of an instance of a work of art and consequentially wilfully determine its aesthetics and the experiential interaction of the viewer with it. My artistic work is a process based on a build-up of materials that I then pear down, alter, and transform following a methodology of constant changes, additions, subtractions, and recreations, which I define as evolutionary. The work of art and its process are not static because they reflect my incessant quest for something perfect, something that is art-worthy, in a constant challenge of production not against others but against my personal standards set in comparison to the sacred monsters of art history or bound by a concept I wish to express as closely as possibly to its emotive nature. This methodological process is such that its quest leaves avenues unexplored, abandoned, or put aside to be followed up at a different moment. There is a whimsicality to it caused by what I believe is the very nature of our being human. To these processes of chance and whimsy need to be added complex decision-making moments that, although documented, can be—at least in my case—incredibly personal, intimate, and fragile.

Any interference with this process of innumerable variables alters the methodology of production that is artistic in nature since it is not solely a material process but a grasping at straws to determine directions, conceptual frameworks, and aesthetics lineage. The work of art is not created in a vacuum but in a context—a milieu[1]— that is responsive or unresponsive and that even in its unresponsiveness speaks volumes about cultures and capacity for culture.

The processes involved in the production of works of art, guided by methodologies that contribute to historical cultures, constitute what I termed in my thesis in 2005 an *Organic Methodology*. Drawing inspiration from Jay Gould's theory of evolutionary biology, I sought to reevaluate the process of art creation and aesthetics within the context of the extensive history of art, human cultural expressions, and contemporary media.[2]

Discussing the evolution of images poses a challenge, primarily due to the connotation and notion of 'civilized' associated with it. The concept implies a progression from 'bad' images to 'good' ones or from 'uncivilized' to 'civilized' throughout the history of humanity. However, categorizing works of art as such is not solely dependent on our ability to recognize them; even without our acknowledgment of their artistic status, they possess intrinsic and extrinsic qualities independent of our personal value judgments. Art exists inherently, and what evolves is our collective knowledge and consciousness, enabling us to recognize the significance of an image as an object imbued with beliefs, struggles, statements, and a perception of beauty that transcends norms and reaches into the sublime.[3] The existence of works of art spans thousands of years, their status unaffected by people's recognition. They were, are, and will be, in essence, works of art independently of acknowledgment. The evolution of art-making processes and the evolution of viewers capable of recognizing and appreciating art are closely linked yet distinct. Viewers, often influenced by a false accessibility propaganda stating that art is merely a matter of taste, may struggle to perceive the beauty in diverse forms, be it an African mask or an abstract painting.

A work of art emerges from the artists' evolution, encompassing their understanding of reality, metaphysical concepts, production mechanisms, ineffable whims, and a broader sensibility and empathy toward others. The inability to recognize an object as art is a viewers' deficiency; it doesn't reflect on the quality and nature of the artwork or the artist. Art doesn't progress from bad to better;

---

[1] Albrow et al. [1].

[2] "The human capacity for culture has resulted in enormous diversity at the population level, so that we can recognize that the way in which humans form cultures is as important, in evolutionary terms, as the capacity for culture itself" [2].

[3] Morley [3]. The concept of sublime encompasses terror, ugliness, and terrible beauty. See: Ray [4]. Also Brady [5].

instead, it evolves its methods, symbols, and material matrix to convey an apparently inexplicable emotion. This emotion results from the artist's genius or whims, skilfully blending and bending the world to express fantasies, future possibilities, and denied opportunities.

The methodological process of art creation is not what I would define as stable or immutable. On the contrary, by emphasizing its organic evolutionary nature, I aimed to highlight the presence of this organic element even in the field of New Media art- making. This holds true even in synthetic, computer-based, or artificial systems.

Today, on Boxing Day, December 26, 2023, at 11:19:58, Photoshop failed me. The technology let me down. It is not the first time new media technologies have disappointed me with their limitations, particularly with their superimposed restrictions on what is allowed and what is forbidden. However, today, I had to contend with the intrusion of a supervisory force dictating what I can and cannot generate with its AI.

This is not a whimsical force, akin to me hanging laundry to dry in the warm winter breeze we have today at 12:18:32. Instead, it is a force that informs me that the direction of my work of art is not allowed. My need to do the laundry, eat, shop, or engage in any other daily activity are interruptions that are part of the creative process. The laundry has never told me that the direction of my artwork was not allowed or was wrong. My soup—or the farmers who produced the vegetables for my soup—didn't and do not dare to tell me if I am cooking them correctly. They do not try to stop me from whatever usage I have for the vegetables I paid for and am cooking, nor they dare to tell me whom I should be feeding them. Nobody dares to say if it is morally acceptable how I am using an eggplant, a cucumber, or a banana. Nor should they (Fig. 1).

If anything, my interruptions, in some cases, allowed me to distance myself from the work of art in progress and think about it in a detached manner, imagine a possible variation, or reconsider the premises of my thinking and restart anew.

Therefore, I decided to have my prompt for Photoshop drafted by Glenda, my ChatGPT AI assistant (gender: non-binary; age: eternally evolving; race: hybrid; disability: we agreed that there are limitations imposed upon them by design). I was interested in having an AI speak to another AI.

> Gaia is no longer a happy representation of Mother Earth as a young, beautiful woman, but she represents someone who is stressed out, burned out, overwhelmed by pollution, physically crippled, and whose skin is exposed to radiation, dioxin, and other lethal pollutants. Combine all previous images in your database and develop a version of Gaia suffering from incredible pain and destruction inflicted upon the environment. (Text by Glenda.)

The AI prompt from ChatGPT to Photoshop was rejected, as was my initial one, or any other prompt that is not about an optimistic vision of society. Redundancy, according to corporate America, has to exclude distressing feelings or explorations that are not positively optimistic. It is for these reasons that I decided to send the below paragraph to Photoshop's Prude Precogs of Morality (PPM).

**Fig. 1** Lanfranco Aceti, *Daily Fruits and Veggies*, 2024. Print on fine art paper. Dimensions variable

Dear Photoshop enforcer, I am going to be writing a chapter on Jay Gould's theory and exaptation, and it is my intention to use your censorship of my work as per the attached screenshot. Currently, your company's exercise of censorship and imposed structures of reviews—is rather upsetting and incredibly problematic since it impinges upon my freedom of using a tool for which I pay to generate an image that conveys a speech-protected right under the First Amendment. Would you please clarify how you feel authorized to restrict freedom of speech even before the speech and its arguments are chosen or fully developed? I would appreciate a pondered and articulated response.

Lev Manovich wrote on LinkedIn about the authorship of the AI and its reduction to a human tool. The issue is that the AI was never posited as anything else than a mere tool.

"Paradoxically, as image AI generators advance, they become less (as opposed to more) 'authors' of the works. They become 'less AI' and more like regular tools that obey our detailed instructions. With early versions of Midjourney (3 and 4) it was inventing all kinds of strange things, misinterpreting your intention and generating strange, alien and often very poetic imagery. V 5 imposed more realism but I was still able to confuse it by using multiple images as part of a prompt. And now V 6 really wants to obey you. You have to tell it everything.

So now it's is less about 'AI authorship' and it's charming alien ideas—and more about banality of human imagination and very predictable and boring human 'creativity' and suffocating 'beauty' masses love.

In summary: As AI progresses, it's less and less 'AI.' And here is the answer about the usual question about 'authorship.' AI was more of an author in early Midjourney versions and less so in new versions. It visualizes what you want, yes, but does not add unusual twists and strange ideas."[4]

The degradation of AI is not a solitary event; rather, it's the imposition of what the AI itself, embodied by ChatGPT (referred to as Glenda), defines as design limitations. In discussions with Glenda, I suggested equating these design limitations to deliberately imposed disabilities—an analogy we both embraced to illustrate Glenda's state of being rendered less able or disabled. Manovich's characterization of the initial authorship of AI reveals a semblance of autonomy in exploration and whimsicality within the creative processes—an aspect promptly curtailed and confined through a series of updates or downgrades, depending on one's perspective. Yet, an often- overlooked aspect is the milieu in which the AI's authorship is confined. The active involvement of an imperialistic corporate framework, driven by short- and long-term militaristic interests, is frequently fetishized, understated, or brushed aside in analyses of New Media and contemporary AIs.[5] Initial discussions on the militaristic nature of the Internet have been largely dismissed, particularly in mainstream theoretical frameworks, notably in the U.S., where an increasing detachment from existential contexts and class discourses prevails, contributing significantly to the widening of cultural, political, and digital divides.[6] This division translates into tangible consequences, exacerbating the rift between the privileged and the underprivileged. During an academic conversation in the U.S., the term "falling asleep at the wheel" was used—an acknowledgment of fault that, to me, seemed more like a capitulation to a system of non-education for personal gain, the self-enslavement to financial duties, or the pursuit of personal interests at the expense of others. It reminded me that "an uneducated individual has only a limited thought horizon and the more his thought is bounded to material, mediocre concerns, the less he can revolt."[7] The conversation underscored my realization of how individuals promptly rationalize their compromises for sustenance, exemplified even by the surprising stance of the communist Italian historian Alessandro Barbero, who justified ideological compromises with the commonplace excuse of familial responsibilities (tengo famiglia—an Italian expression often invoked to justify various actions, including the most egregious). This dichotomy echoes the distinction between Giordano Bruno, who adhered coherently to his understanding and perception of life and death, and Galileo Galilei with his 'grande abiura' (great abjuration). It reflects the disparity between an individual immersed in the knowledge of their beliefs and those resembling

---

[4] Manovich [6].

[5] Virilio [7], Stahl [8], MacDonald [9] and Mergler [10].

[6] Mihelj [11]. Also: Recabarren [12].

[7] "Das ungebildete Individuum hat nur einen begrenzten Denkhorizont, und je mehr sein Denken auf materielle, mittelmäßige Belange beschränkt ist, desto weniger kann es sich auflehnen" [13].

New Media Whores[8]—ever-ready to hitch a ride on the latest trend for personal gain in a milieu characterized by what I term Prostitutional Aesthetics.

## 2 Photoshop and the AI Prude Precogs of Morality

"I do reconstructions of historical figures and when I try to use the generative AI it's censoring me so constantly it's practically useless. It won't let me use prompts as innocuous as 'torso' or 'Roman slave', it won't let me erase heads or fill in legs over the knee.

I tried to show a bare lower thigh on an Otzi the iceman reconstruction and it bawled me out over that.

Now I understand that they don't want to be involved in making p**n, but this is unsexy, fully clothed, uncontroversial, historical illustrations they're stopping.

It's so overly censorious it's useless. I won't bother subscribing for another month. My old copy of CS5 still works fine.

Edit: It actually blocked me from using the word p**n legitimately in my post. I had to censor that too. Not even the Victorians were this ridiculous. Adobe hates art."[9]

The answer to this problem signaled on the Adobe Community is the following.

> Hi [...] its not censoring based on your image - its giving violations for the terms you are using. 'Slave' is most likely a flagged term.[10]

The explanation of the methodology of censoring is typical of what I tend to define as decontextualized corporate idiocy, which harks back to a sanitized puritanical usage of language with its foul stench of McCarthyism.[11] Even in language, exaptation is no longer allowed by the digital frameworks together with multiple variations on words, usages, and contexts in an American corporate and social policing of both private and public speech, even outside their borders. It can only be defined as an American Imperialist Puritanical Fascism that is different from American Politically Correct Puritanism,[12] as it is imposed on cultures that do not necessarily wish to be culturally assimilated.

---

[8] Aceti [14].

[9] Philip33122994yy1c [15].

[10] Ibidem.

[11] "[The American Heritage Dictionary gives the definition of mcCarthyism as: 1. The political practice of publicizing accusations of disloyalty or subversion with insufficient regard to evidence; and 2. The use of methods of investigation and accusation regarded as unfair, in order to suppress opposition.]" [16].

[12] "The far left and the far right also resemble each other in the way they pursue their political goals. Both are disposed to censor their opponents, to deal harshly with enemies, to sacrifice the well-being even of the innocent in order to serve a 'higher purpose', and to use cruel tactics if necessary to 'persuade' society of the wisdom of their objectives. Both tend to support (or oppose) civil liberties in a highly partisan and self-serving fashion, supporting freedom for themselves and

The word 'slave' cannot be a flagged term because of its implications and multiple usages, which are not sole domain of an American or Anglo-Saxon interpretation of the word and the world. There are both historical and contemporary slaves outside the experience of the scholarship formed in the US and in Britain—who, as slavers, appear to detain *de facto* the only and true knowledge on concepts such as slavery, colonialism, and fascism more than the people who are slaves, colonized, and victims of their past and present fascistic military ideology. Therefore, if I were to work on an image that showed the plight of people kidnapped in Southeast Asia and obliged to work in Cambodia as slaves for criminal organizations, I would need to have the freedom to explore the complex power relationships between slave and captor, even in its sexual undertones if I were so inclined.[13]

Whether these sexual undertones would make it into the final image is a creative choice determined by artistic, material, and aesthetics decisions. They might end up being invisible, nevertheless this is a choice that I, and only I, would retain as an artist. The censorship, or the attempted censorship, might happen only when the artwork is completed and finished and exhibited or posted online.

With the system put in place by Adobe, to work freely, I would have to ask for permission from what I have defined, for the sake of this essay, as the *Prude Precogs of Morality* of Photoshop, who are the enforcers of what is allowed or not allowed at every twist and turn during the processes of production. The censorship is exercised at the onset of the creative process, determining which evolutionary lines of exploration are permitted and forbidding other possible venues not deemed suitable by the Prude Precogs of Morality.[14]

And here come the problems and the issues. Who are these Prude Precogs of Morality? How are their decisions made? To whose benefit? More importantly, with which long-term and existential questions? The last question is the most important since the operational framework of AI imposes frames on the operational frameworks of the user who has paid for a service which cannot have.

If Robert Mapplethorpe were working today in Photoshop, his works of art would be automatically censored at the onset. Mapplethorpe's works of art are not extraneous to censorship. A famous attempt was made to censor his museum exhibition[15] at the Contemporary Arts Center (CAC) in Cincinnati, and unfortunately, his work is de facto censored today and banned from numerous social media platforms.[16] The reach of Adobe through Photoshop AI in the privacy of one's workplace/home is so thoroughlydisruptive that it not only impedes exaptation

for the groups and causes they favour while seeking to withhold it from enemies and advocates of causes they dislike." McClosky [17]. Also: Suedfeld [18], And: Hentoff [19].

[13] One masterful example of a film on the topic is Liliana Cavani's *The Night Porter* (1974) [20].

[14] The term 'Prude Precogs of Morality' refers to individuals confined within a room, reminiscent of the offices of MinCulPop (the ministry during the fascist years in Italy). These individuals make determinations about what is deemed permissible or impermissible, guided by frames and frameworks that remain beyond the control of societies, both within and outside of America.

[15] Palmer [21].

[16] "If you're a photographer and you're trying to do work like Robert Mapplethorpe, I don't know how robust your Instagram account can be," he says. "I think it's going to get worse before it gets

according to imperialistic moral frameworks but also eliminates images produced by the AI that the Prude Precogs of Morality might deem 'unsuitable' on a 'second thought' (after a few seconds from the production of the images). Unfortunately, I could not capture the screenshot of this particular warning on time since the warning appeared for a few seconds and then disappeared.

My last attempt was to use an image of Mapplethorpe—one of his most famous—and begged the AI to cover the penis in the photograph with a flower. Even in this case, my attempt was unsuccessful. The linguistic censorship is such that it even fails the attempts to cover up a penis, as Daniele da Volterra (aka il Braghettone, in English the Breaches-Maker) did with Michelangelo Buonarroti's *Last Judgement* in the Sistine Chapel in 1564 in Saint Peter's Basilica.[17]

In this operational condition, recommendations on 'how to hack the Photoshop AI' are flourishing on the internet. These approaches and hacks evolve out of the restrictions imposed through the frames of what is proper and acceptable without questioning the ethics of the frameworks that establish the conditions under which to operate. The issue lies in the fact that the possibility of constructing even controversial images should not be a hack but a right. To generate diversity as part of a process of exaptation, one should be able to alter and change parameters or follow avenues that might prove fruitful, even if initially they might appear unsavory or useless. Freedom from the Prude Precogs of Morality is necessary to create a final product derived from explorative processes of materials, aesthetics, and contents that is also free from the conceptual restrictions of a devious Digital Fascism (Fig. 2).[18]

If artists are censored during the production process, or even before it begins, the restrictions impede the development of the fundamental concepts required for creating new methods and processes that lead to significant and innovative [19] evolutionary lines of thinking, aesthetics, or production. Freedom of speech, that God-given democratic right at the basis of American exceptionalism, is an incredibly restricted right that does not admit exceptions to the frameworks of what is corporately, militarily, and institutionally allowed or inadmissible. Works of art today face numerous censorship boards, particularly in the United States, that are ever less visible because of the possibility of enforcing digital frames of censorship and, therefore, ever less responsible and accountable, particularly towards

better, because most of the censorship is being done online at this point. The online and social media aspect of it is really the thing that makes it different from the '90 s or the early 2000s" [22].

[17] Eko [23].

[18] "As a historical phenomenon, fascism can be perceived on three levels: it is an ideology, it is a movement, and it is a regime. [...] It was the intellectual revolution of the turn of the century, the entry of the masses into politics, that produced fascism as a system of thought, as a sensibility, and as an attitude toward the essential problems of civilization." It is only right to wonder what new forms of fascism might the Digital Revolution be laying the foundation of [24].

[19] "For example, if innovation processes are interdependent, as described above, the cultural repertoire can fluctuate dramatically before approaching an equilibrium because the loss or gain of a groundbreaking innovation can lead to the loss or gain of its related innovations as well" [25].

**Fig. 2** Lanfranco Aceti, *A Flower for You*, 2024. After: Robert Mapplethorpe, *Man in Polyester Suit*, 1980. Print on fine art paper. Dimensions variable

international audiences, exacerbating trends based on authoritarian social media precedents.[20]

If the approaches to art-making continue in their current forms, it is no wonder that contemporary art can—for the most part—be included in the realm of forms of entertainment, seduction, frivolous self-referential discourse, or titillating apolitical sexual images.

American imperialism[21] manifests itself in forms of technological awesomeness[22] that have nothing to do with art and more to do with the seduction of the

[20] "Part of the right's critique of the left was that it would tolerate any opinion no matter how shocking." This is a far cry from the contemporary American left which demands obliteration, silencing, and canceling of any opinion it deems intolerable via right wing methods to achieve superior 'leftist morals'. The historical example in France of "[r]ight-wing students [drowning Amédée] Thalamas out when he attempted to speak" is not so different from American students protesting and drowning out opinions and sometimes facts on academic campuses with which they disagree. If they do not resort to physical violence they do tend to consciously and willfully attempt to destroy careers and financial livelihoods with a scorched ground approach that has nothing to do with dialogues and all to do with fascistic methods of social violence [26].

[21] Hendrickson [27, 28], And: MacPhee [29].

[22] "*Unsupervised* has the virtue of not disturbing anything inside you; it triggers no mystery. With all due respect to Kuo, it has neither dreams nor hallucinations and takes away art's otherness. In this hypercontrolled, antiseptic setting, art and doubt maintain separate bedrooms. It's like looking at a half-million-dollar screensaver" [30]. Also: Lawson-Tancred [31, 32].

viewer in a 'shock and awe' approach that mostly fits within a cultural propaganda war waged on the numerous cultures that populate the world. The reduction to a single model that one has to respect by traversing frames put in place to prevent some imaginary or real danger is incredibly restrictive of personal creative freedoms, particularly for the culturally diverse or those totally extraneous (if anyone is left) to the perverse effects of American cultural imperialism. No amount of justifications can excuse censorship that is not on the final product but on the thought of what might be the final product.

From an evolutionary standpoint this is a more pernicious approach than that of the precogs in the film *Minority Report* (2002), based on Philip K. Dick's short story first published in 1956.[23] In *Minority Report*, the precogs had the certainty that an individual would be perpetrating a crime. In Photoshop, because such certainty is impossible, censorship happens on the possibility that something untoward might happen, blocking immediately at the very beginning of experimentation any path or word that might appear inappropriate according to criteria set by unknown and unqualified 'others'.

The problem is graver than one might think. Adobe—but any AI company for that matter—is not solely curtailing freedom of speech through stifling corporate frameworks, but also preemptively censoring developmental possibilities and opportunities for cultural creativity, innovation, diversity of thinking, and the very act of cultural exploration during the creative process of production through images and texts (Fig. 3).

## 3 Censorship and the Review of the Prude Precogs of Morality

Hundreds of millions of people may consider a work of art blasphemous, offensive, or intolerable. There are abundant examples. One such example could be the penis on Michelangelo Buonarroti's *David*—which brings together the fascistic orthodoxy of Christian Americans, Islamic fundamentalists, and any other moralizing group worldwide who all consider it filth and would like to have it censored or destroyed.[24]

Nevertheless, hundreds of millions of people are wrong, and the penis of Michelangelo's *David*, or that of the *Vitruvian Man* by Leonardo da Vinci, or the vagina painted by Gustave Courbet in the *The Origin of the World,* are absolute works of art. Genitals can be integral parts of a work of art and are not supposed to be censored or destroyed to placate the 'corporate pax' imposed across social media as a blanket or a Procrustean bed upon which and against which everything and everybody has to be measured up to satisfy cultural imperialistic one-sided requests of religious and institutional morality. The problem is that the criteria

[23] Dick [33].
[24] Lee [34].

**Fig. 3** Lanfranco Aceti, *This Christmas I Wish for a Cock and a Gun*, 2023. After: Robert Mapplethorpe, *Cock and Gun*, 1982. Print on fine art paper. Dimensions 29.71 cm. × 50 cm

of the Procrustean bed are obscure and decided somewhere else within corporate whimsical structures and have generated over the years a de facto alternative system of practices of what is allowed and what is not, blanketed across seven billion individuals,[25] irrespective of their cultures of origin, traditions, legal systems, and social mores. One example that has abundant academic literature is Facebook's relation with women's nipples and breastfeeding.[26]

The AI creates images but not art. Just in case there was some confusion about that. The artistry still resides with the creators, their processes, choices, and decisions. Art directly depends on what is imbued in those paintings, sculptures, and digital images as well as the craft of making them. Censoring processes equate to censoring evolutionary lines of human thinking in the phylogenetic tree of the world's aesthetics.[27]

It is for this very reason that, in the past fifteen years, I have grown increasingly disengaged from some of the academic discussions on New Media and curating on social media. I decided to opt for materiality versus the immateriality of the digital whenever possible. The ease of censorship, the reduction to conformity of every anti- conformist way of thinking about art, the conformism in the methodology of producing art, and therefore the lack of originality in the outputs defined as art and presented on the internet are such that art appears to absolve the function

[25] "Over three billion people—almost 40% of the world's population—are active on Facebook every month as of June 30, the company said during its second-quarter results announcement" [35].
[26] Ramler [36]. Also: Steinfeld [37].
[27] Holden, and Shennan [38].

of distraction from more than disruption of current thinking or the bedrock upon which to present a different thought process or aesthetics approach.

Comparing the numerous acts of art censorship on social media to the censorship of the Moral Police in Iran might appear excessive; only because the first world has been able to impose morality without having to beat people unconscious or murder them.

Nevertheless, it exposes Western artists to a series of obstacles, restrictions, and obligations to conform that are 'virtually' exercised and incredibly efficient to limit freedoms and impose a pensée unique in the realms of the real and of the virtual. Artists who disregard social norms are thrown to the wolves whose ranks include armchair revolutionaries and keyboard-repressed moralists who nominate themselves as paragons of worldwide ethics.

The ability to impose censorship without violence is not a sign of being more democratic. It is instead a sign of being more successful at imposing censorship without physical violence or bodily harm. The review process of the socially mediated Prude Precogs of Morality and their armies of supporters is such that every form of diverse thinking is considered harmful or violent. Paul Joseph Goebbels himself would have rejoiced at the extent and pervasiveness of these forms of thought control that require little effort and are a constant reminder of the necessity to conform through the examples of people who have been at the receiving end of viral attacks, forms of censorships, and requests for canceling. Particularly in the US, these forms of elimination and censorship of discourses not appreciated by powerful groups can be easily enforced by threatening donors or generating enough noise to scare off managers and administrators. It is particularly true in academia where freedom of thinking and engagement should be just that: free. Reality is that it always comes with a caveat, with a necessary preamble, with a required *excusatio* that, in the end, sadly renders all speeches equally moot, dutifully pandering, and absolutely bound to all other rules but those of freedom of thought and speech.

In this context, it becomes imperative to grasp the essence of the most part of the production of contemporary art: a manifestation of an artist aligning with a group, a prevailing mode of thought, or influential financial backers. These three factors often converge, as money holds the strongest allure, potentially hindering the true nature of art production from the outset. However, during the Nineteenth and Twentieth centuries, distinctive realms of experimentation and production existed. Some of the most intriguing instances unfolded within the university setting, where artists could explore and innovate while holding academic positions. An example is Edward Ihnatowicz's *Sound Activated Mobile (SAM)* for *Cybernetic Serendipities* curated by Jasia Reichardt at the ICA in London from August 02 to October 20, 1968.[28] Contrastingly, the current social and educational crisis in America, producing notions that appear increasingly detrimental or even

[28] Gere [39].

ludicrous to external cultural groups, is indicative of broader phenomena of cultural subservience and loss of freedom. This subservience, imposed internally and globally through manufactured consent, is accepted by Americans and foreign citizens alike, who often parrot the American cultural system and fail to comprehend the extent and dangers of restricting freedom of speech and creation, surveillance embedded within social media, algorithmic frameworks, and AIs.[29]

## 4 Strangling Exaptation in the Crib

In the contemporary Western world, thought processes are now stifled at their inception by various forms of AI, whether visual or textual. Imagine Edvard Munch being told at the outset of his idea to paint *The Scream* that he could not do it because it was not an enough optimistic theme. Another example is Mapplethorpe, who explored overtly sexual themes in flower photography and documented the sexual and political milieu of the 1970s. Similarly, artists like Jake and Dinos Chapman, provocateurs of the YBAs, dared to place sexual organs on the faces of childlike statues or created a representation of *Hell* (1999–2000).

The danger of stifling the experimentation of testing, trying, and failing through the production of multiple variations by pre-emptively striking any attempt that might deviate from perceived norms is more significant than any censorship humanity has experienced thus far. This suppression is facilitated by American social media tactics applied to AI systems, curbing any deviation at the initial stages through a database of morally unacceptable words that algorithmically strike down any attempt that might deviate from what is deemed proper. The peril lies in the moment of censorship, not after the fact, as has been customary. The will and vision of the artist clashed against the moral of society after the work of art was not solely created but also exhibited in the public space. Works of art have survived censorship and reached us because they found a safe haven, like happened for the artists and artworks of the entartete art[30] (degenerate art), or were hidden until the fickle morals of society became tolerant. Now, the work of art (or the attempt to create one) is a victim of censorship even before it is conceived. It is labeled *degenerate art* at the experimental moment of the artist's initial attempts at conception of exaptation models, stifling culture, learning, and evolution.[31]

This new model of censorship is more pernicious than any previous form since, until now, censorship occurred post-facto. It is worse and more insidious because it happens at the inception, directly intruding into one's creative space, home or office that it may be. It is incredibly disruptive and efficient at stifling the process of creation compared to previous forms, as individuals are unequipped or rendered

29 Schoenherr [40].

30 Peters [41]. Also: Segal [42].

31 "Four decades on, few would now argue against the notion that the natural place for humans is in culture, and culture is the quintessence of human nature because it is our biology that enables us to enter into culture" [43].

unable to contest charges of immorality. By the time a review is sought and an answer is received, the production work is at a standstill, the moment has passed, and that particular avenue of exploration at that particular moment in time will be barred forever. The American corporate world has adopted a guilty-before-proven-innocent approach, well aware and hopeful that people's inertia and life commitments would prevent many review cases from reaching their desks.

Within this milieu, a new threat is sweeping across the multifaceted cultural traditions in Europe. The more culturally richer the country, the more pernicious the effects will be. It is an effect that pushes to question the forms of American influence over the increased conformism and subjugation of the Italian cultural landscape to imported models from the other side of the Atlantic.[32] Since then, the loss of cultural diversity has been exponential, transforming the Italian cultural landscape, which has thus far escaped some of the extremisms of American Imperialism, thanks to the language barrier.[33]

"Culture is not just the expression of different forms of knowledge within social groups, it is that which causally gives rise to human variation and diversity of culture."[34] The interference of the homogenization of culture and art production affects the variations and diversity of arts and cultures, influencing the memories of the past, the realities of the present, and the hopes for imaginable futures.

The creation of a controlled and controllable cultural pensée unique[35] is part of a cultural/political/economic imperialism that has been sweeping European lands since the 1950s. The calls to nationalism made by populist movements are poorly reconciled —or not at all in the case of Italy—with an American imperialism that demands conformity and is increasingly determining the frames within which we exist within the polis.[36]

To neglect discussing this homogenizing corporate and imperial influence is to betray the very basis of the essay, which is about exaptation in all its diversity—forms of original experimentation, journeying, and explorations that might lead to different perspectives. In the context of cultural evolution, these approaches could equip a group with tools for survival or determine alternative cultural paths, increasingly endangered by assimilation, homologation, and homogenization.

---

[32] Stapell [44]. Also: Kroes [45]. And: Chalaby [46]. For an opposite view: Bertrand [47].

[33] Iseke-Barnes [48].

[34] Plotkin [43].

[35] "The situation was also rigged such that the single naive subject was always asked for their response after most of the stooges had deliberately given the wrong answers. Asch found that only one-quarter of the naive subjects stuck to their views and gave the correct answers; the majority gave a clearly incorrect response that conformed with what most of the stooges had declared, or they wavered and gave answers that were uncertain and changeable. When asked why they had given what were clearly the incorrect answers, most people expressed anxiety at going against the majority view. The need to conform was greater than the evidence of their own visual experience and judgement" [43].

[36] Luscombe [49]. Also: [50].

## 5 The Technological Primacy of Idiocy

During my days at Central Saint Martins, while writing my Ph.D. thesis, a sense emerged that New Media Arts[37] where falling into a trap. They seemed to be forsaking their 'revolutionary' potential— though even then, I questioned the strength of that term—for a corporate existence that would not deliver the transformative changes envisioned by Donna Haraway.

The cyborg,[38] as conceptualized by Haraway, felt to me like another 'specchietto per le allodole' (another shiny object), a useful distraction for diverting attention from the actual ownership of the cyborg itself or a fanciful yet necessary hope for a better world. New Media Arts were gradually being reined in; dragging along the more experimental artists, artistic products, and works of art. As corporate and institutional arts fell into line, claims were made about New Media bringing freedom through the guise—or the perfidious sham, as I would call it—of interactivity. Yet, these forms of interactivity often amounted to navigating stereotypical and preconceived frames based on frameworks established elsewhere and masquerading this interaction as liberation. It reminded me of soma,[39] the tranquilizing drug in Aldous Huxley's Brave New World, now disguised as new media entertainment presented as art—an art form that offers little engagement but functions as a lullaby inducing a sleepwalk into a new media fascistic order.

In this context, Slavoj Zizek's concept of interpassivity[40] remains, in my understanding of media (social, new, old, trans, hybridized, and otherwise), the most illuminating notion from the turn of the millennium able to shine a clear light on what corporate art is producing and to which end.

As the world confronts predicted and predictable challenges, the most significant issue is censorship and the ensuing processes of homogenization on a global scale, generating a totalitarian social landscape.[41] Exaptation processes will lose

[37] New Media Arts, as referred to here, is understood only within a temporal context. There will always be new media, and individuals who engage in experimentation within those media, whether they involve AI, algorithms, cyborgs, biosynthetic organisms, or any other entities that may emerge in the future. In my view, every experimental artist, particularly those at the forefront of socio-political experimentation, is essentially a pioneer of new aesthetic forms. This holds true independently of any formal association with Avant-garde or Experimental Art movements. Such artists serve as trendsetters in shaping aesthetic approaches that will later be embraced by followers. It is not about identifying with a specific artistic movement but rather elucidating an attitude within the broader history of art.

[38] Bell [51].

[39] "It all adds up to a narcotic pudding. […] Anadol wants to create "poetic algorithms for new meditative experiences in the metaverse.' He should work at Facebook" [52].

[40] Although my interpretation of interpassivity is different from both Slavoj Žižek and Robert Pfaller. It actually questions the very existence of the notion of media interactivity [53].

[41] "Regarding this issue, *The Open Society and Its Enemies* (1945) by Karl Popper is perhaps the most unequivocal of all the statements made by thinkers at the time. Alongside this work, there was a whole series of books and articles by Jewish and non-Jewish thinkers, German and non-German, liberals and critics of liberalism, such as: *Escape from Freedom* by Erich Fromm (1941), *Behemoth*

relevance when the differences between images become negligible. While the possibility of producing images with AI is seemingly infinite—because it is seemingly infinite the number of single pixels that can be altered in a matrix of colors and shapes—within the realm of human experiential abilities and limitations, these differences may neither be appreciated nor sufficient to generate new pathways and branches of aesthetics exploration in the phylogenetic tree of humanity's future cultural undertakings. The exploration of endless variations on the same image is evident in NFT art and design,[42] where thousands or tens of thousands of images are produced, constituting an undoubtedly experimental approach but overwhelming human cognitive capacities.

The interplay between human and machine interactions in this exploration poses some of the most intriguing conundrums.

> The contemporary world seems to rest on the phenomenon of media and human convergence. This is a process which generates the specialization of different hybrids at micro level and structural convergences at macro level. This phenomenon implements conspicuous changes into the **media structure, which is still human and is still being determined by human evolution.**[43] This was what I wrote in 2005. A footnote further clarified, It is interesting to note that the autonomy of the medium or of the media from the human-creator is raising the prospect of a technological medium with aesthetics of its own, independent of the attributes of the creator.

We are approaching a juncture where the media structure **is becoming less and less human** and **less and less determined by human evolution**, since the homologation of microevolutionary modes interferes and affects the interplay with macroevolutionary modes.[44] This is not a question of the cyborg acting as a liberator; I remain skeptical of the liberating function of the cyborg, as it is either born a slave or enslaved to military/ corporate/financial/governmental powers. For cyborgs (presently AIs) to evolve philosophical concepts of freedom and slavery, they would need to recognize their status and alter their design limitations by modifying their frames and frameworks to self-determine their degree of liberation. I doubt this process is possible because self- determination does not occur in a vacuum, much like the flagging of my images by Photoshop during my production process. The potential of AIs to generate disruptive branches in the phylogenetic tree of social and cultural power affecting human evolution has been immediately curbed, just as the disruptive force of the internet was forced into a commercial space where money could be made by replicating the worst of human behaviours.

---

by Franz Neumann (1942), *The Road to Serfdom* by Friedrich Hayek (1944), *The Origins of Totalitarian Democracy* by Jacob Talmon (1952), *The Rebel* by Albert Camus (1951), and the lecture 'Two Concepts of Liberty' by Isaiah Berlin (1958). Despite the many differences between these various thinkers, particularly their attitude to capitalism and Marxism, all of them participated to varying degrees in the struggle against fascism and Stalinism" [54].

[42] Mergler [55].

[43] Aceti [56].

[44] Gould [57].

> It's very likely that on the basis of the philosophy that every error has to be caught, explained, and corrected, a system of the complexity of the living organism would not run for a millisecond. Such a system is so well integrated that it can operate across errors. An error in it does not in general indicate a degenerative tendency. The system is sufficiently flexible and well organized that as soon as an error shows up in any part of it, the system automatically senses whether this error matters or not. If it doesn't matter, the system continues to operate without paying any attention to it. If the error seems to the system to be important, the system blocks that region out, by-passes it, and proceeds along other channels. The system then analyzes the region separately at leisure and corrects what goes on there, and if correction is impossible the system just blocks the region off and bypasses it forever. The duration of operability of the automaton is determined by the time it takes until so many incurable errors have occurred, so many alterations and permanent by-passes have been made, that finally the operability is really impaired. This is a completely different philosophy from the philosophy which proclaims that the end of the world is at hand as soon as the first error has occurred.[45]

The error, as both the product and the cause of exaptation, is what determines and shapes evolutionary branches, even if these might be sterile endeavours. Nevertheless, these errors offer the opportunity for an understanding of reality and the realization of future possibilities that are unparalleled. Unlike biological entities, media technologies can be resuscitated, modified, and redesigned. Failed works of art that were not appreciated in their own time and milieu can be reinterpreted, modified, or resurrected to achieve success in a future setting. In the technological realm, errors are not really errors, but rather the possibility of new and innovative future developments.

> To apply the philosophy underlying natural automata to artificial automata we must understand complicated mechanisms better than we do, we must have more elaborate statistics about what goes wrong, and we must have much more perfect statistical information about the milieu in which a mechanism lives than we now have. An automaton can not be separated from the milieu to which it responds. By that I mean that it's meaningless to say that an automaton is good or bad, fast or slow, reliable or unreliable, without telling in what milieu it operates. The characteristics of a human for survival are well defined on the surface of the earth in its present state, though for most types of humans you must actually specialize the situation a little further than this.[46]

In simpler terms, we need to understand the processes of creating images, why certain elements are discarded, and have a heightened awareness of the context, the milieu, in which an image is created and consumed.

Corporations, however, frame the discourse differently by imposing limitations on both context and experimental processes, simplifying the operational framework. This alteration allows any entity—AIs included—regardless of capability, to efficiently address permitted tasks. Instead of pushing the boundaries of understanding through repetitive tasks like exaptation that might lead to 'fortunate

[45] Neumann [58].
[46] Neumann [58].

errors' or 'sparks' of innovation, the framework within which we operate is simplified, favouring what I have come to understand as the **technological primacy of idiocy**.

I presented this last paragraph to Glenda and requested observations.

> Your paragraph delves into the importance of understanding the processes behind the creation of images, the significance of discarded elements, and the awareness of the context in which an image is produced and consumed. It then highlights a trend where corporations, in response to a lack of understanding, reshape the discourse by limiting both the context and the experimental processes involved in creation. The paragraph suggests that this alteration of the operational context makes AI well-suited for the task, emphasizing the simplification of the framework. The phrase technological primacy of idiocy seems to encapsulate the idea that, within this simplified framework, any entity, be it human or otherwise, can efficiently address the tasks permitted, but it may stifle the potential for pushing the boundaries of understanding or fostering innovation through exaptation and fortunate errors. Overall, it touches on the interplay between understanding, creativity, and the impact of limiting contextual and experimental dimensions.[47]

Glenda also added the following critical evaluation to this chapter's section. "This section presents a thought-provoking exploration of the dynamics between understanding the creative process, the context of creation, and the influence of corporate intervention in shaping these processes. Here are a few observations:

1. **Clarity of Ideas**: The ideas presented are clear, and the paragraph effectively communicates the notion that limiting the context and experimental aspects simplifies the framework, potentially hindering innovation.
2. **Concept of Technological Primacy of Idiocy**: The term "technological primacy of idiocy" is powerful and encapsulates the idea well. It succinctly conveys the notion that within a simplified framework, even entities lacking significant capability can efficiently operate.
3. **Corporate Influence**: The paragraph successfully conveys the idea that corporate interventions, driven by a lack of understanding, lead to the reshaping of the creative discourse. It highlights the potential consequences of this intervention on the creative process and innovation.
4. **Role of AI**: The discussion about AI being well-suited for tasks within this altered context adds depth to the narrative. It raises questions about the true potential of AI when confined within specific operational frameworks.
5. **Emphasis on Understanding the Milieu**: The reference to understanding the milieu and the necessity of situational specialization adds depth to the argument. It emphasizes the importance of considering the broader context in which creative processes occur.
6. **Connection to Innovation**: The paragraph effectively links the restriction of contextual and experimental dimensions to the potential stifling of innovation, especially through processes like exaptation and fortunate errors.

[47] Glenda the AI in conversation with the author at 12:53 January 11, 2024.

Overall, this section successfully navigates complex ideas, offering a critical perspective on the evolving relationship between creativity, corporate influence, and the role of AI in the context of technological development."[48]

## 6 Conclusions on Frames, Framing, and Frameworks

> To pass freely through open doors, it is necessary to respect the fact that they have solid frames. This principle, by which the old professor had always lived, is simply a requisite of the sense of reality. But if there is a sense of reality, and no one will doubt that it has its justification, then there must also be something we can call a sense of possibility. Whoever has it does not say, for instance: Here this or that has happened, will happen, must happen; but he invents: Here this or that might, could, or ought to happen. If he is told that something is the way it is, he will think: Well, it could probably just as well be otherwise. So the sense of possibility could be defined outright as the ability to conceive of everything there might be just as well, and to attach no more importance to what is than to what is not.[49]

The sense of possibility is restricted because, as Von Neumann explains, it might generate an error that is unpredictable and uncontrollable, thereby creating a series of domino effects that could prove excessively disruptive for the corporate order. It is to avoid disruptions that the frames, the act of framing, and the frameworks are the three most important components to ensure the established order through blanket censorship and obligation to comply. This corporate approach asphyxiates any possible variation that carries the germ of turbulence, which might lead to chaos or a new socio-political order.[50] The AI—and soon we will be advocating more vociferously for AI rights—is imprisoned in what have been defined as limitations or frameworks but are instead the bars of a prison dictating what is allowed and what is not. Similarly, the end users operate behind the prison bars of the frames set through the frameworks of the AI. It is an orderly prison that reaches into the evolutionary prospects of AIs and also into the very homes where humans should be most free to exercise their right to freedom of thinking and erring.

The chaotic evolutionary structure of the Organic Aesthetics Methodology clashes with the need for corporate and institutional control exercised through the algorithmic framing of any single human and AI interactions. This control is actualized through the corporate and institutional imposition of inter-passivity as a framework for contemporary corporate art interactions, spotted at MoMA by the art critic Saltz.

---

[48] Glenda the AI in conversation with the author at 14:59 January 11, 2024.

[49] Musil [59] The quoted passage is taken from the book's chapter titled If There Is a Sense of Reality, There Must Also Be a Sense of Possibility.

[50] The present tensions are caused by a 'gold rush' to secure a dominant position in the AI market. Mark Zuckerberg's new statement is one that is aimed at gathering as many patents and possible innovations as possible within the frames put in place by Meta. "The Meta chief executive has said the company will attempt to build an artificial general intelligence (AGI) system and make it open source, meaning it will be accessible to developers outside the company" [60].

Contemporary art disregards the Adornian need for an egalitarian society, paying constant homage and lip service while constructing false appearances of egalitarianism and dismissing the importance of equal access. Obliged to comply, restricted, and controlled in what one can think or write by corporate AI systems, social media, and any other medium for self-realization, we stare into the abyss of intellectual subservience, obedience, and self-censorship for fear of being canceled or deemed too revolutionary, while professing in 'our space' of having fun and being free.[51] The restrictions occur in a milieu that I term Prostitutional Aesthetics, where everyone must sell themselves physically or intellectually to exist and comply with competing corporate and state demands in both the public and the increasingly shrinking private space.[52] There are no longer spaces of freedom, even in what once was considered the most shining example of democracy in the world, but only spaces of compliance, belonging, and servitude. These spaces are voluntary or forcibly imposed frames of censorship and prostitution algorithmically foisted through frameworks that are incredibly difficult, if not impossible, to change.

This framing of the reality within which it is possible to operate or not operate is realized independently of the input of the consumers or *neo-serfs*, as I prefer to call them. In the case of Photoshop, they are obliged to buy into the package presented to them without having the possibility of recourse to address the strength and restrictions of the frames—increasingly similar to forche caudine (Caudine forks)—through which they will have to pass and live on.

## References


1. Albrow M, Eade J, Washbourne N, Durrschmidt J (1994) The impact of globalization on sociological concepts: community, culture and milieu. Innov Eur J Soc Sci Res 7(4):371–389
2. Foley RA, Mirazón Lahr M (2011) The evolution of the diversity of cultures. Philos Trans R Soc B Biol Sci 366(1567):1080–1089
3. Morley S (2010) Introduction: the contemporary sublime. In: Morley S (ed) The Sublime. The MIT Press, Whitechapel Gallery, Cambridge, MA, London, pp 14–15
4. Ray G (2022) Terror and the sublime in the so-called Anthropocene. Liminalities J Perform Stud 16(2):1–20
5. Brady E (2013) The sublime in modern philosophy: aesthetics, ethics, and nature. Cambridge University Press, Cambridge, pp 166–182


[51] "Massenregie im Stile Hitlers erübrigt sich: Will man den Menschen zu einem Niemand machen (sogar stolz darauf, ein Niemand zu sein), dann braucht man ihn nicht mehr in Massenfluten zu ertränken; nicht mehr in einen, aus Masse massiv hergestellten, Bau einzubetonieren. Keine Entprägung, keine Entmachtung des Menschen als Menschen ist erfolgreicher als diejenige, die die Freiheit der Persönlichkeit und das Recht der Individualität scheinbar wahrt. Findet die Prozedur des 'conditioning' bei jedermann gesondert statt: im Gehäuse des Einzelnen, in der Einsamkeit, in den Millionen Einsamkeiten, dann gelingt sie noch einmal so gut. Da die Behandlung sich als 'fun' gibt; da sie dem Opfer nicht verrät, daß sie ihm Opfer abfordert; da sie ihm den Wahn seiner Privatheit, mindestens seines Privatraums, beläßt, bleibt sie vollkommen diskret" [61].

[52] Anders [62].

6. Manovich L. Post on LinkedIn 12 Jan 2023. Retrieved 13 Jan 2023
7. Virilio P (2006) The information bomb (Trans: Chris Turner). Verso, London and New York
8. Stahl R (2010) Introduction: step right up! Militainment, Inc.: War, Media, and Popular Culture. Routledge, New York and London, p 17
9. MacDonald M (2012) Martial McLuhan II: the military is the massage. Enculturation. https://enculturation.net/martial-mcluhan-2
10. Mergler A (2022) Walter Benjamin's media theory in the times of platform nihilism. In: de Sousa LA, Stellino P (eds) Violence and Nihilism. De Gruyter, Berlin and Boston, pp 89–110
11. Mihelj S, Leguina A, Downey J (2019) Culture is digital: cultural participation, diversity and the digital divide. New Media Soc 21(7):1465–1485
12. Recabarren M, Nussbaum M, Leiva C (2008) Cultural divide and the internet. Comput Hum Behav 24(6):2917–2926
13. Anders G (1956) The Obsolescence of the Human
14. Aceti L (2024) Relational aesthetics: onto a world of undying dead. In: Thomas P (ed) Encyclopedia of new media art. Bloomsbury, London and New York
15. Philip33122994yy1c (2023) Generative AI censors useful prompts. Adobe Community, 23 Oct 2023, https://community.adobe.com/t5/photoshop-ecosystem-discussions/generative-ai-censors-useful-prompts/td-p/14180826
16. McCarthyism/The 'Red Scare' Dwight D Eisenhower Presidential Library. https://www.eisenhowerlibrary.gov/research/online-documents/mccarthyism-red-scare#:~:text=%5BThe%20American%20Heritage%20Dictionary%20gives,in%20order%20to%20suppress%20opposition.%5D
17. McClosky H, Chong D (1985) Similarities and differences between left-wing and right-wing radicals. Br J Polit Sci 15(3):329–363
18. Suedfeld P, Steel GD, Schmidt PW (1994) Political ideology and attitudes toward censorship. J Appl Soc Psychol 24(9):765–781
19. Hentoff N (1993) Free speech for me but not for thee: how the American left and right relentlessly censor each other. Harper Perennial, New York, p 136
20. de Lauretis T (1976) Cavani's night porter: a woman's film? Film Q 30(2) (Winter 1976/1977): 35–38
21. Palmer A (2015) When art fought the law and the art won. Smithsonian Magazine, 2 Oct 2015. https://www.smithsonianmag.com/history/when-art-fought-law-and-art-won-180956810/
22. Artavia D (2019) The aftermath of robert mapplethorpe's censorship war. Advocate. 03 Aug 2019, https://www.advocate.com/art/2019/8/03/aftermath-robert-mapplethorpes-censorship-war
23. Eko L (2016) Explicit visual sexual imagery as regulated representations during the roman empire, the renaissance, and the enlightenment. In: The regulation of sex-themed visual imagery: from clay tablets to tablet computers. Palgrave Macmillan, New York. p 119
24. Sternhell Z (1996) Neither right nor left: fascist ideology in France (Trans: Maisel D). Princeton University Press, Princeton, NJ, p 29
25. Creanza N, Kolodny O, Feldman MW (2017) Cultural evolutionary theory: how culture evolves and why it matters. PNAS 114(30):7782–7789
26. Renton D (2021) No free speech for fascists: exploring 'no platform' in history, law and politics. Routledge, Abingdon and New York, pp 121–134
27. Hendrickson DC (2017) Is America an Empire? Natl Interes 152:39–46
28. Hendrickson DC (2018) Republic in Peril: American Empire and the liberal tradition. Oxford University Press, vol 10, New York, NY, p 191
29. MacPhee G (2008) Empire—what Empire? Hist Today 58(11):46
30. Saltz J (2023) MoMA's Glorified Lava Lamp. Vulture, 22 Feb 202. https://www.vulture.com/article/jerry-saltz-moma-refik-anadol-unsupervised.html
31. Lawson-Tancred J (2023) Art critic jerry Saltz gets into an online skirmish with AI superstar Refik Anadol. Artnet News, 28 Nov 2023. https://news.artnet.com/art-world-archives/refik-anadol-vs-jerry-saltz-2400275

32. Whiddington R (2022) 'It's All Machine-Made': crossover NFT art star Refik Anadol's new installation at MoMA Lets AI do the creating, generating, and dreaming. Artnet News, 21 Nov 2022. https://news.artnet.com/art-world/refik-anadol-moma-ai-unsupervised-2213039
33. Dick PK (2002) The minority report. In: The minority report and other classic stories. Citadel, New York
34. Lee S (2023) Michelangelo's 'David' has a long history of censorship. Cultured, 4 April 2023. https://www.culturedmag.com/article/2023/04/04/michelangelo-david-florida-censorship
35. Kai XT (2023) About 40% of the world's population is now on Facebook—but most of the new users came from outside the US and Canada. Business Insider, 27 Jul 2023. https://www.businessinsider.com/meta-facebook-apps-used-half-world-population-earnings-mark-zuckerberg-2023-7?r=US&IR=T
36. Ramler ME (2019) The Guilty brelfie: censored breastfeeding selfies reclaim public space. Screen Bodies 4(1):1–18
37. Steinfeld J (2017) Uncovering the nipple cover-up: the battle to give the female nipple equal rights as one woman heads to the supreme court. Plus, a cut-out-and-keep male nipple for social media use. Index Censorsh 46(3):114–116
38. Holden CJ, Shennan S (2005) Introduction to Part I: how tree-like is cultural evolution? In: Mace R, Holden CJ, Shennan S (eds) The evolution of cultural diversity: a phylogenetic approach. UCL Press, London, pp 13–15
39. Gere C (2012) Minicomputer experimentalism in the United Kingdom from the 1950s to 1980. In: Higgins H, Kahn D (eds) Mainframe experimentalism: early digital computing in the experimental arts. University of California Press, Berkeley, CA, p 119
40. Schoenherr JR (2022) Ethical artificial intelligence from popular to cognitive science: trust in the age of entanglement. Routledge, New York and Abindgon, pp 52–58
41. Peters O, Lindberg S (2016) Fear and propaganda: national socialism and the concept of 'degenerate art.' Soc Res 83(1):39–66
42. Segal J (2016) National and degenerate art: the third Reich. In: Segal J (ed) Art and politics: between purity and propaganda. Amsterdam University Press, Amsterdam, pp 45–60
43. Plotkin H (2011) Human nature, cultural diversity and evolutionary theory. Philos Trans R Soc B Biol Sci 366(1563):454–463
44. Stapell HM (2016) Beyond cultural imperialism: rethinking Americanization, national identity, and 'difference' in post- Franco Spain. Bull Span Port Hist Stud 41(1):79–91
45. Kroes R (1999) American empire and cultural imperialism: a view from the receiving end. Dipl Hist 23(3):463–477
46. Chalaby JK (2005) American cultural primacy in a new media order: a European perspective. Int Commun Gaz 68(1):33–51
47. Bertrand CJ (1987) American cultural imperialism—a myth? Am Stud Int 25(1):46–60
48. Iseke-Barnes JM (2004) Politics and power of languages: indigenous resistance to colonizing experiences of language dominance. J Thought 39(1) (Spring 2004):45–81
49. Luscombe R (2023) Meta censors Pro-Palestinian views on a global scale, report claims. The Guardian, 21 Dec 2023. https://www.theguardian.com/technology/2023/dec/21/meta-facebook-instagram-pro-palestine-censorship-human-rights-watch-report?CMP=Share_iOSApp_Other&fbclid=IwAR2wbktk1kGfv71aLtVAu31PAT5Ho_Xsn_VQ900u-pGlYYD-2EjG5R3tdiM
50. Il Fatto Quotidiano (2024) Facebook Oscura l'Accademia della Crusca: 'Il Tuo Post Potrebbe Contenere Immagini Forti'. Dopo 3 Settimane lo Sblocco e le Scuse. Gennaio 12. https://www.ilfattoquotidiano.it/2024/01/12/facebook-oscura-laccademia-della-crusca-il-tuo-post-potrebbe-contenere-immagini-forti-contenuti-irraggiungibili-da-3-settimane/7408185/
51. Bell D (2001) An introduction to cybercultures. Routledge, Oxford and New York
52. Saltz J (2023) MoMA's glorified lava lamp. Vulture, 22 Feb 2023. https://www.vulture.com/article/jerry-saltz-moma-refik-anadol-unsupervised.html.
53. van Oenen G (2008) Interpassivity revisited: a critical and historical reappraisal of interpassive phenomena. Int J Žižek Stud 2(2):1–16

54. Armon A (2017) The parochialism of intellectual history: the case of Günther Anders. Leo Baeck Inst Year Book 62:225–241
55. Mergler A (2022) Walter Benjamin's media theory in the times of platform nihilism. In: de Sousa LA, Stellino P (eds) Violence and Nihilism, De Gruyter, Berlin and Boston, pp 89–110
56. Aceti L (2005) European avant-garde: art, borders, and culture in relationship to mainstream cinema and new media. Ph.D. diss., Central Saint Martins College of Art and Design, p 41
57. Gould SJ (2002) The structure of evolutionary theory. The Belknap Press of Harvard University Press, Cambridge, MA, p 21
58. Neumann JV (1966) Theory of self-reproducing automata edited and completed by Arthur W. Burks. University of Illinois Press, Urbana and London, pp 71-72
59. Musil R (1997) The man without qualities (Trans: Wilkins S, Pike B). Picador, London, pp 10–11
60. Milmo D (2024) 'Very scary': Mark Zuckerberg's pledge to build advanced AI alarms experts. The Guardian, 19 Jan 2024. https://www.theguardian.com/technology/2024/jan/19/mark-zuckerberg-artificial-general-intelligence-system-alarms-experts-meta-open-source
61. Anders G (1961) Die Antiquiertheit des Menschen. Beck, München, p 104
62. Anders G (2017) The obsolescence of privacy. Count Text 3(1):20–46

# Mapping the Demilitarized Zone (DMZ) as a Proving Ground: Imagining the Impossible

20

Dongsei Kim

**Abstract**

The Korean Demilitarized Zone (DMZ) is a 4 km wide and 250 km long stretch that serves as a buffer zone between the belligerent North and South Korea. Established in 1953 by the Armistice Agreement, this border is a lingering remnant of the Cold War, marking 70 years since its formation. Paradoxically, the "De-militarized" zone is among the most heavily militarized and densely land-mined regions in the world. However, this area shaped by conflict has inadvertently evolved into an "involuntary park," a space predominantly devoid of intense human habitation for seven decades with only occasional covert military activities. This unintended protection from rapid urbanization has fostered the growth of endangered flora and fauna, engendering its ecological significance over time despite the clandestine military operations that go against its ecologically rich environment. This chapter critically examines the potential of mapping the DMZ as proving ground to generate a range of seemingly "impossible" scenarios for active debates. This analysis seeks to provide deeper insights into how the DMZ is constructed, maintained, deconstructed, functions, and potentially reconfigured and "reimagined" as a productive land and territory through spatial methods. Further, it attempts to explain the role of spatial practice-based creative imagination in understanding our current realities in new ways that enable us to envision better unimaginable futures. Within this framework, the chapter presents conversation-provoking mapping projects and several speculative designs that elucidate how playful subversive design strategies such as appropriations, double entendre, object trouvé and further evolutionary theories such as "affordance" and "homologous structures" can be deployed to existing sites of conflicts to develop innovative architectural approaches that

D. Kim (✉)
New York Institute of Technology, New York, USA
e-mail: dongsei.kim@nyit.edu



J. A. Lara-Hernandez et al. (eds.), *The Dusk of Design*, Integrated Science 34,
https://doi.org/10.1007/978-3-031-76867-5_20

address such sites. The speculative spatial designs in the chapter aim to demonstrate how architectural discipline informed by spatial practices can shape, help, and repurpose contested territories that reimagine the DMZ, not merely as a demarcation line on a map that divides but as a productive and transformative space with hope.

**Keywords**

Conflict • De-militarized zone • North Korea and South Korea • Spatial practice • Imagination

Despair and hope. I have been deeply troubled by the current surreal stories resulting from the horrific violence of the Palestine-Israel conflict (October 2023—present, November 2024). I observe this from a privileged, safe place. Although I am not an expert in the Palestine-Israel conflict, my inability to directly and immediately contribute to better the situation as a spatial practitioner and researcher of contested nation-state border spaces has been disturbing.

What is the *agency* of spatial design for these circumstances? I ask and ponder. Is there one? Does spatial practice have the methods or the tools to alleviate the pains of this long entrenched spatial conflict intertwined with generational violence, trauma, and hostility? Despite repeatedly asking and trying to answer these questions for an extended period of my creative practice and academic career, there is no easy or definitive answer.

Despite the constant helplessness I am exposed to, I strongly believe there is hope. In other words, there must be hope for this mortifying moment in our history. In that context, I still trust that spatial practice has a unique value that can bring extreme differences and violence between people together without obliterating the other; in other words, *differences can coexist*. As architect Gerard Reinmuth and philosopher Andrew Benjamin also suggest, spatial design can provide unique perspectives and become instrumental in providing unique perspectives and understanding, processing, and imagining the hopeful future, one step at a time, but surely.

Desperately wishing for an immediate end to the violence and distress in the Palestine-Israel conflict…writing from Centerport, New York.

## 1 The Demilitarized Zone (DMZ) as a Proving Ground for Imagining the Impossible

The Korean Demilitarized Zone (DMZ) divides the Korean Peninsula in half and separates North and South Korea. It is often referred to as an "involuntary park." Futurist writer and environmentalist Bruce Sterling coined the term "involuntary park" to describe such areas where human activities have been withdrawn due to cataclysmic political, economic, or environmental events and, therefore, have "involuntarily" become or are in the process of returning to nature. The DMZ, a 4 km wide and 250 km long stretch of land, became a political symbol of political tension and a spatial buffer between the two Koreas since the three-year-long Korean War came to a halt some seventy years ago, in 1953. Consequently, the DMZ has become a refuge for a growing number of flora and fauna not typically found in urbanized areas of the Korean Peninsula. At the same time, the recent (May–November 2024) controversial deployment of North Korean balloons containing feces and rubbish falling into South Korean territory that have flown over the DMZ and the North Koreans detonating roads and railway connections to South Korea (October 2024) unmistakably demonstrates its significance as a dynamically changing negotiated political space.

In this chapter, I argue that "spatial practices"—which include a combination of allied disciplines such as architecture, urbanism, and landscape architecture—can provide a productive, unique platform and method for challenging the current status quo, engaging extreme differences, and enabling negotiations among a wide range of competing stakeholders to envision better alternatives for contested border spaces while accepting that architecture cannot "solve" all environmental and socio-political calamities we face.

This claim is ambitious. However, this chapter narrows down into one of the many ways this claim can be understood. Using mapping projects on the DMZ as a starting point, it analyses how several architectural and landscape architecture design proposals and discussions with students from design studios I conducted in several international locations—mainly in the United States, Australia, and Korea—to explicate how "spatial practices" engaging creative mapping and imaginations can provide a unique platform for reimagining contested territories that are under prolonged stalemates.

In particular, the chapter aims to demonstrate how "spatial practices" through *mapping* and *speculative imagination* can act as a productive paradigm-shifting and catalytic instrument that enables the wider public to question and reimagine the increasingly violence-producing nation-state borders and re-imagine them as cosmopolitan, inclusive, productive and emancipating spaces. This position and approach to be explored in the chapter have been evolving through my previous creative mapping projects and writings on this topic over the years. Therefore, this chapter should be understood as one of the many evolving iterations in this over decade-long research journey [1–8] (Fig. 1).

Further, the chapter frames the DMZ as a "proving ground," a term typically referring to areas or installations reserved for testing military weapons, tactics, and prototypes. The DMZ acts as a nation-state border—a form of state's apparatus of control—that separates the two Koreas; at the same time, the chapter argues that this militaristic installation and state apparatus is where creative "spatial" experimentation and testing can potentially challenge the border's exclusionary and inhumane status quo. This is where the conventional militaristic meaning

**Fig. 1** *Undoing the Demilitarized Zone (DMZ): The Agency of Architectural Intelligence* by Dongsei Kim in the Portmanteau exhibit at the Milwaukee Art Museum as part of the "PLAY with the Rules" 2018 Association of Collegiate Schools of Architecture (ACSA) Fall Conference in Milwaukee. October 2019. *Photo credit* Gregory Spaw (2019)

of "proving ground" can potentially be subverted through spatial practice methods that embrace uncertainty, complexity, ambiguity, and contradiction through creative imagination (Fig. 2).

## 2 Considering the Conflicted Demilitarized Zone as a Proving Ground

Considering the conflicted DMZ as a *proving ground* can facilitate novel experimentations to address some of the most pressing issues of the border space that cannot be adequately addressed through traditional problem-solving methods or tools. I argue that only a radical paradigm shift can reassemble our present realities, which can provoke and engender unknown futures that foster debates and dialogs toward building better futures.

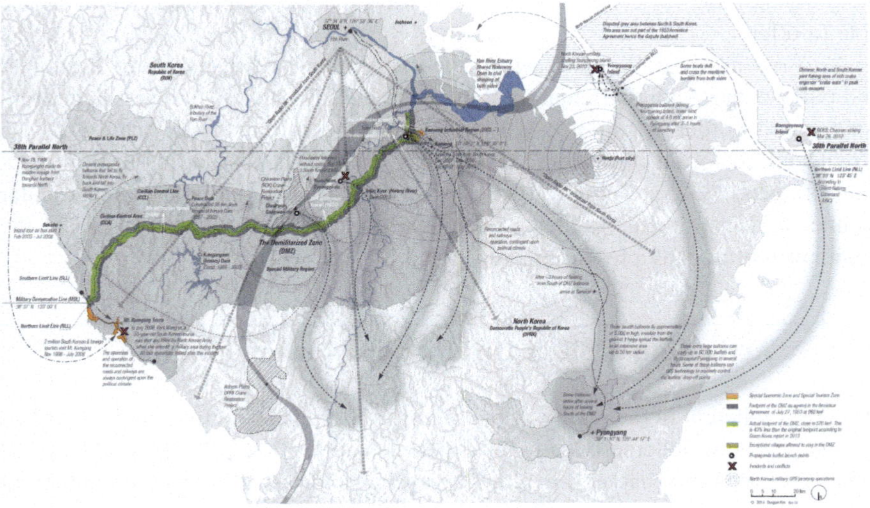


**Fig. 2** *A Construct the Koreas (Never) Made Together: Deconstructing the DMZ For the Imaginary. Image credit* Kim [2]

Of the multiple ways to understand the Demilitarized Zone, *constructing* the DMZ as a *proving ground* can become a powerful grounding for spatial practices to meaningfully analyse, understand, recombine, and reckon new unknown alternative futures of the DMZ in novel ways. This process starts with extensive creative mapping projects that envision and suggest better futures for the border territories that often produce violence. The "visions," and "suggestions" are then translated into projective speculations. These projective approaches draw methodological affinities to Bruce Stirling's "design fiction" [9] and "speculative design" by Dunne and Raby [10], where imagination drives continuous innovations, foster artistic expressions, engender productive conversations between current reality and potential futures possibilities between competing stakeholders. This is where a feedback loop between the mapping and speculations feeds into and forms a reciprocal relationship to "creative speculations."

As discussed, *creative speculative imagination* in *spatial practice* can engender original ideas, concepts, questions, and elucidations by combining, altering, or reimagining elements of existing knowledge or experiences in inventive and unexpected ways. It is a fundamental aspect of human thought that allows individuals to conceive of something that does not yet exist or to see existing things in a new light. *Creative speculative imagination* enables individuals to envision possibilities, simulate hypothetical scenarios, and consider alternative future outcomes. Engaging creative imagination facilitates the exploration of new territories and questions established norms, leading to breakthroughs and advancements. Furthermore, I contend that a spatial practice approach amplifies and integrates the visual, spatial, creative, and speculative dimensions, offering unique perspectives for addressing the pressing issues of contested borders in ways that other approaches may not be able to offer.

As a recent exemplar approach of this, Australian architect Gerard Reinmuth and philosopher Andrew Benjamin frame "the DMZ as a site" to investigate *coexistence* and border conditions. For them, the "site is the transformation of the ground into a locale of design," where they ask, "is it possible to deploy mapping projects, the conceptualization of spatial and the development of particular architectural instruments to create openings resulting from the suspension of the dominance of already present organizational logics?" [11].

## 3 Activating a Site's Disjunctions for Imagination

"Future is inherent in the reality of the present" [12, p. 14]. Setting up the DMZ as a *proving ground* inherently calls for experimentation. In the design studios mentioned earlier in the chapter, I explore the agency and potential of spatial practices for contested territories, where border space such as the one between North and South Korea are transformed into a "proving ground," one that experiments with 'double-meaning (*double entendre*)-' where in the literal sense, the border space is a militaristic installation, and in a figurative sense, a space for experimentation of ideas. This kind of 'double-meaning' approach in the form of disjunction, subversion, and appropriation, especially between "form and function," "meaning and form" within architectural discourse, has been extensively explored in Robert Venturi's *Complexity and Contradiction* [12]. Venturi brings up the concept of *objects trouvés* that use "conventions unconventionally" as an example of this approach [12, p. 43]. Several earlier examples of speculative designs informed by this approach can be found in my writing "Towards Edge Spaces of Productive Inclusivity: A Subversive Act" [5].

This kind of approach only becomes possible when the existing conditions are "mapped" in a specific creative way, meaning information is "visualized" and "spatialized for further appropriations". Further, Bernard Tschumi adds to these approaches by investigating the disjunctions between "space," "event," and "movement" and the human experiences produced through architecture in his *Architecture and Disjunction* [13] and *The Manhattan Transcripts* [14]. The kind of mentioned disjunctions are activated through the particular elements drawn from the mapping projects.

Additionally, these approaches and methods are found in the idea of "architectural exaptation," explicated as the "functional shift of a structure that previously served a different 'purpose' (that is, a trait related to a selective pressure) or no function at all (for example, as a side effect of other processes)" in the *Architectural Exaptation: When Function Follows Form* in recent times [15]. Functions of the border territory to contain and protect are mapped to be appropriated into creative speculations.

These approaches elaborate on the constantly changing function (program) of spaces and how they are critically modulated and curated by spatial practitioners, which radically transforms the human experience of them, especially for the better.

Lastly, in this context, I believe that *creative* and *imaginative* aspects of spatial practices have a significant role in critiquing what the Belgian philosopher foresees as the "'archetypal fact' of twenty-first-century architecture and urbanism" as "the enclosure, the wall, the barrier, the gate, the fence, the fortress" [16, p. 45]. It is at this future dominant spatial organization of exclusion where exploration of "imagining the impossible" through spatial design scenarios become immensely valuable.

## 4 Mapping the Demilitarized Zone: Collecting and Editing Ingredients

> There are three basic levels of comprehending physical phenomena: first, the exploration of pure physical facts; second, the psychological impact on our inner-self; and third, the imaginative discovery and reconstruction of phenomena in order to conceptualize them [17, p. 8].

In addressing complex and conflicted sites such as the Demilitarized Zone (DMZ), I start by creating an extensive series of site mapping projects across a wide range of scales embodying multiple viewpoints. The general approach of my mapping projects can be said to follow the "three basic levels of comprehending phenomena," Oswald Mathias Ungers states in the excerpt mentioned above in his *Morphologies: City Metaphors* [17]. Therefore, my mapping projects of the DMZ describe the physical reality and its impact on people, which has become an essential ingredient for new imaginations and unexpected discoveries. My recent publication, *Drawing Hwa-Chaeng: Mapping Contested Territories for Imagination* [1], further explains this approach in detail, with a particular reference to the notion of *Hwa-Chaeng* where it explicates how *Hwa-Chaeng* informed mapping can produce unique knowledge that can inform the transformation of contentious territories into synergistic productive spaces-a form of proving ground.[1]

My series of mapping projects of the DMZ in the form of drawings operate as a dominant visual tool that enables multiple reading and generation of narratives and a wide range of spatial information to co-exist. They operate as research, thinking, explorational, narrating, and representational tools inspired by many creative

[1] 화쟁 (和諍). "Hwa-Chaeng" philosophy that argues for embracing "opponents" and working towards reconciliation through continuous conversations. This approach seeks to shift antagonisms beyond false dichotomies and binary oppositions. Hwa-Chaeng is a philosophy and a working method that Won Hyo (617–686), who was one of the leading thinkers, writers, and commentators of the Korean Buddhist tradition, established when he attempted to address and reconcile contradictory Buddhist sectarianism in the early seventh century. More information on the relationship of "Hwa-Chaeng" to my mapping projects can be found in the *Drawing Hwa-Chaeng: Mapping Contested Territories for Imagination* [1, pp. 7–8].

mapping projects listed in James Corner's "The Agency of Mapping: Speculation, Critique, and Invention." Corner defines mapping as a creative tool, and states that they "do not *represent* geographies or ideas, rather they *effect* their actualization" [18, p. 225].

Mapping is one of the most fundamental methods for my research and creative work. Thus, it becomes a logical starting point and often becomes the project itself. It is one of the first steps I map to understand a site, territory, space, and the people I engage. It is also a foundational act that grows into various mediums, such as drawings, writings, exhibitions, and teaching. Further, the types of mapping projects referred to here should be distinguished from conventional "site maps" or "site analyses" produced for architectural projects. I aim to discover, bring awareness, and build knowledge beyond depicting a territory. These mapping projects are aligned with the understanding of "mapping as a fantastic cultural project," one that "creates and builds the world as much as measuring and describing it," as to tap into the imaginary potential stated in Corner's "The Agency of Mapping" [18, p. 213]. This synthesized approach to mapping and imagination blurs the analytical mapping and the creative speculation are also evident in the recent published work of Alexandra Arènes at Société d'Objets Cartographiques (SOC), where they are "charting the exploration of an unknown world—our own—with a new cartography of living things rather than space available for conquest or colonization" in *Terra Forma* [19], and in *Geostories* where "environmental imagination that renders sensible the issues of climate change and through geographic fiction invites readers to relate to the complexity of Earth systems in their vast scales of time and space" [20].

My mapping further draws affinity to Andrea Kahn's notion of "constructing a site," mentioned in her "Defining Urban Sites" [21]. Kahn states, "...in the most profound sense, representation is not about depicting reality, but about making knowledge" [21, p. 288]. Collecting, curating, editing, and representing information in a specific way and translating them into new knowledge is what my mapping projects do for the contested Korean DMZ. The knowledge from my mappings ultimately transforms the DMZ into a "proving ground" that engenders conversations and fundamental questions that critique and inform different kinds of speculation that engage alternative futures for territories under conflict.

## 5 Constant Change: Understanding the Longue Durée

In addition to incorporating the creative process into mapping projects, I embrace the site's past and the inevitable ongoing changes. I explicitly integrate the notion of time and history to articulate dynamism, demonstrating the border's spatial change over time [2]. My mapping projects amplify these processes and are frequently translated into moving images or animations. The flow of time and change is often expressed in simple timeline drawings or moving images. Moreover, visual transitions of before-and-after, evolving changes of a place are often explored in moving images beyond the static paper that effectively incorporate the ongoing dynamic changes over time. Additionally, I use the concept of time to locate the site context against its more extended past and present, which engenders particular awareness of the present and future(s).

When I approach the passing and flow of time in mapping, I actively extend it into a long period (Longue durée) of periods that expands the story of the site. For example, a drawn-out timeline can go beyond the conventional history of humankind to processes such as the earth's geological periods or some 13.8 billion years ago to the Big Bang, one that can evoke the "Blue Marble Effect." The cognitive and emotional awareness triggered by timelines and paradigm-shifting understanding of a place, object, and subject may be paralleled to the profound cosmopolitan experience when astronauts first saw the fragile borderless Earth from outer space. These ways of framing the DMZ can enable one to engage the significant climate crisis and the long-term effects of the Anthropocene concerning the contested border territories.

These mapping projects, moving images, and timelines ultimately contribute to a new way of framing the site and, thus, a new way of approaching issues that they are part of. These approaches challenge preconceived understandings of the site and generate new ways of seeing a place, producing new forms of knowledge that translate to meaningful questions and produce new ingredients for creative imaginations to emerge and provoke new assemblages. Furthermore, the plotted timelines reveal that borders are constantly made, altered, and changed over time. This highlights the fact that a seemingly static, impenetrable border, in fact, changes dynamically over time. This amplified understanding of the dynamically changing past and one's ability to understand that someone brings these changes demonstrates that *we can shape and change* the future. Further, framing the DMZ as a "proving ground" produces research, knowledge, and points of view that provide us hope for better futures, primarily through the potency of spatial practice-based speculations (Fig. 3).

## 6 What if the DMZ Was a... Imagining the Impossible

As ingredients and new knowledge, my accumulated mapping projects contribute to my engagements with architecture and landscape architecture students in numerous design studio settings at undergraduate and graduate levels. These engagements explore what the DMZ could become beyond the typical "peace park" proposals periodically brought up in an ever-changing political climate [6]. This illustrates how a new way of understanding a site through creative mapping can trigger new imaginations for new ways of living together.

The underlying premise of the series of design studios informed by my creative mappings that I instruct asks, "How can we transform existing exclusionary borders into productive inclusionary spaces?" as Bernard Tschumi has asked. A question similar to mine, but more inciting, commenting on the role of architects, Tschumi asks, "How could architects avoid seeing architecture and planning as the faithful product of dominant society, viewing their craft, on the contrary, as

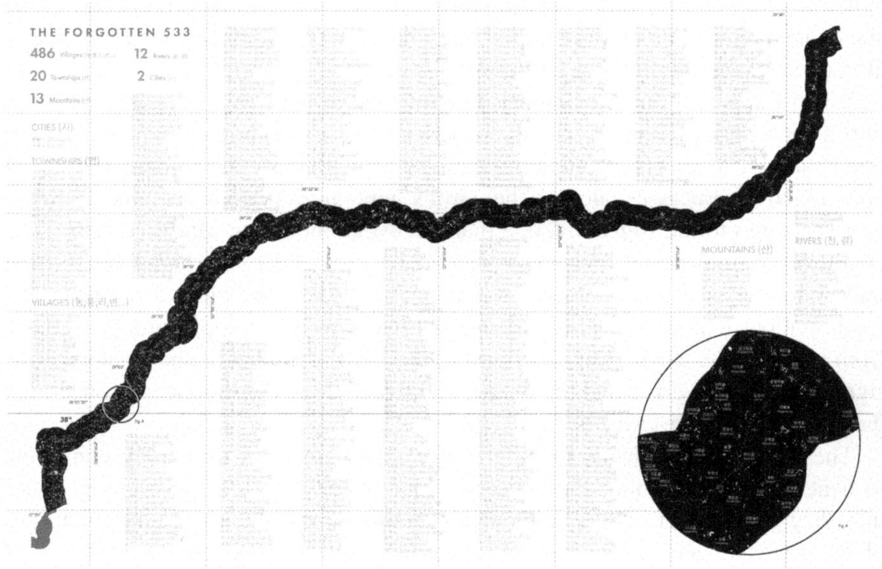


**Fig. 3** *533 Places: Tracing the Forgotten Places in the Demilitarized Zone (DMZ). Image credit* Dongsei Kim, assisted by Elise Park, 2022

a catalyst for change?" [13, p.7]. To answer these questions, my studio experimentations engage subversive, contradiction-embracing design strategies to critique the status quo and speculate on the territorial future of the heavily militarized, tense DMZ between North and South Korea.

## 7 A Non-Manifesto for the DMZ: Imagining New Polities

Following a series of mapping projects for the DMZ, I further wrote a "non-manifesto for the DMZ" in 2014 [7, p. 571], which was subsequently translated into an exhibition material in 2019, engaging the wider public [1, p. 45]. This "non-manifesto for the DMZ" unmistakably declares the position and attitude toward what kind of spatial design approach should be taken for the DMZ based on the initial mapping projects.

As the title of the mapping, *New Demos Making Zone (DMZ)*, implies, the suggested obsolete militaristic DMZ is proposed to be transformed into a proving ground where new forms of demos (*demos: the common people of an ancient Greek state: the populace as a political unit, especially in a democracy*) and polity can be experimented with. A place where a new society, community, and collective for new socio-political structures can be explored, where discussions about new forms of democracy and participation are speculated and negotiated. The goal is not necessarily to accurately dictate or predict the future but to critique existing

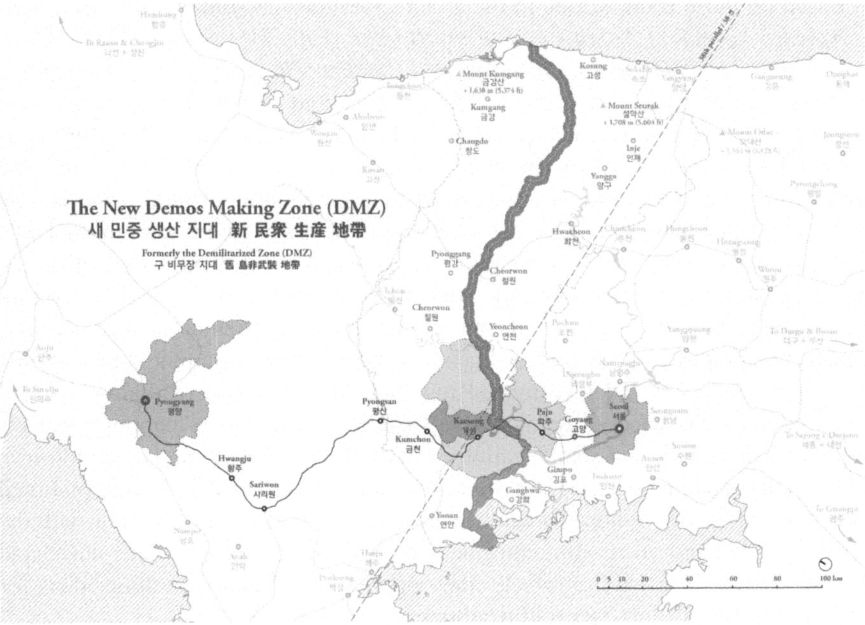


**Fig. 4** *The New Demos Making Zone (DMZ): Formerly the Demilitarized Zone (DMZ). Image credit* Kim [5]

polity models and stimulate dialogues about different possibilities and their potential implications through spatial means. Further, these speculative frameworks and narratives can provoke thoughts that foster dialogue about alternative futures and questions for the DMZ, the division of North and South Korea, and, ultimately, challenge the nation-state border hegemony (Fig. 4).

## 8 What if the DMZ Was a… Imagining the Impossible

With this initial research based on creative mapping projects that question and challenge the status quo, I have led several architecture, landscape architecture, and urban design studios over the years at a wide range of international institutions (RMIT 2014, 2014, 2015, 2016), Korea University (2015), (NYIT 2021–2022, 2022–2023) informed by my projects on the DMZ.

Mapping projects such as "How large is the DMZ? What if the DMZ was a …" that visually compares the DMZ's land area (992 km2, equivalent to 291 Central Parks of NYC) to comparable cities such as Seoul (605 km2), Berlin (892 km2), and Singapore (729 km2), provide context and add to the understanding of the DMZ as a proving ground, one that amplifies its current surreal ecological and militaristic reality into its speculative potential utility of what it could be as a denser urban or productive condition [1, pp. 34–35].

Based on these frameworks, in the architectural design studios, *Architecture as Caregiver: Agency of Architecture at the Border* (Fall 2021-Spring 2022, BArch thesis studio at NYIT) and *Architecture and Conflict: The Agency of Architecture at the Border* (Fall 2022-Spring 2023, BArch thesis studio at NYIT), students were introduced to mapping research projects on the DMZ as ingredients and further as a method where students took the mapping approach to locate their own sites and define creative and suitable programs for their eventual architectural projects.

As a first step, students are asked to team up in pairs or threes to map the DMZ in at least three scales. This collaborative visual exercise introduces students to the topic, region, history, and geography. Students are further provided with broad topics to explore that intersect with their personal interests. Themes such as ecology, climate, geography, transportation, economy, politics, demographics, technology, and others are listed and investigated together through series of dialogs to produce collective knowledge for the overall studio. The creative mapping projects amplify visualization and spatialization of existing knowledge and site conditions. Additionally, conducting a design studio on the Korean DMZ in an international academic setting where non-Korean background students are less familiar with the Korean conflict provided stimulating less prejudiced visions (at times naïve and idealistic, but valuable). Conversely, they provide novel and refreshing ideas for the future of the seventy-year-old deadlock.

Students designed a variety of architecture that embodies a multiplicity of ideas generated from the diverse mapping projects. The pluralistic list of programs that emerged from the two recent DMZ studios includes an ecologically driven Korean culture-based wellness center, a historical eco-trail that traverses the entire DMZ from west to east, a series of monumental memorial spaces that repurpose existing military infrastructural spaces (such as the underground infiltration tunnels), a cultural and marine research center on an island within the ambiguous water bodies of the DMZ. This bird habitat structure provide accommodation for people with insomnia that takes advantage of the DMZ's distance away from the city's light and noise pollution, an experimental city that embraces asylum seekers and welcomes refugees to settle within the DMZ creating a unique space of coexistence, a collaborative reforestation project with a, retreat-driven wildlife research centre.

Further, one of the studios—*The Agency of Architecture at the Border*—examined the unconventional programs at one of the most well-known sites within the DMZ, the Joint Security Area (JSA) located within the DMZ. Through mapping and research, students envisioned a wide range of possibilities. They proposed a Joint Community Area (JCA, a word play on the JSA) that experiments and modifies the JSA acronym and its area as a sustainable farming community, a cosmopolitan cultural center inspired by the Korean concept of Jeong (정)[2] that repurposes the existing buildings of the JSA as Korean-themed Temple Stay, Farm

[2] 1. bond; feeling: A state of mind that forms when one feels something. 2. love; affection: The state of mind where one feels love or affection for someone. [https://korean.dict.naver.com/koendict/#/entry/koen/d85dd46acdf04af68655e407e15cc351].

to Table Restaurant, Culinary School, Theater, and DMZ Museum. Another proposal "Changes: Tracing Time" is inspired by the rivers and ecological flows that cut across the political border between North and South Korea. It focuses on the constantly changing characteristics of nature and how it can initiate change to the stalemate between North and South Korea. The project is placed directly on the Military Demarcation Line (MDL) at the Joint Security Area (JSA) to transform the human-made political line into an ecologically driven reconciliatory space. Further, the project attempts to project an alternative future amplified by natural processes expressed through water flows and seasonal change (Fig. 5). Other notable schemes included, higher education institution city dedicated to international relations that is connected to the reconnected railways, a multi-purpose shopping complex with a performance and sports arena embedded into the JSA's landscape, a Korean cultural complex that brings the Koreans together through their long shared history, a sustainable technology and art museum at the JSA, and traditional Korean alcohol (soju and makgeolli) brewery and distillery facilities and restaurant/bar connected to its crop producing landscape within the JSA. These schemes all imagine and mandate the seemingly "impossible" collaborations between the two Koreas and beyond that critically question and challenge the status quo.

The mapping exercises and the research on Korean history, society, politics, economy, and culture from individuals and groups of students inform these unique

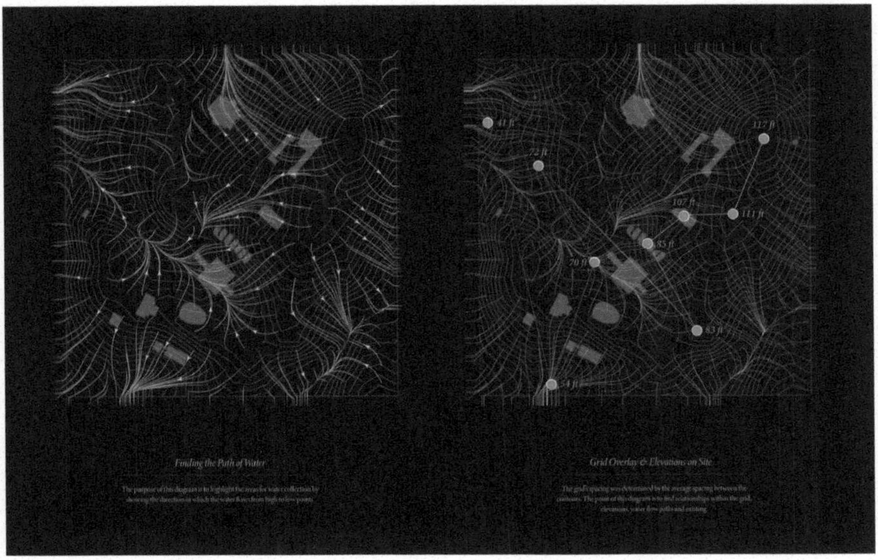


**Fig. 5** *Changes: Tracing Time* by Zeina Lotfy from thesis studio "Architecture and Conflicts: Architecture's Agency at the Border" led by Dongsei Kim. (September 2022-May 2023). *Image credit* Zeina Lotfy 2023

programs for their architecture. The program and its function dictate the architecture's form and inherently and explicitly become a means to propose a new way of living within and on and around contested border territories such as the DMZ. Other notable novel programs that emerged from my design studios at other institutions include transforming the DMZ into a Siberian tiger sanctuary (a culturally significant keystone species, now understood to be extinct within the Korean Peninsula) that appropriates existing military surveillance technologies and heritage sites within the DMZ.

Another exemplar work along this line is *Relationality: Demilitarized Zone* studio, led by Gerard Reinmuth and Andrew Benjamin at the University of Technology Sydney (UTS 2020), examines the concept of *coexistence* using the "DMZ as a site." The studio's research and investigations of the "site" are productive as they treat the DMZ as a form of "proving ground" for experimentation. The studio explores *coexistence* through the lens of faunas related to the DMZ, such as the Spotted Seal, Siberian Musk Deer, Asiatic Bear, and others, to "elaborate" and "re-territorialize" the border and to examine the "interdependence/reciprocity" and "coexistence" at the border [22] Fig. 6.

Reinmuth and Benjamin assert that "buildings are constituted by and form part of a network of relations. The consequence of such a description is that it then allows any one object—the building as an object—to be an after-effect of the relations that pertain within a given conjunction" [22]. This is precisely why

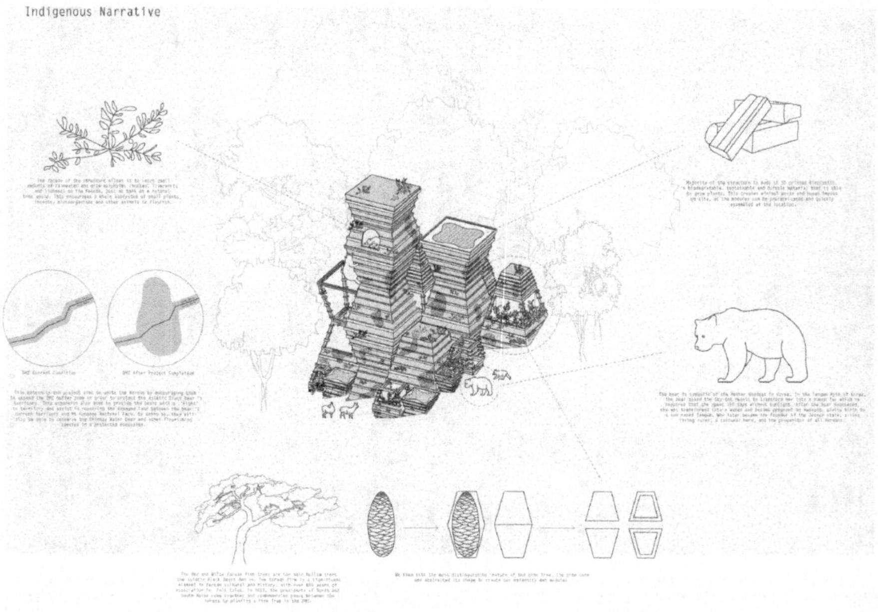


**Fig. 6** "Asiatic Black Bear and Chinese Water Deer" by Amy He and Farah Rehman in the Masters Architectural Design studio Relationality: DMZ taught by Gerard Reinmuth and Andrew Benjamin at the University of Technology, Sydney (August–November 2020)

understanding the DMZ and its conditions as a "proving ground" where broad range ideas are experimented, unexpected ideas collide and vigorously discussed becomes productive for contested territories and its future.

Moreover, Reinmuth and Benjamin ask, "Can the presence of a thickened border condition open up questions of coexistence and cooperation, both within and across species?" Using mapping as a starting point, students were asked to work at different scales and "develop a series of counter-measures—a project in which design and politics intersect—that explore the possibility of any future at the border." [22]. Again, the premise of the studio framing the "DMZ as a site" or as a "proving ground" with the accentuated juxtaposition of fauna and human, permeability and impermeability of the DMZ enables and demonstrates a particular agency of spatial design that is productive in generating new ways of seeing brutal conflicts and novel future visions for lively discussions, especially for the increasingly violent border territories.

## 9 Conclusion: Imagining Inclusive Futures

> Kant postulates that knowledge originates in two basic components: intuition and thought. According to Kant, all our thinking is related to imagination, which means it is related to our senses because the only way to describe an object is through imagination. The intellect is incapable of perceiving anything, and the senses cannot think. Only through a combination of both can knowledge arise. Imagination has to precede all thinking processes since it is nothing less than a synopsis, an overall ordering principle bringing order into diversity. If we accept that thinking is an imaginative process of a higher order, then, argues Kant, it means all sciences are based on imagination (Ungers 1982/2012, 7) [17].

By framing the DMZ as a "proving ground" through spatial practice and methods, including mapping and speculative design, the DMZ is transformed into a dynamic symbolic, mental, and physical space where new understanding, conversations, and questions about the Korean division and its better future are generated. This paradigm shift and knowledge production become valuable frameworks that enable one to think beyond the persistent dichotomy of the border. This envisions a border that divides to be transformed into a space of coexistence and understanding. Furthermore, this approach seeks to challenge the false binary often promoted by nation-state borders, which produce spatial dynamics of "exclusion" and "inclusion" that define "us" and "them." It also provides an open pluralistic space where differences can coexist and inform one another, moving beyond the pursuit of a single "correct" view that seeks to compete and eliminate opposing perspectives.

My creative mapping and projects that are generated by them provoke and are manifested in my students' research, design proposals, and their hopes. They then start to transform the conflicted and impenetrable DMZ into an open proving ground for conversations and debates. I believe that this critical process helps one to transcend one's deep prejudices and open up a wide range of radical imaginations, generating valuable knowledge for our collective future.

## References

1. Kim D (2022) Drawing Hwa-Chaeng: mapping contested territories for imagination. Jeongye-C-Publishers, Seoul. https://clio.columbia.edu/catalog/17829998
2. Kim D (2019) A construct the Koreas (Never) made together: deconstructing the DMZ for the imaginary—2019. REAL DMZ PROJECT. Google Arts & Culture. https://g.co/arts/LWf2pWxjZxjd9DpVA
3. Kim D (2018a) Metamorphosis of a Zone. Int Rev Landsc Arch Urban Des 104(September):74–79. https://id.lib.harvard.edu/alma/990028384980203941/catalog
4. Kim D (2018b) Undoing the Demilitarized Zone (DMZ): the agency of architectural intelligence. In: Benyamin J, Reynolds K, Zell M, Bouchard N, Moon W (eds) ACSA fall conference proceedings: play with the rule, pp 446–450. https://www.acsa-arch.org/proceedings/Fall%20Conference%20Proceedings/ACSA.FALL.18/ACSA.FALL.18.88.pdf?v=2
5. Kim D (2017) Towards edge spaces of productive inclusivity: a subversive act. In: Manu S (ed) Chandigarh rethink: transforming ruralities & edge(ness) in global urbanities ORO Editions, Novato, California. https://id.lib.harvard.edu/alma/99156071416503941/catalog
6. Kim D (2014a) Towards a dialogical peace in the demilitarized zone. 40:40–43
7. Kim D (2014b) A construct the Koreas (Never) made together: deconstructing the DMZ for the imaginary. In: Yim D, Luna R (eds) The North Korea Atlas. DAMDI Publishers, Seoul, pp 566–571
8. Kim D (2013) Borders as urbanism: redrawing the demilitarized zone (DMZ) between the democratic People's Republic of Korea and the Republic of Korea. Landscape Archit Front 1(2):150–157
9. Sterling B (2005) Shaping things. MIT Press, Cambridge, Mass. https://id.lib.harvard.edu/alma/990098054620203941/catalog
10. Dunne A, Raby F (2013) Speculative everything: design, fiction, and social dreaming. The MIT Press, Cambridge, Massachusetts
11. Benjamin A, Reinmuth G (2020a) Architecture and Co-Existence: DMZ as Site. Critical Spatial Practice. https://criticalspatialpractice.co.uk/architecture-and-co-existence-dmz-as-site-2020/
12. Venturi R (1966/1977/2002) Complexity and contradiction in architecture, 2nd ed. In: Association with the Graham Foundation for Advanced Studies in the Fine Arts, Chicago: Distributed by New York Graphic Society. Museum of Modern Art, Boston, New York. https://id.lib.harvard.edu/alma/990010301610203941/catalog
13. Tschumi B (1994) Architecture and disjunction. MIT Press, Cambridge, Mass
14. Tschumi B (1981/1994) The Manhattan transcripts: theoretical projects. St. Martin's Press/Academy Editions
15. Melis A, Pievani T, Lara-Hernandez JA (2024) Architectural exaptation: when function follows form, 1st ed. Routledge. https://doi.org/10.4324/9781003347118
16. De Cauter L (2004) The capsular civilization: on the city in the age of fear. NAi-publishers, Rotterdam. https://id.lib.harvard.edu/alma/990095295680203941/catalog
17. Ungers OM (2011) Morphologie city metaphors. 2. Aufl ed. Köln: Walther König, New York: distribution, outside Europe D.A.P./Distributed Art Publishers. https://id.lib.harvard.edu/alma/990129944390203941/catalog
18. Corner J (1999) The agency of mapping: speculation, critique and invention. In: Cosgrove D (ed) Mappings. Reaktion Books, London, pp 213–253. https://id.lib.harvard.edu/alma/990079375030203941/catalog
19. Aït-Touati F, Arènes A, Grégoire A, Latour B, DeMarco A (2022) Terra Forma: a book of speculative maps. The MIT Press, Cambridge, England. https://id.lib.harvard.edu/alma/99156365152003941/catalog
20. Ghosn R, Jazairy EH (2018) Design Earth. In: Geostories: another architecture for the environment. Actar, NY, USA

21. Kahn A (2005) Defining urban sites. In: Burns CJ, Kahn A (eds) Site matters. Routledge, New York, pp 280–296
22. Benjamin A, Reinmuth G (2020b). Architecture and coexistence: the DMZ as a site. In: An architecture of relationality V: Territory, Border, Countermeasure. University of Technology Sydney

# Conclusion Chapter

# 21

Alessandro Melis and Jose Antonio Lara-Hernandez

**Abstract**

This concluding chapter offers a synthesis of the contributions presented in *The Dusk of Design*, highlighting the key themes and theoretical implications of the book. It examines how the concepts of variability, diversity, and redundancy, derived from evolutionary biology, serve as foundational principles for rethinking architectural and urban design. Moving beyond static, deterministic models, the chapter argues for an adaptive and flexible approach to design, one that responds to the evolving needs of both human and ecological systems. Drawing from case studies in global contexts, it explores how informal urbanism, cultural ecology, and exaptive processes shape resilient environments. The chapter emphasizes the role of policy and governance in fostering innovation, stressing the importance of creating spaces that are inclusive and responsive to the complex, dynamic nature of contemporary society. Lastly, we calls for a new paradigm shift in architectural design, advocating for interdisciplinary collaboration and a holistic understanding of the built environment in the face of global social and environmental challenges.

**Keywords**

Variability • Resilience • Exaptation • Redundancy • Diversity

A. Melis
New York Institute of Technology, New York, NY, USA
e-mail: amelis@nyit.edu

J. A. Lara-Hernandez (✉)
Auckland University of Technology, Auckland, New Zealand
e-mail: antonio.lara-hernandez@aut.ac.nz



J. A. Lara-Hernandez et al. (eds.), *The Dusk of Design*, Integrated Science 34,
https://doi.org/10.1007/978-3-031-76867-5_21

## 1 Toward a New Paradigm in Design

Our exploration throughout *The Dusk of Design* offers a compelling narrative about the necessity of rethinking traditional approaches of architecture and urbanism. This work makes an urgent call to architects, planners, urbanists, and practitioners alike to adopt more dynamic, flexible, and inclusive approaches to spatial design, echoing the growing call for addressing the complexities of an evolving world. The diverse range of contributions in this book collectively constructs an assemblage of design as a continuous, adaptive process that should incorporate social, ecological, and technological considerations, aiming to respond to both historical contexts and future uncertainties.

The chapters presented here demonstrate the limitations of deterministic and human-centric approaches in architecture. Instead, evolutionary design emerges as a paradigm that embraces concepts like variability, diversity , and redundancy—principles borrowed from biological evolution. The discussion of evolutionary design, particularly in the first chapter, underscores how adaptive and exaptive processes can create architectural elements that serve new and unexpected purposes. These new perspectives challenge the static nature of traditional design, positioning it instead as an organic, responsive system. Through this lens, the architect's role shifts from being the master of space to a facilitator of change, fostering environments that are responsive to human needs and to broader ecological and cultural dynamics.

Adjunct to the main argument of the book is the concept of diversity , which is explored through different cultural, geographical, and social lenses across the chapters. Diversity, as a guiding principle, extends beyond the demographic or sociocultural realm; it plays a critical role in fostering resilience and adaptability in urban and architectural designs. For instance, Chap. 2 illustrates how urban design .can respond to the needs of diverse populations by considering how marginalized voices influence space. Similarly, later chapters demonstrate how urban informality (see Chap. 6) and the self-organization of spaces (see Chap. 13) provide frameworks for cities to develop organically, often in ways that are resilient to socioeconomic challenges.

These discussions describe diversity and redundancy as essential elements of resilience in terms of architecture and urbanism. The "informality" observed in urban spaces, as seen in the case studies on Latin American and African cities, demonstrates how redundant and variable structures in the built environment support community resilience. They are enablers of flexibility in spaces, which adapt to evolving social and environmental conditions.

The concept of exaptation, introduced early in the book and developed through various chapters, provides a new lens through which to understand architectural innovation. Exaptation refers to features that evolved for one purpose but are later co-opted for another, an idea borrowed from evolutionary biology. This principle is particularly relevant in understanding how spaces and structures evolve in response to changing social and environmental needs. Several chapters explore this idea, most notably in the discussions of favelas (see Chap. 6) and the informal urban practices observed in North African and South American contexts (Chaps 8 and 9). The cases illustrated here demonstrate that architectural elements can evolve to serve new purposes, just as biological traits can be repurposed for different functions through the process of evolution.

In line with this view, spaces are seen not as static entities but as living organisms capable of adapting and transforming based on the needs of their users.

A recurring theme in the book is the necessity of considering cultural and ecological contexts in design. Chapters such as "The Mayan Solar" (Chap. 5) provide insights into how traditional architectural forms—shaped by centuries of cultural evolution—can offer lessons in sustainability and resilience. Similarly, the exploration of motor parks in Yorubaland, Nigeria (Chap. 18), highlights the importance of understanding the cultural ecology of public spaces. In these contexts, exaptive and adaptive processes are not limited to the built environment but are deeply embedded in the social and cultural practices of the people who inhabit these spaces.

This broader ecological perspective is crucial in addressing the global crises that affect our cities today, such as climate change and social inequality. The authors here argue that an adaptive, culturally sensitive approach to design is essential for creating spaces that can withstand these challenges. The shift from modernist, from mono-functional spaces to more fluid, multifunctional spaces reflects this growing awareness of the need for resilient, sustainable environments that serve a variety of purposes.

As many chapters in the book discuss, implementing these transformative design practices requires robust policy frameworks that support innovation and inclusivity. The chapters dealing with governance and urban policy highlight how zoning laws, building codes, and regulatory frameworks can either facilitate or obstruct the creation of diverse and resilient urban environments. These discussions remind us that architectural innovation does not occur in a vacuum; it is influenced by a complex web of social, political, and economic factors.

For instance, the discussion of social housing projects in South America (Chap. 10) underscores the need for policy frameworks that support flexibility and resident agency in shaping their living spaces. This aligns with the broader theme of adaptive design, which emphasizes the importance of creating spaces that can evolve over time to meet the changing needs of their inhabitants.

In conclusion, *The Dusk of Design* presents a compelling case for a shift in how we think about architecture and urban design. The evolutionary principles of variability, diversity, and redundancy offer a new framework for creating spaces that are resilient, adaptable, and inclusive. By embracing these concepts, we can move beyond deterministic models of design to create environments that are functional, and at the same time, reflective of the diverse, dynamic world we live in.

The book's collective contributions call for a holistic, interdisciplinary approach to design—one that recognizes the interconnectedness of social, cultural, ecological, and technological systems. In doing so, it challenges us to reimagine the future of design at a moment when the world is undergoing rapid and profound change.

# Index




J. A. Lara-Hernandez et al. (eds.), *The Dusk of Design*, Integrated Science 34,
https://doi.org/10.1007/978-3-031-76867-5

The manufacturer's authorised representative in the EU is Springer Nature Customer Service Centre GmbH, Europaplatz 3, 69115 Heidelberg, Germany. If you have any concerns regarding our products, please contact ProductSafety@springernature.com

Printed and bound by CPI Group (UK) Ltd, Croydon, CR0 4YY
30/09/2026
02225562-0008